Previously published

Biodiversity Monitoring and Conservation: Bridging the Gap between Global Commitment and Local Action
Edited by Ben Collen, Nathalie Pettorelli, Jonathan E.M. Baillie and Sarah M. Durant
ISBN: 978-1-4443-3291-9 Hardcover;
ISBN: 978-1-4443-3292-6 Paperback; April 2013

Biodiversity Conservation and Poverty Alleviation: Exploring the Evidence for a Link
Edited by Dilys Roe, Joanna Elliott, Chris Sandbrook and Matt Walpole
ISBN: 978-0-470-67478-9 Paperback;
ISBN: 978-0-470-67479-6 Hardcover; December 2012

Applied Population and Community Ecology: The Case of Feral Pigs in Australia
Edited by Jim Hone
ISBN: 978-0-470-65864-2 Hardcover; July 2012

Tropical Forest Conservation and Industry Partnership: An Experience from the Congo Basin
Edited by Connie J. Clark and John R. Poulsen
ISBN: 978-0-4706-7373-7 Hardcover; March 2012

Reintroduction Biology: Integrating Science and Management
Edited by John G. Ewen, Doug. P. Armstrong, Kevin A. Parker and Philip J. Seddon
ISBN: 978-1-4051-8674-2 Paperback;
ISBN: 978-1-4443-6156-8 Hardcover; January 2012

Trade-offs in Conservation: Deciding What to Save
Edited by Nigel Leader-Williams, William M. Adams and Robert J. Smith
ISBN: 978-1-4051-9383-2 Paperback;
ISBN: 978-1-4051-9384-9 Hardcover; September 2010

Urban Biodiversity and Design
Edited by Norbert Müller, Peter Werner and John G. Kelcey
ISBN: 978-1-4443-3267-4 Paperback;
ISBN: 978-1-4443-3266-7 Hardcover; April 2010

Wild Rangelands: Conserving Wildlife While Maintaining Livestock in Semi-Arid Ecosystems
Edited by Johan T. du Toit, Richard Kock and James C. Deutsch
ISBN: 978-1-4051-7785-6 Paperback;
ISBN: 978-1-4051-9488-4 Hardcover; January 2010

Reintroduction of Top-Order Predators
Edited by Matt W. Hayward and Michael J. Somers
ISBN: 978-1-4051-7680-4 Paperback;
ISBN: 978-1-4051-9273-6 Hardcover; April 2009

Recreational Hunting, Conservation and Rural Livelihoods: Science and Practice
Edited by Barney Dickson, Jonathan Hutton and Bill Adams
ISBN: 978-1-4051-6785-7 Paperback;
ISBN: 978-1-4051-9142-5 Hardcover; March 2009

Participatory Research in Conservation and Rural Livelihoods: Doing Science Together
Edited by Louise Fortmann
ISBN: 978-1-4051-7679-8 Paperback; October 2008

Bushmeat and Livelihoods: Wildlife Management and Poverty Reduction
Edited by Glyn Davies and David Brown
ISBN: 978-1-4051-6779-6 Paperback; December 2007

Managing and Designing Landscapes for Conservation: Moving from Perspectives to Principles
Edited by David Lindenmayer and Richard Hobbs
ISBN: 978-1-4051-5914-2 Paperback; December 2007

Conservation Science and Practice Series

Elephants and Savanna Woodland Ecosystems: A Study from Chobe National Park, Botswana

Edited by

Christina Skarpe, Johan T. du Toit and Stein R. Moe

LIVING CONSERVATION

WILEY Blackwell

Library of Congress Cataloging-in-Publication Data

Elephants and savanna woodland ecosystems : a study from Chobe National Park, Botswana / edited by
Christina Skarpe, Johan T. du Toit and Stein R. Moe.
   pages cm
 Includes bibliographical references and index.
 ISBN 978-0-470-67176-4 (cloth)
1. Elephants – Botswana – Chobe National Park. 2. Grassland ecology – Botswana – Chobe National Park.
3. Chobe National Park (Botswana) I. Skarpe, Christina, 1946- editor of compilation. II. Du Toit,
Johan T., editor of compilation. III. Moe, Stein R., 1960- editor of compilation.
 QL737.P98E443 2014
 599.67096883 – dc23

                                                                           2013046029

Cover image: Front cover: Elephants *Loxodonta africana* browsing on *Croton megalobotrys*.
         Photo: Stein R. Moe.
         Back cover: Alerted impala ram and a group of bull elephants, Mababe, Botswana.
         Photo: Christina Skarpe.
Cover design by Design Deluxe

Typeset in 9.5/11.5 pt MinionPro by Laserwords Private Limited, Chennai, India
Printed and bound in Singapore by Markono Print Media Pte Ltd

1   2014

Contents

List of Contributors

Per Arild Aarrestad Norwegian Institute for Nature Research, Trondheim, Norway

Kathy A. Alexander Department of Fisheries and Wildlife Conservation, Virginia Tech, Blacksburg, USA

Harry P. Andreassen Faculty of Applied Ecology and Agricultural Sciences, Hedmark University College, Evenstad, Norway

Roger Bergström Gropgränd 2A, Uppsala, Sweden

Simon Chamaillé-Jammes Centre d'Ecologie Fonctionnelle et Evolutive, Montpellier, France

Kjell Danell Department of Wildlife, Fish, and Environmental Studies, Swedish University of Agricultural Sciences, Umeå, Sweden

Johan T. du Toit Department of Wildland Resources, Utah State University, Logan, Utah, USA

Øystein Flagstad Norwegian Institute for Nature Research, Trondheim, Norway

Hervé Fritz Laboratoire de Biométrie et Biologie Evolutive, Villeurbanne, France

Peter G.H. Frost Science Support Service, Wanganui, New Zealand

Duncan J. Halley Norwegian Institute for Nature Research, Trondheim, Norway

Håkan Hytteborn Department of Plant Ecology and Evolution, Evolutionary Biology Centre, Uppsala University, Uppsala, Sweden

Department of Biology, Norwegian University of Science and Technology, Realfagbygget, Trondheim, Norway

Craig Jackson Department of Biology, Norwegian University of Science and Technology, Realfagbygget, Trondheim, Norway

Thor Larsen Norwegian University of Life Sciences, Aas, Norway

Hillary Madzikanda Scientific Services, Zimbabwe Parks and Wildlife Management Authority, Causeway, Harare, Zimbabwe

Shimane Makhabu Department of Basic Sciences, Botswana College of Agriculture, Gaborone, Botswana

Gaseitsiwe Masunga Okavango Research Institute, University of Botswana, Maun, Botswana

Stein R. Moe Department of Ecology and Natural Resource Management, Norwegian University of Life Sciences, Aas, Norway

Rapelang Mojaphoko Ministry for Environment, Wildlife and Tourism, Botswana

Sekgowa S. Motsumi Department of International Environment and Development Studies, Norwegian University of Life Sciences, Aas, Norway

Department of Environmental Affairs, Gaborone, Botswana

Gosiame Neo-Mahupeleng Poso House, Gaborone, Botswana

Norman Owen-Smith Centre for African Ecology, School of Animal, Plant and Environmental Sciences, University of the Witwatersrand, Johannesburg, South Africa

A. H. M. Raihan Sarker Department of Biology, Norwegian University of Science and Technology, Realfagbygget, Trondheim, Norway

Susan Ringrose PO Box HA 65 HAK Maun, Botswana

Tuulikki Rooke Research and Assessment Department, Swedish Environmental Protection Agency, Stockholm, Sweden

Eivin Røskaft Department of Biology, Norwegian University of Science and Technology, Realfagbygget, Trondheim, Norway

Lucas Rutina Okavango Research Institute, University of Botswana, Maun, Botswana

Thato B. Sejoe Department of International Environment and Development Studies, Norwegian University of Life Sciences, Aas, Norway

P.O. Box 1826, Gaborone, Botswana

Christina Skarpe Faculty of Applied Ecology and Agricultural Sciences, Hedmark University College, Evenstad, Norway

Sigbjørn Stokke Norwegian Institute for Nature Research, Trondheim, Norway

Jon E. Swenson Department of Ecology and Natural Resource Management, Norwegian University of Life Sciences, Aas, Norway

Cyril Taolo Department of Wildlife and National Parks, Gaborone, Botswana

Marion Valeix Laboratoire de Biométrie et Biologie Evolutive, Villeurbanne, France

Mark E. Vandewalle CARACAL, Kasane, Botswana

Märtha Wallgren Forestry Research Institute of Sweden (Skogforsk), Uppsala Science Park, Uppsala, Sweden

Per Wegge Department of Ecology and Natural Resource Management, Norwegian University of Life Sciences, Aas, Norway

Foreword

Norman Owen-Smith

Centre for African Ecology, School of Animal, Plant and Environmental Sciences, University of the Witwatersrand, South Africa

After a long journey through the dry Kalahari sand woodlands stretching over northern Botswana, a spectacular sight confronts one on arrival at the Chobe River. The emerald landscape of meadows and water on the associated floodplain is thronged by elephants, buffaloes and hippos, alongside numerous smaller ungulates and birds. This is a prime example of a megaherbivore-dominated ecosystem, lacking only the rhinos that were once also there. However, the very largest of these herbivores has been disrupting the structure and diversity of the riparian woodland, to the consternation of wildlife managers, tourist operators and visitors. The Chobe River front has attained notoriety for the woodland devastation wrought by the elephant concentrations there. Gaunt trunks of dead trees stand amongst battered shrubs above a sparse herbaceous cover. Should not something be done about the elephants to rectify this situation? Along with Tsavo National Park in Kenya, the state of woodland destruction at Chobe is commonly invoked as justification for culling elephants to alleviate the vegetation transformation and its ramifications for biodiversity.

This was the context for the BONIC project, established as an institutional collaboration involving the Norwegian Institute for Nature Research, the Norwegian University of Life Sciences, and the Botswana Department of Wildlife and National Parks. Its aim was to advance local capacity to address this and other problems in the management of Botswana's rich wildlife resource, through research and training. The expectation was that participating wildlife ecologists from the far north, untainted by the polarised standpoints about the management of elephants that prevail within African countries, would be openly receptive to the prospect of culling elephants if ecologically justified. After all, Norway has persisted in harvesting of whales, despite international criticism. The crucial research needed was to interpret the impacts that elephants were obviously having on riparian woodlands within the context of other ecosystem components and processes. Hence studies were focused not merely on the elephants, but also on their effects on soil properties and tree regeneration, the consequences of woodland transformation for other browsers and grazers, and ramifying influences on predators, small mammals and even some groups of birds. The chapters of this book document the findings from this comprehensive suite of studies, summarising or elaborating numerous papers that have appeared in scientific journals as well as the contents of several unpublished theses.

Surprisingly to some, these studies did not find adverse consequences of the obvious elephant impacts on vegetation for any animal species, apart from bushbuck, which have declined in abundance from earlier times when the bush cover was much thicker. The stumbling block for restoration of the woodlands lies in the high local abundance of impala, a much smaller herbivore that is a mixed grazer-browser. Exclosure plots demonstrated that their browsing of tree seedlings is so thorough that very few of these seedlings have much chance of escaping towards tree height. Moreover, rather than competing for browse, elephants favour different woody species from those utilized by browsing ruminants. This finding reinforces the suggestion that megaherbivore extinctions largely through human hunting had negative consequences for post-Pleistocene large mammal diversity worldwide (Owen-Smith, 1987, 1989). The disturbing impacts that mammoths, mastodons and other very large herbivores must have had on woodlands and forests would have brought more browse within the reach of smaller browsers, while increasing the extent of meadows for grazers. Hence the largest herbivores can facilitate the coexistence of smaller species through these mechanisms, rather than competing with them for shared vegetation resources.

Nevertheless, the extent of the woodland destruction along the Chobe River front is regrettable negative scenically. However, happenings in the more distant past have probably contributed to this situation. After the extirpation of the elephants, browsing antelope became decimated by rinderpest, allowing unfettered growth by trees. Are the Chobe woodlands merely reverting to the messy state that had prevailed when herbivores from elephants size down had all been hugely abundant? But expanded human pressures are an exacerbating influence, funnelling elephants into a narrow section of river front between Kasane town and villages within the Kachikau enclave, and blocking movements across the international border into Namibia.

Fundamental questions remain about how riparian woodlands can withstand the elephant concentrations that develop along Africa's "pristine" rivers during the dry season. Seasonal elephant densities even greater than those near the Chobe River have become established along the Linyanti River to the west, following the drying of the Savuti Channel and other water sources in the interior. Will the vegetation transformation there progress towards the state prevalent along the Chobe? Or be averted by the wider scope that elephants have for movement in the Linyanti region? Research is in progress to assess this trend and where it might eventually lead.

Elephants and their impacts on trees have drawn most of my attention. But the chapters in this book encompassing broader ecosystem ecology will be more widely valuable as a counterpoint to the renowned studies undertaken in the grassland ecosystem prevalent in the Serengeti region of Tanzania. Most of Africa is very different in aspect from Serengeti, having vast areas of fairly well wooded savanna occupying nutrient deficient soils of granitic or Aeolian origin. In these parts the very largest grazers and browsers assume dominance of the herbivore biomass rather than the "plains game" typical of the East African plateau. Botswana is an amazingly diverse country, with dry savanna woodlands juxtaposed with the wetlands of the Okavango Delta and abutting Kalahari semi-desert in the south. Interspersed are localities where the concentration of herbivores and predators rival those in the Serengeti ecosystem. There are huge challenges in managing Botswana's rich wildlife legacy, not least because of the continuing expansion of the elephant population, approaching 150,000 animals at the time of writing. A major contribution of the BONIC programme was its fostering of local wildlife scientists equipped with the qualifications to take responsibility for this custodianship. Regional planning is well advanced for Chobe National Park to become a component of the vast Kavango-Zambezi Transfrontier Conservation Area stretching from Botswana through adjoining parts of Namibia, Zimbabwe, Zambia and Angola. The findings from the BONIC studies will help inform this ambitious development in its aim of promoting the coexistence of people and wildlife.

References

Owen Smith, N. (1987) Late Pleistocene extinctions: the pivotal role of megaherbivores. *Paleobiology* 13, 351–362.

Owen Smith, N. (1989) Megafaunal extinctions: the conservation message from 11 000 years BP. *Conservation Biology* 3, 405–412.

Preface

The common image of an African savanna, held by people living far from savanna environments, is a landscape of short-cropped grass and scattered *Acacia* trees that is teeming with medium-sized grazing ungulates such as plains zebra, *Equus quagga*, and blue wildebeest, *Connochaetes taurinus*. This is an image based largely on the Serengeti-Mara ecosystem in Tanzania and Kenya, well known from extensive popular science publications, films and TV programs, as well as seminal scientific publications. The woodland savannas of Chobe National Park in Botswana, the focus for this book, present a very different image. Whereas mean annual rainfall is about the same in large areas of Serengeti and Chobe, most other ecological conditions are different. Instead of the nutrient-rich volcanic soils of the Serengeti savannas, the Chobe woodlands grow on nutrient-deficient Kalahari sand and instead of scattered fine-leafed *Acacia* trees, the Chobe woodlands consist primarily of large, broad-leafed trees. The abundant medium-sized grazers in the Serengeti are replaced in the Chobe woodlands by a dominance of larger-bodied species such as African buffalo, *Syncerus caffer*, and elephant, *Loxodonta africana*, with more than half the herbivore biomass in Chobe National Park contributed by elephants alone.

Elephants were virtually exterminated from the Chobe ecosystem by an intense bout of commercial ivory hunting in the late 19th century. Over the following decades, woodlands established on the previously open narrow strip with alluvial soil close to the Chobe River. These woodlands were different from those on the Kalahari sand in most of the Chobe National Park. Once the elephant population eventually began recovering, elephants killed the trees in these newly established woodlands by debarking them, and since the 1960s managers and conservationists have been concerned about the destruction of the scenic woodlands along the Chobe River. This 'Chobe elephant problem' was the rationale for the Botswana–Norway Institutional Cooperation Project (BONIC), a research and capacity building project run in cooperation between the Botswana Department of Wildlife and National Parks (DWNP) and two Norwegian research institutions: the Norwegian Institute for Nature Research (NINA) and the Norwegian University for Life Sciences (UMB), with funding from the governments of Botswana and Norway. BONIC operated from 1997 to 2003 and encompassed studies on diverse aspects of the Chobe ecosystem, all with a focus on the ecological implications of the increasing elephant population. At the termination of the project, a workshop was held in Kasane, adjacent to the study area, over 13–15 March 2003 including about 50 people from DWNP, NINA, UMB, University of Botswana, some non-governmental organisations and three specially invited experts: Patrick Duncan,

Norman Owen-Smith and Anthony R.E. Sinclair. The workshop resulted in the first compilation of preliminary results from the project in a volume of proceedings, edited by Mark Vandewalle and published by the Government Printer, Gaborone, Botswana in 2003. It was followed in 2004 by an overview article about the project in the journal *AMBIO* by C. Skarpe and 26 co-authors from the project.

The workshop in 2003 gave rise to the idea of an edited book drawing from the main results from the project. First, over the ten years following the completion of the project, the results of various sub-projects were written up as graduate theses and peer-reviewed journal publications. Finally, this book synthesizes the ecological research conducted under the auspices of BONIC. The book's aim is to present results from the project related to the effects of elephants on ecosystem dynamics and heterogeneity, and finally to discuss the extent to which there is an 'elephant problem' in Chobe. The book compiles information from a nutrient-poor and elephant-rich savanna to allow comparison with other African savannas, for example: the nutrient-rich Serengeti-Mara ecosystem as presented in the three books edited by A.R.E. Sinclair and colleagues (University of Chicago Press 1979, 1995, 2008); the Kruger National Park on mixed soil types, as described in the book edited by J.T. du Toit, K.H. Rogers and H.C. Biggs (Island Press 2003); the nutrient-poor and elephant-free Nylsvley savanna described by R.J. Scholes and B.H. Walker (Cambridge University Press, 1993).

Most studies in the BONIC project are included in this book, some constituting individual chapters. We refer as far as possible to published data from the project, although data from PhD and MSc theses are also referred to along with some previously unpublished data, which are included without references. Four staff members from DWNP completed their PhD projects within BONIC and all of them contributed as authors to this book: Shimane W. Makhabu, Gaseitsiwe S. Masunga, Lucas P. Rutina and Cyril L. Taolo. Eight staff members from DWNP completed their MSc studies within BONIC: Kingsley M. Leu, Itani Mathumo, Thato B. Morule, David K. Mosugelo, Sekgowa S. Motsumi, Elsie T. Mvimi, Gosiame Neo-Mahupeleng and Claudia S. Zune. Their work has directly and indirectly contributed to this book.

Neither the BONIC project nor this book could have come into existence without DWNP, being the project leader, and the Ministry of Environment, Wildlife and Tourism of Botswana and their staff. Apart from staff members appearing as authors in the book, we particularly thank Jan Broekhuis, Joe Matlhare, Sedia Modise, Dan Mughogho, Bolt Othomile and Botshabelo Othusitse. Further, Thatayaone Dimakatso, Frederick Dipotso, Wilson Marokane, Moses Mari, Mpho Ramotadima, Ditshoswane Modise (now deceased), Zenzele Mpofu (now deceased), Lettie Sechele and many others helped as counterparts and field assistants in data collection and research. Abraham Modo (now deceased), then District Coordinator for Ngamiland, contributed by allowing his staff to participate in the project as field assistants.

The Norwegian Embassy in Gaborone provided valuable support to the project, and we are particularly indebted to Jan Arne Munkeby, who was the Norwegian Chargé d'Affaires at the time of project inception. Britt Hilde Kjølås, Embassy Secretary, also provided invaluable assistance during the initial phase of the project.

The chapters of this book were improved by the critical comments of independent expert reviewers and the following are thanked for their valuable contributions: George Batzli, Jane Carruthers, David H. M. Cumming, Hervé Fritz, Jacob R. Goheen, Ricardo Holdo, R. Norman Owen-Smith, Steward T.A. Pickett, Robert J. Scholes, Peter Scogings, Anthony R.E. Sinclair, Izak Smit, Marion Valeix and George Wittemyer. We sincerely thank the editing team at Wiley-Blackwell, particularly Ward Cooper, Kelvin Matthews and Carys Williams for encouraging and efficient collaboration as well as great patience. We further thank Ola Diserud and Andreas Brodén for contributions to Chapters 5 and 9, respectively; Marit Hjeljord for drawing many of the figures; and Lin Cassidy for drawing the map in Figure 1.3.

Finally, the entire text of this book was copyedited by Peter Frost who applied his writing skills, understanding of the English language, broad knowledge of African savannas and, above all, his meticulous professionalism, to substantially enhance the final product. For his valued contributions the editors owe Peter a large debt in gratitude.

References

Sinclair, A.R.E. & Northon-Griffiths, N. (1979) *Serengeti: Dynamics of an Ecosystem.* University of Chicago Press.

Scholes, R.J. & Walker, B.H. (1993) *An African Savanna: Synthesis of the Nylsvley Study.* Cambridge University Press.

Sinclair, A.R.E. & Arcese, P. (1995) *Serengeti II: Dynamics, Management, and Conservation of an Ecosystem.* University of Chicago Press.

du Toit, J.T., Rogers, K.H. & Biggs, H.G. (2003) *The Kruger Experience: Ecology and Management of Savanna Heterogeneity.* Island Press.

Sinclair, A.R.E., Packer, C., Mduma, S.A.R. & Fryxell, J.M. (Eds.) (2008) *Serengeti III: Human Impacts on Ecosystem Dynamics.* University of Chicago Press.

Part I
The Chobe Ecosystems

Elephants and Savanna Woodland Ecosystems: A Study from Chobe National Park, Botswana,
First Edition. Edited by Christina Skarpe, Johan T. du Toit and Stein R. Moe.
© 2014 John Wiley & Sons, Ltd. Published 2014 by John Wiley & Sons, Ltd.

Introduction

Christina Skarpe[1] and Stein R. Moe[2]

[1] Faculty of Applied Ecology and Agricultural Sciences, Hedmark University College, Norway
[2] Department of Ecology and Natural Resource Management, Norwegian University of Life Sciences, Norway

The basis for this book was laid by ivory hunters and cattle herders operating more than a 100 years ago south of the Chobe River in what today is northern Botswana. In the book we explore how the virtual extinction of the elephants, *Loxodonta africana*, by ivory hunters in the second half of the 19th century and the simultaneous reduction in many ungulate populations caused by rinderpest, a viral disease spread with cattle, initiated a dramatic history of ecosystem perturbations, many of which are still ongoing. The dynamics involved many aspects of the ecosystem, quantitative as well as qualitative, including plants, soil and animals varying in multiple spatial and temporal scales. To analyse this complex heterogeneity, we adopted the framework presented by Pickett *et al.* (2003). A framework or a model is a simplification and generalisation providing a context, in this case allowing us to identify origin, spatial and temporal scale and pattern of variation. The framework is hierarchical in nature, allowing shifting between scales to assess different degrees of detail.

For a pattern of heterogeneity to exist in time and space there must be a substrate exhibiting the heterogeneity and an agent creating the heterogeneity by its actions on the substrate. For the pattern to be functionally meaningful there must also be some organism or process that responds to the pattern of the substrate, and generally there are also one or more factors controlling the interactions between agent, substrate and responder (Pickett *et al.*, 2003). Therefore, to understand the function of ecological

Elephants and Savanna Woodland Ecosystems: A Study from Chobe National Park, Botswana,
First Edition. Edited by Christina Skarpe, Johan T. du Toit and Stein R. Moe.
© 2014 John Wiley & Sons, Ltd. Published 2014 by John Wiley & Sons, Ltd.

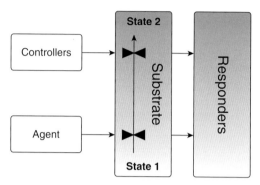

Figure 1.1   The conceptual model of the Chobe ecosystem dynamics. Agents, for example, elephants, create or maintain heterogeneity on specific scales by acting on a substrate, in our example, the vegetation. Substrates change or are maintained in a certain state by action of the agent (or the absence thereof); in our example the vegetation changed from woodland (state 1) to shrubland (state 2). Controllers affect the action of the agent on a substrate or the resultant transition (and reversibility) between states of a substrate, for example, impala preventing regeneration of the woodland once elephants have killed the large trees. Responders are the variables influenced by the change or state transition of the substrate, for example, other herbivores. Drawing by Marit Hjeljord. (Source: Adapted from Pickett *et al.*, 2003.)

heterogeneity in a certain ecosystem, we need to specify what substrate exhibits the variation, what agent works on the substrate to cause the variation, what organisms or processes respond to the variation in the substrate and what factors control the interactions (Pickett *et al.*, 2003; Figure 1.1). For example, on the alluvial soil along the Chobe River the increasing population of elephants, functioning as agents, have altered the vegetation, the substrate, by debarking and killing most of the large trees. The resulting tree-less state of the vegetation is being maintained in turn by seedling-eating impala, *Aepyceros melampus*, which are acting as controllers by preventing regeneration of the trees (Moe *et al.*, 2009; Chapter 10). The change in the state of the substrate has led to adjustments in population size and/or behaviour of a number of herbivore and carnivore species, which thus act as responders. So have the increasing openness of the vegetation resulting from the elephant and impala activities been a disadvantage for the thicket-preferring bushbuck, *Tragelaphus scriptus*, but might have favoured the puku, *Kobus vardonii*, preferring open plains (Dipotso and Skarpe, 2006; Dipotso *et al.*, 2007; Chapter 13).

Agents such as elephants can create, maintain or transform states of the substrate upon which they are acting. They are agents of change only as long as they themselves change, for example in density or behaviour. When the substrate has ceased exhibiting directed change, the role of the agent may be to maintain the acquired state (Skarpe,

1992; Pickett *et al.*, 2003). In the Chobe situation, the earliest mentioning of the alluvial flats above the flood plain, where the elephants now have killed the large trees, described a very open vegetation (Selous, 1881), which have been maintained in that state by the activities of large herbivores acting as agents and controllers (Chapters 4 and 12). A change in the agent and controller, such as the decline of elephants and ungulates following the ivory hunt and the rinderpest panzootic, respectively, initiated profound changes in the substrate by providing a window of opportunity for the establishment on the alluvial flats of the woodlands, which were later reduced with the recovery of the elephants constituting an agent of change (Chapters 4 and 10).

The heterogeneity framework described can be seen as a hierarchy of spatial and temporal scales, where small-scale pattern and processes may be contained within a such larger scale. Biotic and abiotic agents create heterogeneity in all ecosystems and at all temporal and spatial scales, from global climate governing continental-scale variation in plant and animal communities over decades, centuries and millennia to selective grazing by buffalo, *Syncerus caffer*, creating patchiness in grassland vegetation, termites modifying soil around their nest or twig-browsing greater kudu, *Tragelaphus strepsiceros*, causing differentiation of shoot growth within a tree canopy. Such differences in scale mean that an agent in one scale can be a responder or a controller at another, or sometimes have different roles at the same scale (Pickett *et al.*, 2003). In a local scale on the Chobe flood plains buffalo are agents, increasing heterogeneity in the grass sward by creating and maintaining grazing lawns. Responders to this change in the substrate are other grazing species such as impala and puku utilising the lawns. In another scale buffalos are responders facilitated by elephant-induced changes in the vegetation, and controllers influencing the spatial distribution of elephant grazing (Taolo, 2003; Chapter 11). Other examples of multiple roles are the few remaining live trees on the alluvial flats. These trees are the substrate for elephant activity, but are also agents themselves in a smaller scale, causing local changes in soil properties and microclimate to which many organisms, plants and animals, respond (Campbell *et al.*, 1994; Pickett *et al.*, 2003). The book is mainly structured according to a spatial and temporal scale to which many human activities such as conservation and management relate, where vegetation is the main responder and elephants the agent creating, modifying or maintaining heterogeneity.

References

Campbell, B.M., Frost, P., King, J.A., Mawanza, M. & Mhlanga, L. (1994) The influence of trees on soil fertility in two contrasting semi-arid soil types at Matopos, Zimbabwe. *Agroforestry Systems* 28, 159–172.
Dipotso, F.M. & Skarpe, C. (2006) Population status and distribution of puku in a changing riverfront habitat in northern Botswana. *South African Journal of Wildlife Research* 36, 89–97.

Dipotso, F.M., Skarpe, C., Kelaeditse, L. & Ramotadima, M. (2007) Chobe bushbuck in an elephant-impacted habitat along the Chobe River. *African Zoology* 42, 261–267.

Moe, S.R., Rutina, L.P., Hytteborn, H. & du Toit, J.T. (2009) What controls woodland regeneration after elephants have killed the big trees? *Journal of Applied Ecology* 46, 223–230.

Pickett, S.T.A., Cadenasso, M.L. & Benning, T.L. (2003) Biotic and abiotic variability as key determinants of savanna heterogeneity at multiple spatiotemporal scales. In: du Toit, J.T., Rogers, K.H. & Biggs, H.C. (eds.) *The Kruger Experience. Ecology and Management of Savanna Heterogeneity*. Island Press, Washington, DC, pp. 22–40.

Selous, F.C. (1881) *A Hunter's Wanderings in Africa*. Richard Bentley & Son, London, UK.

Skarpe, C. (1992) Dynamics of savanna ecosystems. *Journal of Vegetation Science* 3, 293–300.

Taolo, C.L. (2003) Population ecology, seasonal movement and habitat use of African buffalo (*Syncerus caffer*) in Chobe National Park, Botswana. PhD Thesis, Norwegian University of Science and Technology, Trondheim, Norway.

2

The Chobe Environment

Christina Skarpe[1] and Susan Ringrose[2]

[1] Faculty of Applied Ecology and Agricultural Sciences, Hedmark
University College, Norway
[2] PO Box HA 65 HAK, Maun, Botswana

Environmental factors set the conditions for living organisms and ecological processes in all spatial and temporal scales. At the largest scales continental drift has determined what genetic material is available for evolution, and is a reason for the largely different floras and faunas of different continents. Variation in geology and climate, topographic relief and hydrology creates environmental heterogeneity which promotes diversity of plant and animal communities and of ecosystems. If these environmental factors are seen as having bottom-up effects on species and communities, others such as fire, herbivory and human activities, for example, forestry, agriculture and livestock grazing, might be seen as having top-down effects. Which factors form the environment for ecological processes and which are interactive components of the ecosystem depends on the scale of observation, and for example fire and herbivory are important factors in savanna ecology, but constitute interactive parts of ecosystems in all but the smallest scales (Skarpe, 1992).

Geomorphology

The Chobe ecosystem is part of the dissected Southern African plateau, formed over time by intermittent uplift of the region following the fragmentation of Gondwana, some 180 million years ago. The Kalahari upland basin, which is inset into the plateau, is one of the largest inland sedimentary basins of Africa. During the Jurassic-Cretaceous periods it received considerable deposition (Karoo deposition) which now form the

Elephants and Savanna Woodland Ecosystems: A Study from Chobe National Park, Botswana,
First Edition. Edited by Christina Skarpe, Johan T. du Toit and Stein R. Moe.
© 2014 John Wiley & Sons, Ltd. Published 2014 by John Wiley & Sons, Ltd.

host sediment for the Kalahari sands. The sands which are generally up to 250 m thick, underlie most of Botswana including the Chobe area. The Kalahari sands however thin out over the Chobe area where basalts are exposed in the uplands south of the river. Early drainage dissected the original Southern African plateau and included the proto-Zambezi, Kwando and Okavango Rivers (Moore and Larkin, 2001).

The Chobe area has developed as a result of palaeoenvironmental shifts which have influenced the courses of the original Okavango, Kwando and Zambezi rivers. The three initial proto-rivers were truncated by epeirogenic flexuring (minor uplift) along the Ovambo-Kalahari-Zimbabwe axis, which includes the area to the south of the Makgadikgadi Pans (Moore et al., 2009). This caused the rivers to drain into an early extensive Makgadikgadi-Okavango-Zambezi depression, producing a large palaeolake which may have covered most of northern Botswana and the adjacent Caprivi about 60 million years ago (McCarthy and Rubidge, 2005). This palaeolake became smaller over time as the East African Rift system began to extend south-westward, leading to a fault controlled drainage diversions which formed the Zambezi and Kwando river courses (Modisi et al., 2000). For instance, the original N-S draining Kwando river was diverted north-eastwards and now drains into and beyond a mini-delta on Kalahari sediments abutting the still active NE–SW trending Linyanti fault. Although some water continues to drain through the Linyanti River and on eastwards as the Itenge River, in most years there is barely enough flow to reach the Chobe area. The Chobe River is flooded mainly by back-flow from the Zambezi River during the annual floods. Nevertheless, the steepness of the river banks in the Chobe National Park and the abundance of calcrete in the river cliffs all testify to earlier, more vigorous stream development and the effect of continuous uplift of the entire Southern African plateau.

During the last 400,000 years, embracing the later Pleistocene and Holocene periods, the climate of the southern hemisphere has shifted as documented by the Vostok ice core in Antarctica (Petit et al., 1999) with extensive cold and dry glacial periods being interspersed by warm, wet interglacials. These alternations of cold-dry and warm-wet intervals have strongly influenced palaeo-climatic change hence landform development, throughout northern Botswana (e.g. Partridge et al., 1999; Thomas and Shaw, 2002; Ringrose et al., 2005; Huntsman-Mapila et al., 2006). Within Chobe National Park, evidence depicting cooler-dry intervals includes extensive systems of fossil sand dunes which are still visible over much of the centre and east of the Park. Evidence of warm-wet periods with higher water levels includes the remnants of several strandlines which stand at least 20 m above the present Mababe depression (Figure 2.1). The Mababe depression is still linked with overflow systems from the Kwando and Okavango, to this day. The higher strandlines indicate former (Late Pleistocene) increased water inflow attributable in part to the past expanded flow down the Zambezi, Okavango and Kwando rivers (Burrough and Thomas, 2008; Cruse et al., 2009). This later diminished due to palaeoclimatic change and due to continued

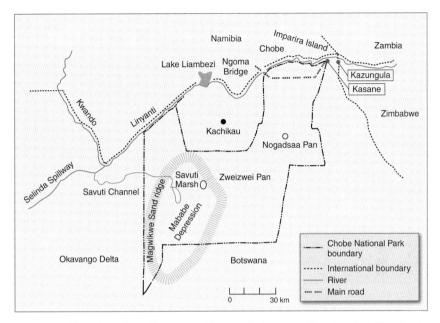

Figure 2.1   **Map of northern Botswana with some features mentioned in text. Drawing by Marit Hjeljord.**

tectonic shifts due through the later south-west propagation of the East African Rift system. This fault controlled tectonic activity is prevalent to this day leading to generally low magnitude earthquakes mostly south of the Kwando-Linyanti area.

Apart from the low undulating topography of fossil dunes and sand ridges, much of the Chobe National Park area is flat with elevations from 1120 m above mean sea level in the north-east and dropping to 920 m in west towards the Mababe depression and the Chobe river (Figure 2.1). Evidence of Holocene or earlier drainage likely lies with the numerous short south-bank tributaries of the Chobe River, which now form dry valleys (e.g. Kalwizikalkanga; Figure 2.2). The Mababe depression, also received relatively recent palaeodrainage from the NE. These drainage lines are now characterised by small pans and dry valleys, for example, the Ngwezumba and Gautumbi valleys and the Nogadsaa and Zweizwe pan areas in Chobe National Park (Figure 2.1; Thomas and Shaw, 1991). Within more recent times, as in the past, cyclic change and the inherent variability of the river related systems are the norm. Recent (2011–2012) high flood levels have been experienced in all the northern rivers (Zambezi, Kwando, Okavango). This has influenced recent inflow events in the Mababe depression, including the Savuti Marsh and its channel, which had been dry for about 30 years. The Savuti channel

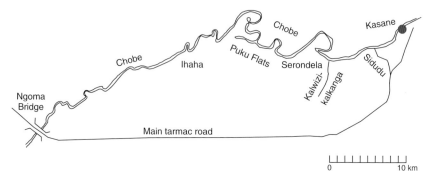

Figure 2.2   **Map of the core investigated area in northern Chobe National Park. Drawing by Marit Hjeljord.**

was infilled via the Linyanti swamps and the Okavango system through the Selinda spillway. Southern Okavango drainage also overflowed into the southern part of the Mababe depression. While to a much lesser extent than the flooding which took place in the Pleistocene-Holocene, these recent flood events are a reflection of the former expansive drainage networks throughout the Chobe area.

The water level fluctuations in the Chobe River depend on contributions from different sources, primarily Zambezi and Kwando Rivers, and is out of phase with the local rainy season (October–April). Consequently, there are four distinct seasons in the floodplains along the river: (i) a low water rainy season from October to March; (ii) a high water (floodplains inundated) rainy season from March to April; (iii) a high water dry season from April to June and (iv) low water dry season from June to October. The Chobe River is the only (natural) permanent water in the region, except during periods when the Savuti Marsh, about 130 km to the southeast, contains water. This makes it a vital resource for water-dependant fauna in the ecosystem. The Chobe River gives its name to the northernmost district in Botswana and to the Chobe National Park, which forms much of its southern bank.

Soils

The soils in Chobe National Park are mostly deep to very deep, well to excessively drained arenosols, developed from the Kalahari sands (Figure 2.3; de Wit and Nachtergaele, 1990). The sand consists of quartz with minor feldspars and mica. In the dunes the grains are coated with iron oxide, colouring the sand brown to deep red, while in seasonally wet, reductive environments the soil is grey to almost white (Leistner, 1967; de Wit and Nachtergaele, 1990). The bottoms of the former lake basins, the Mababe depression and the surroundings of Nogadsaa and Zweizwe pans are characterised by calcareous, fine-textured and compact alluvial luvisols and gleysols of lacustrine and riverine origin (Blair Rains and McKay, 1968; de Wit and Nachtergaele, 1990; Chapter 9). These areas have many pans, shallow, poorly drained depressions that hold water for varying periods after rain. Riverine alluvial soils also make up the recent and uplifted fossil floodplains along the Chobe River. Unlike Kalahari sand, which is poor in minerals and plant nutrients, these alluvial soils are calcareous and moderately fertile. The sands have a cation exchange capacity (CEC) of around $2\,cmol\,kg^{-1}$, compared with about $6\,cmol\,kg^{-1}$ in fossil alluvial soils (luvisols) along the Chobe River and more than $30\,cmol\,kg^{-1}$ in the sodic recent alluvium (gleysols) on the floodplains (Aarrestad *et al.*, 2011; Chapter 9).

The material in the eroding vegetated sand dunes and ridges is partly sorted by wind and water action, which transport finer particles such as silt and clay down towards the bottom of the slope. Nutrients may drain the same way, although lateral movement of water in the sand is probably limited. Such catenas are a characteristic feature of savanna landscapes. The long sand ridge south of the Chobe River rises about 100 m above the river surface, forming an interrupted catena 0.4–2.5 km wide (Simpson, 1974), ending with an abrupt border between the sand and the alluvial deposits of the fossil floodplain.

Climate

The climate in the Chobe region is primarily controlled by the subtropical high pressure belt that dominates the atmospheric circulation of southern Africa. During winter this high pressure belt excludes moist air, creating a stable atmosphere and dry, cold weather. In summer the high pressure belt moves southwards, with moist tropical air from the Inter-Tropical Convergence Zone as well as tropical cyclones formed over the Indian Ocean entering the region. This creates atmospheric instability, warm weather and rain. This is the typical savanna climate with cool dry winters and moist hot summers (Tyson and Crimp, 1998).

The climate of Chobe National Park is semi-arid to sub-humid with an average annual rainfall of about 550 mm in the south-west and almost 700 mm in the north-east (Bhalotra, 1987). Rainfall is highly variable in space and time both within and between years, and the interannual coefficient of variation is about 30%, increasing with decreasing average. More than 90% of annual precipitation falls from October to the end of April, with December and January being the wettest months. Potential evapotranspiration is almost twice of the annual rainfall leading to an annual water deficit of 400–600 mm (Schulze and McGee, 1978). Under present climatic conditions, rain water penetration is thought to be limited to the capillary zone of about 6 m (Boocock and van Straten, 1962; Foster *et al.*, 1982), so that lateral movement of water

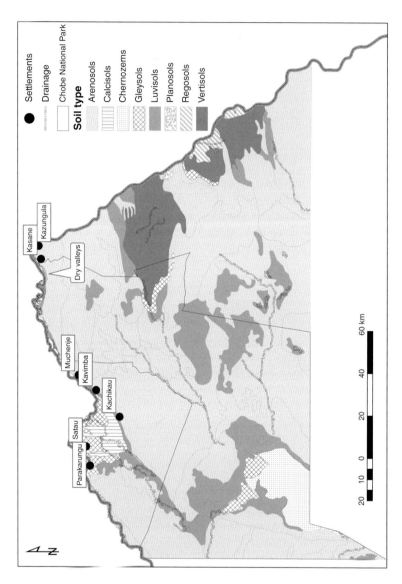

Figure 2.3 Distribution of soils in the Chobe area (UNDP-FAO classification system). Drawing by Dr Lin Cassidy.

is limited to areas with shallow sand. Nevertheless, based on isotope studies, Mazor (1982) suggests temporally and spatially irregular deeper drainage of water, leading to periodic recharge of ground water.

Mean annual temperature in Chobe National Park is about 22 °C and mean daily maximum temperature in October (warmest month) about 36 °C and mean daily minimum in July (coldest month) about 7 °C. Periods with frost are not uncommon (Chafota and Owen-Smith, 2009). The sparse cloud cover means that the Chobe area receives almost 70% of potential sun shine hours over the year (Bhalotra, 1987).

Flora and vegetation in the Chobe savanna

The savanna biome, to which the Chobe savannas and woodlands belong, covers about 40% of Africa and 20% of the world's land area (Scholes and Walker, 1993). Savannas are characterised by an often grass dominated field layer vegetation and a more or less prominent woody component. Although savanna-like ecosystems may have appeared earlier (Prasad *et al.*, 2005), savannas as we know them today have existed in Africa for only about 2.5 million years. From about this time grasses with C_4 photosynthesis are known to have spread in Africa, coincident with rapid speciation among ungulates, including the appearance of the first specialised grazers (Janis, 2008), and a mammalian fauna resembling that of present day African savannas. These savannas today constitute three separate but contiguous regional centres of plant endemism (White, 1983). The Sudanian region lies between the equatorial forests and the Tropic of Cancer, adjoining the Somali-Masai region in the east. This in turn extends from north-east Africa, through East Africa to the Central African plateau, where it is replaced by the Zambezian regional centre, which covers much of the plateau between the equatorial forests and the Tropic of Capricorn. Northern Botswana forms part of this Zambezian centre of endemism, one characterised by many widespread species of grass and trees, among the latter of which members of the family *Fabaceae* are prominent (Werger, 1978; White, 1983).

Across the Kalahari, floristic and functional aspects of the vegetation change along the gradient of increasing rainfall from south-west to north-east (Skarpe, 1986; Scholes *et al.*, 2002). Above-ground potential net primary production of herbaceous vegetation in savanna increases linearly with rainfall up to about 700 mm annual precipitation (Walter, 1970; Rutherford, 1980), although the slope of the relationship varies with soil nutrient conditions (Olff *et al.*, 2002).

With increasing rainfall the importance of soil nutrient resources for vegetation composition and primary production increases. In the Chobe ecosystem, soil types control much of the vegetation structure, directly by bottom-up effects on plants and indirectly by influencing herbivores acting as agents and controllers that mainly exert a top-down effect on vegetation. (Chapters 1, 4, 5 and 9: floristic nomenclature in this chapter follows that in Chapter 5.) The vegetation pattern and dynamics within the approximately 400 km² core study area along Chobe River are described in Chapters 4 and 5, and are mentioned here only briefly, together with some notes on the vegetation south of the core study area. Some characteristics of common woody species are shown in Table 2.1. Along the Chobe River the seasonally inundated floodplain forms an open grassland dominated by *Cynodon dactylon* on fine sodic alluvium (Plate 1). The raised fossil alluvium adjacent to the river is occupied by *Capparis tomentosa – Flueggea virosa* shrubland with a mixture of low forbs and grasses in the field layer (Plate 2). Further from the river, in the narrow ecocline between the alluvial soils and Kalahari sand, is dense woody vegetation dominated by *Combretum mossambicense – Friesodielsia obovata* (Chapter 5; Plate 3). On deep Kalahari sand, *Baikiaea plurijuga – Combretum apiculatum* woodlands occur, often with *Baikiaea plurijuga* as the sole dominant but also with small trees, for example, *Combretum apiculatum*, and shrubs such as *Baphia massaiensis* and *Combretum* spp. The herbaceous layer is dominated by tall forbs and tufted grasses, including *Brachiaria nigropedata*, *Panicum maximum*, *Urochloa* sp. and *Schmidtia pappophoroides* (Chapter 5; Plate 4). South of our main study area are woodlands with *Terminalia sericea*, *Philenoptera* (previous *Lonchocarpus*) *nelsii* and *Colophospermum mopane* with a shrub layer that includes *Combretum apiculatum* and *Grewia* spp. On fine, compacted sand *Colophospermum mopane* may form almost pure stands with a sparse field layer of mainly grasses (Werger and Coetzee, 1978). The recent lacustrine deposits in Savuti have been dry and covered by grassland with scattered clumps of trees or shrubs since the 1980s, but now, once again, forms a wetland, the Savuti Marsh, following floodwaters flowing through from the Linyanti River in 2010.

The mammal community

The biogeography of the Sub-Saharan African mammal fauna resembles the floristic one. Most authors recognise a northern and a southern savanna zone joined by the East African high plateau, and often identify a somewhat variably defined south western arid zone, including a smaller or larger section of the Kalahari (Davis, 1962; Bigalke, 1968). The African continent has about one quarter of all mammal species in the world, but about the same number as the ecologically comparable South America; about 840 species in Sub-Saharan Africa and 810 in South America (Bigalke, 1968, 1978). What makes the African mammal fauna unique is the remarkable richness of large-bodied species, primarily ungulates. Africa houses about 93 wild ungulate species, which is almost twice that of Asia, which is the second richest region. The reason for the extensive speciation, particularly in African Bovidae, during the last about 2.5 million years is debated, but exploiting the 'new' grazing niche may have contributed to this diversification, more than offsetting the contemporaneous decline in browsing species (Cerling *et al.*, 1998; Janis, 2008). Also repeated isolation and

Table 2.1 Traits of common woody species in the study area in Chobe National Park. The table is an extract from the Botswana Tree Data Base compiled by Roger Bergström, Christina Skarpe and Kjell Danell (unpublished) and based on a wide range of literature. All information is for the species generally and is not based solely on data from Chobe National Park. Growth form is more often based on architecture (shrubs are typically multistemmed, trees single-stemmed) than on height. Types of spines are given as prickles (≤1 cm; often hooked) and spines (>1 cm; generally straight). Leaf fall patterns are deciduous (D; i.e. leaves last <1 year) and evergreen (E; i.e. leaves last >1 year). Growth rates are generally subjectively stated, and are likely to be relative.

Species	Family	Growth form	Height (m)	Spinescence	Type of spines	Leaf fall pattern	Leaf type	Leaf length (cm)	Leaf width (cm)	Growth rate	As herbivore food
Acacia fleckii	Mimosoideae	Shrub/tree	10	1	Prickles	D	Compound	4	–	–	–
Acacia nigrescens	Mimosoideae	Tree	16	1	Prickles	D	Compound	–	–	Slow	Used by a variety of browsers, such as elephant and giraffe
Acacia tortilis	Mimosoideae	Tree	11	1	Spines and prickles	D	Compound	3	–	Slow	Leaves and pods valuable fodder
Albizia harveyi	Mimosoideae	Tree	9.5	0	–	D	Compound	15	–	Slow	Leaves utilised by browsers; elephants eat all parts of the tree
Baikiaea plurijuga	Caesalpinioideae	Tree	12	0	–	D	Compound	–	–	–	–
Baphia massaiensis	Papilionideae	Shrub	3	0	–	D	Simple	5.5	4	–	–

(continued overleaf)

Table 2.1 (Continued)

Species	Family	Growth form	Height (m)	Spinescence	Type of spines	Leaf fall pattern	Leaf type	Leaf length (cm)	Leaf width (cm)	Growth rate	As herbivore food
Bauhinia petersiana	Caesalpinioideae	Shrub/tree	3.5	0	–	D	Lobed	4.8	5.5	–	–
Berchemia discolor	Rhamnaceae	Tree	13.5	0	–	D	Simple	5.5	2.5	Slow	Fruits and leaves browsed by several species
Boscia albitrunca	Capparaceae	Tree	7	0	–	E	Simple	4.5	1.2	–	Leaves nourishing forage for game and stock
Canthium glaucum*	Rubiaceae	Shrub/tree	5	1	Spines	D	Simple	3	1.7	–	–
Capparis tomentosa	Capparaceae	Tree	10	1	Spines	E	Simple	5.8	2	–	Leaves palatable for cattle; reported to be poisonous
Colophospermum mopane	Caesalpinioideae	Tree	10	0	–	D	Compound	8	3.5	Slow	Important fodder for many animals; ranks high in diet of elephant
Combretum apiculatum	Combretaceae	Tree	6.5	0	–	D	Simple	6.5	3.5	Slow	–
Combretum celastroides	Combretaceae	Shrub/tree	5.5	0	–	D	Simple	8	4.5	–	–

Species	Family	Form									Notes
Combretum elaeag-noides	Combretaceae	Tree	5	0	–	D	Simple	9	2.4	Fast	–
Combretum engleri	Combretaceae	Shrub/tree	3.5	0	–	D	Simple	3	1.8	–	–
Combretum hereroense	Combretaceae	Tree	4	0	–	D	Simple	3	2	Slow	The leaves browsed by cattle kudu, steenbok, elephant and giraffe
Combretum molle	Combretaceae	Tree	10	0	–	D	Simple	8	5	Fast	–
Combretum mossambi-cense	Combretaceae	Tree	6	1	Spines	D	Simple	6	3	–	High nutritional value for all browsers
Croton gratissimus	Euphorbiaceae	Shrub	6	0	–	D	Simple	7.5	3	–	Sporadically browsed by elephant and kudu
Croton mega-lobotrys	Euphorbiaceae	Tree	15	0	–	D	Simple	10.3	7.2	Fast	Elephants use the tree extensively; leaves eaten by browsers
Dichrostachys cinerea	Mimosoideae	Shrub	5.5	1	Spines	D	Compound	5	–	Slow	Leaves and pods eaten by stock and game
Diplorhynchos condylcar-pon	Apocynaceae	Shrub/tree	7	0	–	D	Simple	5	3	Fast	Elephants very fond of the species
Erythrophleum africanum	Caesalpinioideae	Tree	8	0	–	–	Compound	–	–	–	Leaves and roots poisonous

(*continued overleaf*)

Table 2.1 (*Continued*)

Species	Family	Growth form	Height (m)	Spinescence	Type of spines	Leaf fall pattern	Leaf type	Leaf length (cm)	Leaf width (cm)	Growth rate	As herbivore food
Erythroxylum zambesiacum	Erythroxylaceae	Shrub/tree	5	0	–	D	Simple	4.3	2.7	–	–
Faidherbia albida	Mimosoideae	Tree	30	1	Spines	D	Compound	13	–	Fast	Eaten by a variety of wild animals
Flueggia virosa[†]	Euphorbiaceae	Shrub	2.5	0	–	D	Simple	2.5	1.5	–	–
Friesodielsia obovata	Annonaceae	Shrub/tree	7	0	–	E	Simple	8	5	–	–
Garcinia livingstonei	Clusiaceae	Tree	6	0	–	D	Simple	8	4	Slow	Elephants eat root and bark, giraffe the leaves
Guibourtia coleosperma	Caesalpinioideae	Tree	13	0	–	E	Compound	6.5	3	–	–
Markhamia zanzibarica[‡]	Bignoniaceae	Tree	7	0	–	D	Compound	–	–	Slow	Leaves apparently only used by elephants
Markhamia obtusifolia	Bignoniaceae	Shrub	1.5	0	–	–	Compound	–	–	–	–
Ochna pulchra	Ochnaceae	Tree	5	0	–	D	Simple	7.5	2.7	Slow	–

Species	Family	Habit									Notes
Philenoptera nelsii§	Papilionideae	Tree	4	0	–	D	Simple	12	5.5	–	Leaves are excellent fodder
*Philenoptera violacea*Ω	Papilionideae	Tree	10	0	–	D	Compound	18	9	Slow	Leaves heavily browsed by game and stock
Pterocarpus angolensis	Papilionideae	Tree	16	0	–	D	Compound	30	–	Slow	Leaves eaten by kudu; elephants push over and use the trees
Strychnos madagascariensis	Loganiaceae	Tree	6	0	–	D	Simple	6	3.5	Fast/slow	–
Strychnos potatorum	Loganiaceae	Tree	10	0	–	D	Simple	10.5	6	Fast	Elephants uproot trees to get roots
Trichilia emetica	Meliaceae	Tree	14	0	–	E	Compound	50	–	Fast	–
Vangueria infausta	Rubiaceae	Tree	5	0	–	D	Simple	14.5	9.4	Slow	Leaves eaten by game
Vitex mombassae	Verbenaceae	Shrub	2.5	0	–	D	Compound	17	–	–	–
Xeroderris stuhlmannii	Papilionideae	Tree	10	0	–	D	Compound	37	–	Slow	Leaves seldom eaten by browsers

*Canthium frangula.
†Securinega virosa.
‡Markhamia acuminata.
§Lonchocarpus nelsii.
ΩLonchocarpus capassa.

merging of populations of small and medium sized habitat specialist ungulates, as a result of climatic fluctuations, may have resulted in the evolution of new species (Vrba, 1992; du Toit, 2003). Altogether 165 species of mammals were described from Botswana by Smithers (1971), including 36 species of bats, 33 species of rodents and 31 of ungulates. Smithers (1971) recognised a faunal boundary running roughly east-west through the Kalahari in level with the northern parts of the Makgadikgadi pans, that is, just south of Chobe National Park, representing the northern limit for some typical Kalahari species such as springbok, *Antidorcas marsupialis*, (nomenclature follows Skinner and Chimimba, 2005) and red hartebeest, *Alcelaphus buselaphus*, and the southern boundary for some northern species, for example, sable, *Hippotragus niger* and roan antelope, *Hippotragus equinus*. Chobe National Park has about 54 species of wild mammals weighing more than 2 kg (Table 2.2) and many smaller species (Smithers, 1971, Skinner and Chimimba, 2005).

The elephant *Loxodonta africana* is the only large herbivore using all of Chobe National Park (Plate 5). Other species have more restricted ranges, and the small area of alluvial soils has by far the highest density of species, individual animals and biomass (including elephants: Chapter 13). Floodplain specialists include lechwe, *Kobus leche*, and puku, *Kobus vardonii*, although puku also uses other open vegetation close to water (Chapter 13; Plate 6). Impala, *Aepyceros melampus* and bushbuck, *Tragelaphus scriptus*, are largely confined to the vegetation on alluvial soils whereas, for example, greater kudu, *Tragelaphus strepsiceros*, and giraffe, *Giraffa camelopardalis*, use habitats both on the alluvium and on Kalahari sand. Sable and roan antelope are mainly found further from water, often on Kalahari sand (Chapter 13). Large predators select habitat where catchability of prey is highest or where risk of intraguild predation from other carnivores is low. In the main study area, lion prides are distributed along the Chobe River in the shrublands on alluvial soil where prey density is high (Chapter 13), whereas hyena, *Crocuta crocuta*, cheetah, *Acinonyx jubatus* and wild dog, *Lycaon pictus* seem to select habitats largely to minimise risk of being killed by lions (Chapter 15).

Human impact

The oldest sign of human presence in Chobe National Park are some Middle and Late Stone Age tools and rock paintings in the hills around Savuti (Robbins, 1987; Figure 2.1). However, some 100 km further west, at Tsodilo hills, Robbins and Murphy (2000) have found some of the oldest evidence of human presence in Botswana, dating back at least 100,000 years. The area along the Chobe River is likely to have had a similar long history of human occupancy. Until less than 2000 years ago the only people in southern Africa were San hunter-fisher-gatherers and Khoe pastoralists. Both groups lived by hunting, fishing and gathering, with Khoe engaged also in livestock husbandry; neither group tilled the land nor planted crops (Tlou and Campbell, 1984).

Table 2.2 Traits of mammal species more than 2 kg known by the authors to occur or have recently occurred (e.g. black and white rhinoceros, *Diceros bicornis* and *Ceratotherium simum*, respectively) in Chobe National Park. Information is taken from Skinner and Chimimba (2005) unless stated otherwise. Values for body mass are for mature animals, and therefore are higher than average for whole populations, which include young animals. Preferential grazers and preferential browsers are described as grazers and browsers, respectively. In some species, described as gregarious, old males tend to be solitary.

Scientific name	English name	Mass (kg)		Feeding type	Gregariousness	Diurnal/nocturnal
		Male	Female			
Orycteropodidae						
Orycteropus afer	Aardvark	53.3	51.4	Ants, termites	Solitary	Nocturnal
Elephantidae						
Loxodonta africana	African savanna elephant	5750*	2650*	Grazer/browser	Breeding herds	Diurnal/nocturnal
Leporidae						
Lepus saxatilis	Scrub hare	2.0	2.4	Grazer	Solitary	Nocturnal
Hystricidae						
Hystrix africaeaustralis	Cape porcupine	11.7	12.6	Bark, tubers	Solitary	Nocturnal
Thryonomyidae						
Thryonomys swinderianus	Greater cane rat	4.5	3.6	Grazer	Solitary	Nocturnal
Pedetidae						
Pedetes capensis	Springhare	3.1	2.8	Grazer	Colonies	Nocturnal
Cercopithecidae						
Papio hamadryas	Chacma baboon	28.8	13.9	Omnivore	Troops 10–100	Diurnal
Cercopithecus pygerythrus	Vervet monkey	5.5	4.1	Fruit, leaves, seed	Troops 10–35	Diurnal
Manidae						
Manis temminckii	Ground pangolin	13.3	7.4	Ants, termites	Solitary	Nocturnal

(continued overleaf)

Table 2.2 (*Continued*)

Scientific name	English name	Mass (kg) Male	Mass (kg) Female	Feeding type	Gregariousness	Diurnal/nocturnal
Hyaenidae						
Proteles cristatus	Aardwolf	8.9	8.7	Termites	Solitary	Nocturnal
Parahyaena brunnea	Brown hyaena	40.2	38.4	Scavenger, predator	Solitary/clans < ca. 10	Nocturnal
Crocuta crocuta	Spotted hyaena	57.8	64.8	Predator, scavenger	Clans 4–18	Nocturnal
Felidae						
Acinonyx jubatus	Cheetah	53.9	43.0	Predator	Solitary/family groups	Diurnal
Panthera pardus	Leopard	59.7	31.5	Predator	Solitary	Nocturnal
Panthera leo	Lion	190.0	126.0	Predator	Prides 2–10; up to ca. 30	Nocturnal/diurnal
Caracal caracal	Caracal	13.8	11.9	Predator	Solitary	Nocturnal
Felis silvestris	African wild cat	5.1	4.2	Predator	Solitary	Nocturnal
Leptailurus serval	Serval	11.1	9.7	Predator	Solitary	Nocturnal
Ververridae						
Civettictis civetta	African civet	10.9	11.6	Predator	Solitary	Nocturnal
Herpestidae						
Herpestes ichneumon	Large grey mongoose	3.4	3.1	Predator	Solitary/family groups	Nocturnal/diurnal
Ichneumia albicauda	White-tailed mongoose	4.5	4.1	Predator	Solitary/family groups	Nocturnal
Atilax paludinosus	Marsh mongoose	3.4	3.4	Predator	Solitary	Nocturnal
Canidae						
Otocyon megalotis	Bat-eared fox	4.0	4.1	Insectivorous	Family groups	Nocturnal/diurnal
Lycaon pictus	African wild dog	28.0	24.0	Predator	Packs 2–40	Diurnal/(nocturnal)
Vulpes chama	Cape fox	3.0	2.9	Predator, fruit	Solitary/pairs	Nocturnal

Canis adustus	Side-striped jackal	9.4	8.3	Omnivore	Solitary/family groups	Nocturnal
Canis mesomelas	Black-backed jackal	7.9	6.6	Omnivore	Solitary/pairs	Nocturnal/diurnal
Mustelidae						
Aonyx capensis	African clawless otter	12.3	14.3	Piscivore	Solitary/family groups	Nocturnal/diurnal
Lutra maculicollis	Spotted-necked otter	5.7	4.3	Piscivore	Family groups	Diurnal
Mellivora capensis	Honey badger	11.7	11.5	Predator/omnivore	Solitary	Nocturnal
Rhinocerotidae						
Ceratotherium simum	White rhinoceros	2200	1600	Grazer	Solitary/family groups	Diurnal/nocturnal
Diceros bicornis	Black rhinoceros	852	884	Browser	Solitary	Diurnal/nocturnal
Equidae						
Equus quagga	Plains zebra	313	302	Grazer	Large herds	Diurnal
Suidae						
Potamochoerus larvatus	Bushpig	72.3	68.9	Omnivore	Groups of 4–6	Nocturnal
Phacochoerus africanus	Common warthog	79.6	56.5	Grazer	Family groups	Diurnal
Hippopotamidae						
Hippopotamus amphibius	Hippopotamus	1546	1385	Grazer	Herds 2–20	Nocturnal
Giraffidae						
Giraffa camelopardalis	Giraffe	1192	828	Browser	Loose herds	Diurnal
Bovidae						
Syncerus caffer	African buffalo	631	435	Grazer	Large herds	Diurnal/nocturnal
Tragelaphus strepsiceros	Greater kudu	228	157	Browser	Groups of 6–14	Diurnal/nocturnal
Tragelaphus scriptus	Bushbuck	60†	42†	Browser	Solitary	Nocturnal

(*continued overleaf*)

Table 2.2 (*Continued*)

Scientific name	English name	Mass (kg)		Feeding type	Gregariousness	Diurnal/nocturnal
		Male	Female			
Tragelaphus spekii	Sitatunga	98[†]	54[†]	Grazer	Groups of ca. 6	Diurnal/nocturnal
Tragelaphus oryx	Eland	696[†]	394[†]	Browser	Herds	Diurnal
Connochaetes taurinus	Blue wildebeest	252	215	Grazer	Large herds	Diurnal
Damaliscus lunatus	Tsessebe	130[†]	108[†]	Grazer	Small herds	Diurnal/nocturnal[†]
Hippotragus equinus	Roan	280[†]	260[†]	Grazer	Small herds	Diurnal/nocturnal[†]
Hippotragus niger	Sable	235[†]	220[†]	Grazer	Small herds	Diurnal/nocturnal[†]
Oryx gazella	Gemsbok	240	210	Grazer	Small herds	Diurnal/nocturnal[†]
Sylvicapra grimmia	Common duiker	18.7	20.7	Browser	Solitary	Diurnal/nocturnal
Redunca arundinum	Southern reedbuck	51.8	38.2	Grazer	Solitary/family groups	Diurnal/nocturnal
Kobus ellipsiprymnus	Waterbuck	236[†]	186[†]	Grazer	Small herds	Diurnal/nocturnal[†]
Kobus leche	Lechwe	118	74	Grazer	Small herds	Diurnal/nocturnal[†]
Kobus vardonii	Puku	77.3	61.3	Grazer	Small herds	Diurnal
Ourebia ourebi	Oribi	14.0	14.2	Grazer	Solitary/family groups	Diurnal[†]
Raphicerus campestris	Steenbok	10.9	11.3	Browser	Solitary	Diurnal
Aepyceros melampus	Impala	54.5	40.9	Grazer/browser	Herds	Diurnal

*From Owen-Smith (1988).
[†]From Estes (1991).

Bantu-speaking Iron Age farmers, growing crops and herding livestock, first settled in the northern part of Botswana less than 2000 years ago. Excavations near Serondela in the Chobe National Park show that farmers lived there between about AD 600 and AD 750 (Tlou and Campbell, 1984; Figure 2.2). From about AD 1000, stronger organised societies and kingdoms developed in the wider region. Trade expanded, including the increasing export of ivory, fur, cattle and other products from northern Botswana (Tlou and Campbell, 1984). This trade, with shifting routes, has continued into modern times.

Thus, many groups of people have lived in what is now Chobe National Park: fishing, hunting, gathering, farming and breeding livestock. Land use has been in a dynamic interaction with tsetse fly (*Glossina* spp.), which spread the disease *nagana* in cattle and sleeping sickness in humans, strongly influencing where people settled and their livelihood activities (Tlou and Campbell, 1984). Towards the end of the 19th century, when European travellers brought back written accounts from the region, human settlements in what is now the Chobe National Park were scattered, mainly along the Chobe and Linyanti rivers (Livingstone, 1857; Selous, 1881; Figure 2.1). In the second half of the 19th century, hunting elephants for ivory intensified, resulting in elephants becoming almost extinct from the Chobe region by the end of the century (Chapter 6). At the same time, rinderpest, a disease brought to Africa with livestock and reaching Botswana in 1896, killed a large proportion of wild and domestic even-toed ungulates. This left the area with much reduced herbivore populations, which had profound ecological consequences (Prins and van der Jeugd, 1993; Sinclair *et al.*, 2007; Holdo *et al.*, 2009; Spinage, 2012; Chapters 4 and 6). From the beginning of the 20th century, human and livestock populations both increased in the area (Child, 1968). The main settlements were around Kachikau and in Mababe, with seasonal and permanent cattle posts developed along the Chobe and Linyanti (Figure 2.1) rivers and around pans in the interior. In addition, cattle being exported from Botswana to Zambia were driven through the present-day Chobe National Park, and were in some cases fattened there before being swum over the river to markets. The absence of the large wild browsing animals, most notably elephant, provided a window of opportunity for woody species to establish on the previously tree-less raised alluvium (Chapter 4). This in turn favoured tsetse fly, the spread of which resulted in gradual abandonment of the cattle posts from the 1940s onwards (Child, 1968).

Timber was extracted from the Chobe area, mainly east of Chobe National Park, between 1936 and 1941, and south of Serondela and Ihaha (Figure 2.2) in what is now Chobe National Park between 1944 and 1955–1956. The main timber species were *Pterocarpus angolensis* and *Baikiaea plurijuga* (Chapter 4). A railway and a saw mill were established in Serondela, together with a cattle post, and crops were periodically intensively cultivated (Child, 1968). Large areas were burnt more or less regularly, in connection both with timber exploitation and with livestock rearing. Timber extraction from what is now Chobe National Park finally ended shortly after the Chobe Game

Reserve was established in 1962. A no-burning policy was adopted at the same time. In 1968, the reserve was extended and re-designated as the Chobe National Park.

Closing remarks

Today, after a century and a half with turbulent ecosystem dynamics, fluctuating wildlife populations, exploitation of ivory, grazing by livestock and timber extraction, the Chobe economy again depends on wild animals, as it has done for thousands of years previously, but now exploited for tourism rather than for meat and hides. Another difference from the past is that wild animals and their environments are increasingly controlled by humans. Wildlife based tourism is an important and rapidly growing revenue earner in many African countries, including Botswana, and this generates an incentive, at least among those directly or indirectly favoured by this business, to protect wildlife areas such as the Chobe National Park from for example encroachment of other land uses and from poaching of animals. However, there are new anthropogenic threats to protected areas and their communities of plants and animals including rapid climatic change and globally increasing levels of atmospheric carbon dioxide. These are basic environmental factors, not interacting within natural ecosystems in any scale of interest for management or research. It is anticipated that the increase in carbon dioxide will induce considerable changes in savanna vegetation structure and composition as well as in nutrient fluxes and herbivore community composition (Gordon and Prins, 2008). In this chapter, we have seen that fluctuating climate and habitat fragmentation in the past could have contributed to the high species richness of ungulates in African savanna through alternating isolation and reuniting of populations. Today, areas set aside for wildlife generally form poorly connected patches in a matrix of landscapes under human landuse, limiting the potential to maintain ecologic resilience by mobility. In an effort to increase connectivity and promote conservation and economic development of shared wildlife resources, five southern African countries have jointly established the Kavango-Zambezi Transfrontier Conservation Area, one of the world's largest transfrontier conservation areas, linking conservation areas in Angola, Botswana, Namibia, Zambia and Zimbabwe, and central in the area is the Chobe National Park.

References

Aarrestad, P.A., Masunga, G.S., Hytteborn, H., Pitlagano, M.L., Marokane, W. & Skarpe, C. (2011) Influence of soil, tree cover and large herbivores on field layer vegetation along a savanna landscape gradient in northern Botswana. *Journal of Arid Environments* 75, 290–297.

Bhalotra, Y.P.R. (1987) *Climate of Botswana Part II: Elements of Climate.* Department of Meteorological Services, Ministry of Works and Communication, Gaborone, Botswana.

Bigalke, R.C. (1968) Evolution of mammals on southern continents. III. The contemporary mammal fauna of Africa. *The Quarterly Review of Biology* 43, 265–300.

Bigalke, R.C. (1978) Mammals. In: Werger, M.J.A. (ed.) *Biogeography and Ecology of Southern Africa Volume 2*. Dr W Junk Publisher, The Hague, pp. 981–1048.

Blair Rains, A. & McKay, A.D. (1968) *The Northern State Lands, Botswana*. Land Resource Study No 5. Land Resource Division, Directorate of Overseas Surveys, Tolworth, Surrey, England.

Boocock, C. & van Straten, O.J. (1962) Notes on the geology and hydrology of the central Kalahari region, Bechuana Protectorate. *Transactions of the Geological Society of South Africa* 65, 125–176.

Burrough, S.L. & Thomas, D.S.G. (2008) Late Quaternary lake-level fluctuations in the Mababe Depression: middle Kalahari palaeolakes and the role of Zambezi inflows. *Quaternary Research* 69, 388–403.

Cerling, T.E., Ehrlinger, J.R. & Harris, J.M. (1998) Carbon dioxide starvation, the development of C4 ecosystems, and mammalian evolution. *Philosophical Transactions of the Royal Society of London B* 353, 159–171.

Chafota, J. & Owen-Smith, N. (2009) Episodic severe damage to canopy trees by elephants: interactions with fire, frost and rain. *Journal of Tropical Ecology* 25, 341–345.

Child, G. (1968) *An Ecological Survey of North-eastern Botswana*. Food and Agriculture Organization of the United Nations, Rome.

Cruse, A.M., Atekwana, E.A., Gamrod, J., Atekwana, E.A., Ringrose, S., Teter, K. & Huntsman-Mapila, P. (2009) Environmental change driven by tectonic processes and climate shifts as recorded in the sedimentary record of Paleolake Mababe, northern Botswana. Geological Society of America Annual Meeting, October 2009, Portland, Oregon, USA.

Davis, D.H.S. (1962) Distribution patterns of southern African Muridae, with notes on some of their fossil antecedents. *Annals of the Cape Provincial Museums* 2, 56–76.

de Wit, P.V. & Nachtergaele, F.O. (1990) *Explanatory Note on the Soil Map of the Republic of Botswana*. FAO, Rome.

du Toit, J.T. (2003) Large herbivores and savanna heterogeneity. In: du Toit, J.T., Rogers, K.H. & Biggs, H.C (eds.) *The Kruger Experience – Ecology and Management of Savanna Heterogeneity*. Island Press, Washington, DC, pp. 292–309.

Estes, R.D. (1991) *The Behavior Guide to African Mammals: Including Hoofed Mammals, Carnivores, Primates*. University of California Press, Berkeley, USA.

Foster, S.S.D., Bath, A.H., Farr, J.L. & Lewis, W.J. (1982) The likelihood of active groundwater recharge in the Botswana Kalahari. *Journal of Hydrology* 55, 113–136.

Gordon, I.J. & Prins, H.H.T. (2008) Grazers and browsers in a changing world. In: Gordon, I.J. & Prins, H.H.T. (eds.) *The Ecology of Grazing and Browsing*. Springer Verlag, Berlin, pp. 309–322.

Holdo, R.M., Sinclair, A.R.E., Dobson, A.P., Metzger, K.L., Bolker, B.M., Richie, M.E. & Holt, R.D. (2009) A disease-mediated trophic cascade in the Serengeti and its implications for ecosystem C. *PLoS Biology* 7(9), e1000210. doi:10.1371/journal.pbio.1000210.

Huntsman-Mapila, P., Ringrose, S., Mackay, A.W., Downey, W.S., Modisi, M., Coetzee, S.H., Tiercelin, J.-J., Kampunzu, H. & Vanderpost, C. (2006) Use of the geochemical and biological sedimentary record in establishing palaeo-environments and climate change in the Lake Ngami basin, NW Botswana. *Quaternary International* 148, 51–64.

Janis, C. (2008) An evolutionary history of browsing and grazing ungulates. In: Gordon, I.J. & Prins, H.H.T. (eds.) *The Ecology of Browsing and Grazing*. Springer, Berlin, pp. 21–46.

Leistner, O.A. (1967) *The Plant Ecology of the Southern Kalahari*. Botanical Survey of South Africa, Memoir 38. The Government Printer, Pretoria.

Livingstone, D. (1857) *Missionary Travels and Researches in South Africa*. John Murray, London, UK.

Mazor, E. (1982) Rain recharge in the Kalahari – a note on some approaches to the problem. *Journal of Hydrology* 55, 137–144.

McCarthy, T. & Rubidge, B. (2005) *The Story of Earth and Life, A Southern African Perspective on a 4.6 Billion Year Journey*. Struik Publishers, Cape Town, South Africa.

Modisi, M.P., Atekwana, E.A., Kampunzu, A.B. & Ngwisanyi, T.H. (2000) Rift kinematics during the incipient stages of continental expansion: evidence from the nascent Okavango rift basin, northwest Botswana. *Geology* 28, 939–942.

Moore A.E. & Larkin, P. (2001) Drainage evolution in south-central Africa since the breakup of Gondwana, *South African Journal of Geology* 104, 47–68.

Moore A., Blenkinsop, T. & Cotterill, F., (2009) Southern African topography and erosion history: plumes or plate tectonics, *Terra Nova*, 12 (4), 310–315.

Olff, H., Ritchie, M.E. & Prins, H.H.T. (2002) Global environmental controls of diversity in large herbivores. *Nature* 415, 901–904.

Owen-Smith, R.N. (1988) *Megaherbivores: The Influence of Very Large Body Size on Ecology*. Cambridge University Press, Cambridge, UK.

Partridge, T.C., Scott, L. & Hamilton, J.E. (1999) Synthetic reconstructions of southern African environments during the Last Glacial Maximum (21–28 kyr) and the Holocene Altithermal, (8–6 kyr). *Quaternary International* 57/58, 207–214.

Petit, J.R., Jouzel, J., Raynaud, D., Barkov, N.I., Barnola, J.-M., Basile, I., Benders, M., Chappellaz, J., Davis, M., Delaygue, G., Delmotte, M., Kotlyakov, V.M., Legrand, M., Lipenkov, V.Y., Lorius, C., Pepin, L., Ritz, C., Saltzman, E. & Stievenard, M. (1999) Climate and atmospheric history of the past 420 000 years from the Vostok ice core, Antarctica. *Nature* 399, 429–436.

Prasad, V., Strömberg, C.A.E., Alimohammadian, H. & Sahni, A. (2005) Dinosaur coprolites and the early evolution of grasses and grazers. *Science* 310, 1177–1180.

Prins, H.H.T. & van der Jeugd, H. (1993) Herbivore population crashes and woodland structure in East Africa. *Journal of Ecology* 81, 305–314.

Ringrose, S., Huntsman-Mapila, P., Kampunzu, A.B., Downey, W., Coetzee, S., Vink, B., Matheson, W. & Vanderpost, C. (2005) Sedimentological and geochemical evidence for palaeo-environmental change in the Makgadikgadi subbasin, in relation to the MOZ rift depression, Botswana. *Palaeogeography, Palaeoclimatology, Palaeoecology* 217, 265–287.

Robbins, L. (1987) Stone age archaeology in the Northern Kalahari, Botswana: Savuti and Kudi-akam Pan. *Current Anthropology* 28, 567–569.

Robbins, L.H. & Murphy, M.L. (2000) Archaeology, palaeoenvironment, and chronology of the Tsodilo Hills White Paintings rock shelter, Northwest Kalahari Desert, Botswana. *Journal of Archaeological Science* 27, 1085–1113.

Rutherford, M. C. (1980) Annual plant production–precipitation relations in arid and semi-arid regions. *South African Journal of Science* 76, 53–56.

Scholes, R.J. & Walker, B.H. (1993) *An African Savanna: Synthesis of the Nylsvley Study*. Cambridge University Press, Cambridge.

Scholes, R.J., Dowty, P.R., Caylor, K., Parsons, D.A.B., Frost, P.G.H. & Shugart, H.H. (2002) Trends in savanna structure and composition along an aridity gradient in the Kalahari. *Journal of Vegetation Science* 13, 419–428.

Schulze, R.E. & McGee, O.S. (1978) Climatic indices and classifications in relation to the biogeography of southern Africa. In: Werger, M.J.A. (ed.) *Biogeography and Ecology of Southern Africa 1*. Dr W. Junk Publishers, The Hague, pp. 19–54.

Selous, F.C. (1881) *A Hunter's Wanderings in Africa*. Richard Bentley & Son, London, UK.

Simpson, C.D. (1974) Ecology of the Zambezi Valley bushbuck *Tragelaphus scriptus ornatus* Pocock. PhD Thesis, Texas A & M University, College Station, Texas.

Sinclair, A.R.E., Mduma, S.A.R., Hopcraft, J.G.C., Fryxell, J.M., Hilborn, R. & Thirgood, S. (2007) Long-term ecosystem dynamics in the Serengeti: lessons for conservation. *Conservation Biology* 21, 580–590.

Skarpe, C. (1986) Plant community structure in relation to grazing and environmental changes along a north–south transect in the western Kalahari. *Vegetatio* 68, 3–18.

Skarpe, C. (1992) Dynamics of savanna ecosystems. *Journal of Vegetation Science* 3, 293–300.

Skinner, J.D. & Chimimba, C.T. (2005) *The Mammals of the Southern African Subregion*. Cambridge University Press, Cambridge, UK.

Smithers, R.H.N. (1971) *The Mammals of Botswana*. Museum Memoir, 4. The Trustees of the National Museums of Rhodesia, Salisbury [Harare, Zimbabwe].

Spinage, C.A. (2012) *African Ecology – Benchmarks and Historical Perspectives*. Springer-Verlag, Berlin.

Thomas, D.S.G. & Shaw, P.A. (1991) *The Kalahari Environment*. Cambridge University Press, Cambridge, UK.

Thomas, D.S.G. & Shaw, P.A. (2002) Late quaternary environmental change in central southern Africa: new data, synthesis, issues and prospects, *Quaternary Science Reviews* 21, 783–797.

Tlou, T. & Campbell, A. (1984) *History of Botswana*. Macmillan Botswana Publishing Co., Gaborone, Botswana.

Tyson, P.D. & Crimp, S.J. (1998) The climate of the Kalahari Transect. *Transactions of the Royal Society of South Africa* 53, 93–112.

Vrba, E.S. (1992) Mammals as a key to evolutionary theory. *Journal of Mammalogy* 73, 1–28.

Walter, H. (1970) *Vegetationszonen und Klima*. Eugen Ulmer, Stuttgart.

Werger, M.J.A (1978) Biogeographical division of southern Africa. In: Werger, M.J.A. (ed.) *Biogeography and Ecology of Southern Africa 1*. Dr W. Junk Publishers, The Hague, pp. 145–170.

Werger, M.J.A. & Coetzee, B.J. (1978) The Sudano-Zambezian region. In: Werger, M.J.A (ed.) *Biogeography and Ecology of Southern Africa 1*. Dr W. Junk Publishers, The Hague, pp. 301–462.

White, F. (1983) *The Vegetation of Africa. A Descriptive Memoir to Accompany the UNESCO/AETFAT/UNSO Vegetation Map*. Natural Resources Research XX, UNESCO, Paris.

Elephant-Mediated Ecosystem Processes in Kalahari-Sand Woodlands

Johan T. du Toit[1], Stein R. Moe[2] and Christina Skarpe[3]

[1]Department of Wildland Resources, Utah State University, USA
[2]Department of Ecology and Natural Resource Management, Norwegian University of Life Sciences, Norway
[3]Faculty of Applied Ecology and Agricultural Sciences, Hedmark University College, Norway

Depending on location and context, the African elephant *Loxodonta africana* has been classified as either a keystone species or an ecosystem engineer, sometimes both (reviewed by Caro, 2010). These classifications arise from the abilities of elephants to break, fell and uproot trees in the course of their feeding activities, with the result that, in combination either with fire or with other browsing mammals (Chapter 10), they can transform the structure and function of savanna ecosystems. In this chapter we explore the functional significance of elephants in the Kalahari-sand woodlands, woodlands that occur on a template of aeolian sands and parallel dune-trough topography (Chapter 2). This template provides the conditions for an interaction between the ecosystem's keystone and foundation species that, we hypothesize, is fundamental to the maintenance of biodiversity in the region. Unlike keystone species, foundation species are at the bottom of the food web, are locally abundant and regionally common, and maintain stable environmental conditions for other species while also modulating ecosystem processes. They are typically the dominant trees in a forested ecosystem (Ellison *et al.*, 2005). The foundation species of the Kalahari-sand woodlands is Zambezi teak, *Baikiaea plurijuga* (family Fabaceae, subfamily Caesalpin-ioideae), a deep-rooted, slow-growing, long-lived tree growing to 25 m or more above ground with a dense, rounded canopy. It dominates deciduous woodlands on deep

Elephants and Savanna Woodland Ecosystems: A Study from Chobe National Park, Botswana,
First Edition. Edited by Christina Skarpe, Johan T. du Toit and Stein R. Moe.
© 2014 John Wiley & Sons, Ltd. Published 2014 by John Wiley & Sons, Ltd.

Kalahari sands in Botswana, Zimbabwe, Namibia, Angola and Zambia. Valued for its extremely dense hardwood, Zambezi teak is commercially logged in various parts of its distribution and even in some areas that are now protected, like Chobe, a rarity of large specimens bears testimony to timber extraction in the past.

Large herbivore biomass density and the contribution of elephants

The wildlife populations of the Chobe ecosystem have not been monitored in any coordinated, long-term, landscape-scale programme such as, for example, that conducted in Kruger since the 1970s (Mabunda *et al.*, 2003). The elephant population of northern Botswana has been regularly surveyed from the air, (Chapter 6) but estimating all other wildlife population sizes is methodologically challenging in the closed-canopy woodland that covers most of this region. Nevertheless, knowing the elephant density in Chobe, one can approximate the total biomass density of the large herbivore community by extrapolating from the known total biomass densities of large herbivores (including elephants) in other savanna ecosystems.

In general, large mammal biomass varies as a function of mean annual rainfall across the African savanna biome (Coe *et al.*, 1976), with areas of higher rainfall producing more vegetation for herbivores and, consequently, more prey for carnivores (East, 1984). The nature of the relationship is influenced by an interaction with soil nutrient availability, so there is a family of curves each with a separate shape and slope for each of the major soil types (Bell, 1982). Fritz and Duncan (1994) found that in regions of 'low' soil nutrient availability, such as on Kalahari sands, large herbivore biomass density (B_t, in kg km^{-2}) varies with mean annual precipitation (P, in mm yr^{-1}) in a highly predictable relationship ($r^2 = 0.85$, $p < 0.001$, $n = 24$):

$$\log B_t = 1.96(\log P) - 2.04 \qquad (3.1)$$

Hence, for the Kalahari-sand woodlands of northern Botswana, with a mean annual rainfall of 632 mm in Kasane, this regression predicts a large herbivore biomass density of 2814 kg km^{-2}. A similar estimate of 2680 kg km^{-2} [not used by Fritz and Duncan (1994) in their regression model] was derived from detailed ground-based counts in the 1960s by Dasmann in the adjacent Hwange National Park (reported by Botkin *et al.*, 1981), which also largely comprises Kalahari-sand woodlands. Using an estimated equilibrium density of 0.907 elephants km^{-2} across northern Botswana (Junker *et al.*, 2008) and unit mass of 1725 kg for elephant (Coe *et al.*, 1976), the elephant population alone contributes about 1565 kg km^{-2}, which is 56% of the large herbivore biomass density predicted by Fritz and Duncan's (1994) model. This is almost identical to the 55% suggested by Bell (1986), using data from Kasungu National Park, Malawi, as the contribution of elephants to the large herbivore biomass density in 'moist-oligotrophic' savanna woodlands. Across a wide range of African national parks, Owen-Smith (1988) found megaherbivores (species in which individuals exceed 1000 kg body mass) as a group constituted 40–70% of the total large herbivore biomass.

From this evidence, all the large herbivores in an 'average' square kilometre of northern Botswana have a combined biomass of about 2800 kg. Of that, about 1600 kg is contributed by elephants alone, with the remaining 1200 kg made up of all other species, grazers and browsers combined. By comparison, Ngorongoro Crater of Tanzania receives the same mean annual rainfall as Chobe (630 mm) but with its rich volcanic soils it supports a large herbivore biomass density of 11,000 kg km^{-2}, of which almost all is contributed by grazers; the ecologically insignificant elephant population has always consisted of just a few itinerant old males (Runyoro et al., 1995). At the scale of the African savanna biome, northern Botswana's densely wooded savanna on dystrophic Kalahari sands supports a comparatively modest biomass of large herbivores as a whole, but its capacity for elephant production is impressive. It is particularly impressive in the dry season in areas near water, such as along the Chobe riverfront, where the local elephant density is at least 4 animals km^{-2} (Gibson et al., 1998), which equates to about 6900 kg km^{-2}. In the riparian fringe of the nearby Linyanti River, local elephant densities of more than 20 animals km^{-2} have been reported in the dry season (Teren and Owen-Smith, 2010). Furthermore, at the global scale, the estimated 1200 kg km^{-2} of non-elephant herbivore biomass in Chobe is significant in itself, being about equivalent to that of Yellowstone in North America, for example. The biomass density of large herbivores in Yellowstone's 1554 km^2 Northern Range is estimated to have been about 1200 kg km^{-2} in the late 19th century, after which it climbed to what was arguably an unsustainable 2700 kg km^{-2} in the 1990s (Wagner, 2006). It has been declining since then as a result of the effects of restoring the large predator guild, with wolves, *Lupus lupus*, and grizzly bears, *Ursus arctos*, preying on the elk, *Cervus elaphus*, population (Hamlin et al., 2009).

How can a dystrophic ecosystem support so many elephants?

From the middle of the 19th century, when European explorers and hunters first began describing the interior of south-central Africa, word spread about impressively high densities of elephants in the wooded savannas occuring on Kalahari sands and granitic soils (Chapter 6). Unsustainable commercial hunting for ivory soon followed, but the original abundance of elephants in broad-leafed wooded savannas contrasts with the absence of elephants from the grassy plains of the Serengeti or the shrublands of the Karoo, where the indigenous large herbivore communities are dominated by ruminants. Bell (1982, 1986) drew attention to this pattern and explained that, whereas large herbivore biomass density increases with rainfall across African savannas, the composition of that biomass changes, with elephants dominating in savannas with higher rainfall, lower soil nutrient availability, or both. Bell's hypothesis was that vegetation growing in soils with comparatively low nutrient availability can be abundant – if rainfall is sufficient – but is of low quality to herbivores as a result of a low ratio of digestible nutrients to structural material (wood and fibre), and the presence of plant secondary compounds. Because dietary tolerance increases along the herbivore body-size axis (Bell, 1971; Jarman, 1974), larger-bodied herbivores like elephants can survive on fibrous food that is indigestible to ruminants, for example. Elephants therefore have a competitive advantage in dystrophic wooded savannas where they have exclusive use of a major fraction of the primary productivity.

Bell's hypothesis is supported by empirical evidence from the composition of large herbivore communities when compared on a continental scale (Fritz, 1997). It also conforms to the hypothesis that resource availability is linked to plant antiherbivore defences, according to which browsing ruminants shun the foliage of slow-growing, chemically defended woody plants adapted to low-nutrient soils (Coley et al., 1985; Bryant et al., 1989). However, it is a gross over-simplification to infer from these hypotheses that elephants achieve high biomass densities in dystrophic wooded savannas by unselectively eating the abundant woody browse that smaller-bodied herbivores cannot tolerate. In fact elephants in Kalahari-sand woodlands eat green grass as long as it is available and then feed selectively on woody plants, often with different preferences to those of browsing ruminants (Makhabu, 2005; Chapter 13). In general, they seldom feed on slow-growing tree species (e.g. *Baikiaea plurijuga* and *Burkea africana*) but mainly on the faster-growing species (e.g. *Acacia* and *Combretum* spp.; Stokke, 1999; Chapters 7 and 12) that they also feed on in eutrophic savannas. Clearly, explaining the success of elephants in Kalahari-sand woodlands requires closer consideration of the processes of nutrient cycling through soils, plants, and herbivores.

An elephant ecosystem

We hypothesize that elephants are the keystone species (Power et al., 1996) in Kalahari-sand woodlands because of their comparatively large individual body size and population biomass, which cause the consumption of more woody vegetation by elephants than by all other large herbivore species combined. This facilitates the flow of nutrients from a slow and deep tree-mediated nutrient cycle into a fast and shallow cycle that supports a higher productivity and diversity of plants and animals than would occur in the absence of elephants. Support for this hypothesis, first suggested by Botkin et al. (1981), comes from subsequent studies that have clarified some of the ecological interactions among elephants, plants and soils in dystrophic savannas.

Large herbivore populations in African savannas are limited by nutritional constraints in the dry season, but the abilities to subsist on low quality food and endure extended periods of resource limitation are both size-dependent. Elephants, being large-bodied mixed feeders, build up body reserves by consuming green grass in the wet season and then subsist on browse through the dry season. Grasses require water and nutrients within their shallow rooting zone, whereas woody plants in such ecosystems tend to be deep-rooted and slow-growing, and are chemically defended (Bryant et al., 1989; Chapter 12). Deep-rooted trees bring nutrients from the lower levels of the soil profile to the surface and then return them to the soil in the form of leaf litter laden with phenolic secondary compounds. Such chemicals serve to protect living plant tissues from pathogens and herbivores while also inhibiting microbial decomposition of leaf litter. Slow litter decomposition is theoretically an adaptive trait of long-lived slow-growing trees on leachable soils (Zucker, 1983) because such trees cannot recover microbially released nutrients as rapidly as can grasses and forbs. Hence, nutrient cycling in woodlands operates on a different temporal scale to that in grasslands; the slow and deep tree-mediated cycle conserves nutrients during pulsed episodes of rainfall and leaching, and excludes fast-growing herbaceous plants. This is probably typical of broad-leafed African savannas, where these two distinct nutrient cycling pathways were first suggested by the results of a 9-year study of a dystrophic savanna dominated by *Burkea africana* trees in Nylsvley Nature Reserve, South Africa (Frost, 1985). For Kalahari-sand woodlands we hypothesize that elephants facilitate the flow of nutrients that 'leak' from the slow and deep cycle and become incorporated into the fast and shallow cycle (Figure 3.1).

The surface topography of the sheet of Kalahari sand overlying vast expanses of south-western Africa is characterized by distinctive east-west dune ridges and troughs, strikingly visible on remotely sensed images (e.g. Google Earth). Within the sand sheet there is commonly a substratum of compacted sand forming a hard layer that impedes drainage near the surface (1–1.5 m depth) in the clayey soils of the dune troughs but is either absent or occurs below the rooting depth (~10 m) of *Burkea plurijuga* in the dune ridges (Childes and Walker, 1987). This landscape of catenas drives heterogeneity in the vegetation, with deep-rooted *Baikiaea* woodland on the ridges, grassy *vleis* in the troughs, and shallow-rooted mixed woodland and scrubland in the transition zone along the slopes (Holdo and Timberlake, 2008; Chapter 5). Catenary processes, especially when associated with an underlying hard layer, result in fertile patches, or zones, in dystrophic savannas where lateral water transport brings nutrients into the rooting depth of shrubs and grasses in the bottomlands (Ben-Shahar, 1990; Scholes, 1990). Woody plants in these fertile zones have faster growth rates and lower concentrations of carbon-based plant secondary compounds (Bryant et al., 1989), thus constituting key resources for elephants in the dry season (Chapter 7). Spatial patterns generated by elephants in savanna vegetation show that browsing pressure and associated treefall are highest in mixed woodlands and scrublands along

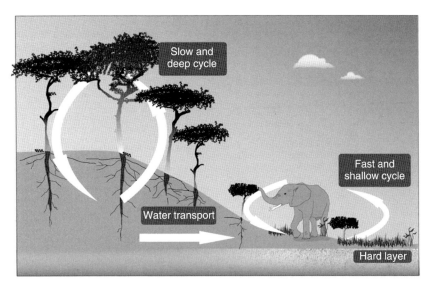

Figure 3.1 This hypothetical cartoon represents a cross section of the catena in Kalahari-sand woodland, with deep-rooted trees (dominated by *Baikiaea plurijuga*) grow-ing on aeolian sand dune ridges and disturbed scrubland and grassland in the dune troughs. A slow and deep nutrient cycle occurs in the upper catena, mediated by phenolic-laden leaf litter that inhibits microbial action. Water transport moves soil nutrients laterally, above a hard layer that impedes drainage, to the lower catena where fast-growing palatable plants are consumed by elephants that maintain a fast and shallow nutrient cycle. The catena (not drawn to scale) is about 200 m from ridge to trough, across which the elevation changes by 5–10 m. Drawing by Marit Hjeljord. (Source: Holdo and Timberlake, 2008).

the lower slopes of the catena (Asner and Levick, 2012). Indeed, in northern Botswana, stem breakage by elephants is almost twice as high in 'rugged' terrain – where the gradient changes sharply at the transition from bottomlands to slopes – compared with 'flat' terrain (Nellemann *et al.*, 2002). Elephants select and maintain nutrient-rich patches (Holdo, 2007; Pretorius *et al.*, 2011) and convert the tissues from the plants growing there (e.g. *Acacia* and *Combretum* spp.; Chapter 7) into dung, urine, meat, and bone. Those nutrients then become available – through defecation, urination, death and decomposition – for quick uptake by grasses, forbs and shrubs in the wet season when the upper soil profile is moist. Elephant-impacted vegetation is selected by other large herbivores, apparently because of enhanced food availability and quality, together with structural changes to the vegetation that improve visibility and thereby reduce the risk of predation (Makhabu *et al.*, 2006; Valeix *et al.*, 2011). In this way, we hypothesize that the elephant population facilitates nutrient transfer from a slow and deep tree-mediated nutrient cycle into a fast and shallow cycle (Figure 3.1) that supports herbaceous vegetation, palatable woody plants, and ultimately the impressive herbivore biomass of the ecosystem.

To summarize, elephants dominate the biomass of large herbivores in northern Botswana just as predicted by Bell (1982) for dystrophic savannas, but not because they consume the abundant low quality woody vegetation of the Kalahari sand woodlands. Rather, they focus their feeding on comparatively fertile patches and zones within the folds of the landscape, where they graze when green grass is available and then selectively browse on fast growing and palatable woody plants in the dry season. They exert top-down control on the woody vegetation in these areas by pruning, breaking, debarking, and uprooting woody plants in ways that other herbivores cannot. Such actions are expected to facilitate the flow of nutrients that 'leak' from a slow and deep tree-mediated cycle into a fast and shallow cycle in which an important nutrient pool is likely to be the elephant population itself, as first hypothesized by Botkin et al., (1981). Quantifying the extent to which elephants contribute to the diversion of nutrients from the slow and deep nutrient cycle is a topic for further research.

Interactions between keystone and foundation species maintain regional biodiversity

The distinctive dune-trough topography underlying the Kalahari-sand woodlands provides a template for heterogeneity at the spatial scale of vertebrate habitats and, at this scale, such heterogeneity is a driver of biodiversity in African savanna ecosystems (du Toit et al., 2003). From research elsewhere, however, it has become widely accepted that when elephant densities exceed about 0.5 animals km^{-2} woodlands become converted to shrubland and grassland, with associated declines in regional biodiversity (Cumming et al., 1997). So, with the current overall density of elephants across northern Botswana estimated at 0.907 animals km^{-2} (Junker et al., 2008) and in Chobe itself at 2.5–3.0 animals km^{-2} (Chase, 2011), are major ecological changes to be expected, not only near the Linyanti and Chobe rivers, but across the Kalahari-sand woodlands as a whole?

In their detailed study of the woody vegetation of the Kalahari sands, Childes and Walker (1987) found that elephants are insignificant agents of change in the Baikiaea woodlands on deep sands, but logging can trigger a sequence of changes. Opening the woodland canopy promotes grass growth and frost damage to the foliage on coppicing tree stumps, which increases the fuel load and thereby enables hot fires. With their large rootstocks these trees can coppice repeatedly (Holdo, 2006a,b) but die back from chronic exposure to fire and frost, allowing the emergence of a scrub phase. Slow regeneration to the woodland phase is inevitable after a series of years with either infrequent fires or a sequence of only mild fires and frosts (Chapter 4). The important issue for managers to understand is that the woodland on deep sands (i.e. on the dune ridges) is not vulnerable to elephant impacts because the chemically defended tree foliage is unpalatable and, with their very deep roots, the trees cannot be uprooted. Logging of *Baikiaea plurijuga* for its valuable hardwood is what triggers structural and functional changes to this ecosystem, as with all forested ecosystems that lose their foundation species to logging (Ellison *et al.*, 2005). Elephants concentrate their feeding in the scrubland on shallower soils (i.e. the dune troughs), which cannot support woodland because of a hard layer that limits rooting depth (Holdo and Timberlake, 2008). Consequently, with parallel bands of woodland and scrubland alternating on deep and shallow sands, respectively, the dune-trough topography of this vast ecosystem provides a template upon which the foundation and keystone species maintain stability. Comparatively high elephant densities seem to be a natural feature of Kalahari-sand woodlands and it is only on the shallow sands and alluvial soils along the major river systems that elephants are important agents of ecosystem change. Although the effects are localized, they are strikingly visible and therefore raise concerns among managers, politicians, and environmental commentators. It was those concerns that triggered the research underlying this book. Nevertheless, bear in mind that Kalahari-sand woodlands cover a vast swathe of south-western Africa across which elephants naturally dominate the herbivore community and stability has long prevailed.

References

Asner, G.P. & Levick S.R. (2012) Landscape-scale effects of herbivores on treefall in African savannas. *Ecology Letters* 15, 1211–1217.

Bell, R.H.V. (1971) A grazing ecosystem in the Serengeti. *Scientific American* 225, 86–93.

Bell, R.H.V. (1982) The effect of soil nutrient availability on community structure in African ecosystems. In: Huntley, B.J. & Walker, B.H. (eds.) *Ecology of Tropical Savannas*. Springer-Verlag, Berlin, pp. 193–216.

Bell, R.H.V. (1986) Carrying capacity and off-take quotas. In: Bell, R.H.V. & McShane-Caluzi, E. (eds.) *Conservation and Wildlife Management in Africa: The Proceedings of a Workshop Organized by the U.S. Peace Corps at Kasungu National Park, Malawi*. U.S. Peace Corps, Washington, DC, pp. 145–181.

Ben-Shahar, R. (1990) Soil nutrients distribution and moisture dynamics on upper catena in a semi-arid nature reserve. *Vegetatio* 89, 69–77.

Botkin, D.B., Mellilo, J.M. & Wu, L.S-Y. (1981) How ecosystem processes are linked to large mammal population dynamics. In: Fowler, C.W. & Smith, T.D. (eds.) *Dynamics of Large Mammal Populations*. John Wiley & Sons, Inc., New York, pp. 373–387.

Bryant, J.P., Kuropat, P.J., Cooper, S.M., Frisby, K. & Owen-Smith, N. (1989) Resource-availability hypothesis of plant antiherbivore defence tested in a South African savanna ecosystem. *Nature* 340, 227–229.

Caro, T. (2010) *Conservation by Proxy: Indicator, Umbrella, Keystone, Flagship, and Other Surrogate Species*. Island Press, Washington, DC.

Chase, M. (2011) *Dry Season Fixed-wing Aerial Survey of Elephants and Wildlife in Northern Botswana*. Elephants Without Borders, Kasane, Botswana; Department of Wildlife and National Parks, Botswana; and Zoological Society of San Diego, USA. [online] http://www.elephantdatabase.org/population_submission_attachments/102

Childes, S.L. & Walker, B.H. (1987) Ecology and dynamics of the woody vegetation on the Kalahari sands in Hwange National Park, Zimbabwe. *Vegetatio* 72, 111–128.

Coe, M.J., Cumming, D.H.M. & Phillipson, J. (1976) Biomass and production of large African herbivores in relation to rainfall and primary production. *Oecologia* 22, 341–354.

Coley, P.D., Bryant, J.P. & Chapin, F.S. III, (1985) Resource availability and plant antiherbivore defense. *Science* 230, 895–899.

Cumming, D.H.M., Fenton, M.B., Rautenbach, I.L., Taylor, R.D., Cumming, G.S., Cumming, M.S., Dunlop, J.M., Ford, G.S., Hovorka, M.D., Johnston, D.S., Kalcounis, M.C., Mahlanga, Z. & Portfors, C.V. (1997) Elephants, woodlands and biodiversity in southern Africa. *South African Journal of Science* 93, 231–236.

du Toit, J.T., Rogers, K.H. & Biggs, H.C. (2003) *The Kruger Experience: Ecology and Management of Savanna Heterogeneity*. Island Press, Washington, DC.

East, R. (1984) Rainfall, soil nutrient status and biomass of large African savanna mammals. *African Journal of Ecology* 22, 245–270.

Ellison, A.M., Bank, M.S., Clinton, B.D., Colburn, E.A. Elliott, K., Ford, C.R., Foster, D.R. Kloeppel844, B.D. Knoepp, J.D., Lovett, G.M., Mohan, J. Orwig, D.A., Rodenhouse, N.L., Sobczak, W.V., Stinson, K.A., Stone, J.K., Swan, C.M., Thompson, J., Von Holle, B. & Webster, J.R. (2005) Loss of foundation species: consequences for the structure and dynamics of forested ecosystems. *Frontiers in Ecology and the Environment* 3, 479–486.

Fritz, H. (1997) Low ungulate biomass in west African savannas: primary production or missing megaherbivores or large predator species. *Ecography* 20, 417–421.

Fritz, H. & Duncan, P. (1994) On the carrying capacity for large ungulates of African savanna ecosystems. *Proceedings of the Royal Society of London B* 256, 77–82.

Frost, P.G.H. (1985) Organic matter and nutrient dynamics in a broadleafed African savanna. In: Tothill, J.C. & Mott, J.J. (eds.) *Ecology and Management of the World's Savannas*. Australian Academy of Sciences, Canberra, pp. 200–206.

Gibson, D.C., Craig, C.G., & Masogo, R.M. (1998) Trends of the elephant population in northern Botswana from aerial survey data. *Pachyderm* 25, 14–27.

Hamlin, K.L., Garrott, R.A., White, P.J. & Cunningham, J.A. (2009) Contrasting wolf-ungulate interactions in the Greater Yellowstone Ecosystem. In: Garrott, R.A., White, P.J. & Watson, F.G.R. (eds.) *The Ecology of Large Mammals in Central Yellowstone*. Academic Press, San Diego, pp. 541–577.

Holdo, R.M. (2006a) Elephant herbivory, frost damage, and topkill in Kalahari sand woodland savanna trees. *Journal of Vegetation Science* 17, 509–518.

Holdo, R.M. (2006b) Tree growth in an African woodland savanna affected by disturbance. *Journal of Vegetation Science* 17, 369–378.

Holdo, R.M. (2007) Elephants, fire and frost can determine community structure and composition in Kalahari woodlands. *Ecological Applications* 17, 558–568.

Holdo, R.M. & Timberlake, J. (2008) Rooting depth and above-ground community composition in Kalahari sand woodlands in western Zimbabwe. *Journal of Tropical Ecology* 24, 169–176.

Jarman, P.J. (1974) The social organisation of antelope in relation to their ecology. *Behaviour* 48, 215–267.

Junker, J., van Aarde, R.J. & Ferreira, S.M. (2008) Temporal trends in elephant *Loxodonta africana* numbers and densities in northern Botswana: is the population really increasing? *Oryx* 42, 58–65.

Mabunda, D., Pienaar, D.J. & Verhoef, J. (2003) The Kruger National Park: a century of management and research. In: du Toit, J.T., Rogers, K.H. & Biggs, H.C. (eds.) *The Kruger Experience: Ecology and Management of Savanna Heterogeneity*. Island Press, Washington, DC, pp. 3–21.

Makhabu, S.W. (2005) Resource partitioning within a browsing guild in a key habitat, the Chobe Riverfront, Botswana. *Journal of Tropical Ecology* 21, 641–649.

Makhabu, S.W., Skarpe, C. & Hytteborn, H. (2006) Elephant impact on shoot distribution on trees and on rebrowsing by smaller browsers. *Acta Oecologica* 30, 136–146.

Nellemann, C., Moe, S.R. & Rutina, L.P. (2002) Links between terrain characteristics and forage patterns of elephants (*Loxodonta africana*) in northern Botswana. *Journal of Tropical Ecology* 18, 835–844.

Owen-Smith, R.N. (1988) *Megaherbivores: The Influence of Very Large Body Size on Ecology*. Cambridge University Press, Cambridge, UK.

Power, M.E., Tilman, D., Estes, J.A., Menge, B.A., Bond, W.J., Mills, L.S., Daily, G., Castilla, J.C., Lubchenco, J. & Paine, R.T. (1996) Challenges in the quest for keystones. *BioScience*, 46, 609–620.

Pretorius, Y., de Boer, W.F., van der Waal, C., de Knegt, H.J., Grant, R.C., Knox, N.M., Kohi, E.M., Mwakiwa, E., Page, B.R., Peel, M.J.S., Skidmore, A.K., Slowtow, R., van Wieren, S.E. & Prins, H.H.T. (2011) Soil nutrient status determines how elephants utilize trees and shape environments. *Journal of Animal Ecology* 80, 875–883.

Runyoro, V.A., Hofer, H., Chausi, E.B. & Moehlman, P.D. (1995) Long-term trends in herbivore populations of the Ngorongoro Crater, Tanzania. In: Sinclair, A.R.E. & Arcese, P. (eds.) *Serengeti II: Dynamics, Management, and Conservation of an Ecosystem*. University Chicago Press, Chicago, pp. 146–168.

Scholes, R.J. (1990) The influence of soil fertility on the ecology of southern African dry savannas. *Journal of Biogeography* 17, 415–419.

Stokke, S. (1999) Sex differences in feeding-patch choice in a megaherbivore: elephants in Chobe National Park, Botswana. *Canadian Journal of Zoology* 77, 1723–1732.

Teren, G. & Owen-Smith N. (2010) Elephants and riparian woodland changes in the Linyanti region, northern Botswana. *Pachyderm* 47, 18–25.

Valeix, M., Fritz, H., Sabatier, R., Murindagomo, F., Cumming, D. & Duncan, P. (2011) Elephant-induced structural changes in the vegetation and habitat selection by large herbivores in an African savanna. *Biological Conservation* 144, 902–912.

Wagner, F.H. (2006) *Yellowstone's Destabilized Ecosystem: Elk Effects, Science, and Policy Conflict*. Oxford University Press, New York.

Zucker, W.V. (1983) Tannins: does structure determine function? An ecological perspective. *The American Naturalist* 121, 335–365.

Part II
The Substrate

The concept of a 'substrate' has a central role in the model by Pickett *et al.* (2003) that forms the framework for this book. The substrate is the entity that changes state or is maintained in a state by the action of an agent, and that in turn influences responders, which are sensitive to the state of the substrate. Both the action by the agent on the substrate and the resultant effect on the substrate can be influenced by one or more controllers. Changes in the substrate can occur across multiple spatio-temporal scales and a substrate can be abiotic or biotic. Soil is the substrate for the activity of nest-building termites, creating spatial variation in soil physical and chemical properties, and *Acacia* trees can be the substrate for browsing giraffe, enhancing variation in morphology and chemistry of shoots in the canopy. In this section of the book we describe the vegetation as a substrate for actions by elephants under the influence of controllers such as soil type. We show that elephants create, enhance or maintain heterogeneity in the vegetation at scales varying from individual shoots in a browsed tree canopy to populations of trees to vegetation structure and composition in a landscape. For example, on the nutrient rich alluvial soil close to the Chobe River tree vegetation, including the knob thorn, *Acacia nigrescens*, a preferred browse species for elephants, has varied inversely with elephant density over the last 150 or so years (Drawing by Marit Hjeljord).

Elephants and Savanna Woodland Ecosystems: A Study from Chobe National Park, Botswana,
First Edition. Edited by Christina Skarpe, Johan T. du Toit and Stein R. Moe.
© 2014 John Wiley & Sons, Ltd. Published 2014 by John Wiley & Sons, Ltd.

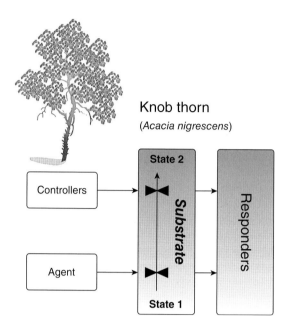

Knob thorn
(*Acacia nigrescens*)

References

Pickett, S.T.A., Cadenasso, M.L. & Benning, T.L. (2003) Biotic and abiotic variability as key determinants of savanna heterogeneity at multiple spatiotemporal scales. In: du Toit, J.T., Rogers, K.H. & Biggs, H.C. (eds.) *The Kruger Experience. Ecology and Management of Savanna Heterogeneity*. Island Press, Washington, Covelo, London, pp. 22–39.

4

Historical Changes of Vegetation in the Chobe Area

Christina Skarpe[1], Håkan Hytteborn[2,3], Stein R. Moe[4] and Per Arild Aarrestad[5]

[1] Faculty of Applied Ecology and Agricultural Sciences, Hedmark University College, Norway
[2] Department of Plant Ecology and Evolution, Evolutionary Biology Centre, Uppsala University, Sweden
[3] Department of Biology, Norwegian University of Science and Technology, Norway
[4] Department of Ecology and Natural Resource Management, Norwegian University of Life Sciences, Norway
[5] Norwegian Institute for Nature Research, Norway

The co-dominance of trees and grasses that characterises savannas, and the relative ease with which the dominance shifts between the two components, has long intrigued managers and researchers (Walter, 1954; Eagleson and Segarra, 1985; Skarpe, 1992; Sankaran *et al.*, 2008). Changes in vegetation structure between typical savanna, with co-dominance of woody plants and grasses, and open grassland and woodland at the two extremes, have been described as variations around an equilibrium, transitions between two or more stable states, or fluctuations in event-driven systems (Eagleson and Segarra, 1985; Ellis and Swift, 1988; Dublin *et al.*, 1990). Whereas woody growth is constrained by water availability up to a mean annual precipitation of about 700 mm, allowing the existence of climatically determined savannas, tree and grass coexistence under higher rainfall is generally unstable and depends on, for example, fire or

herbivory. These often modify the tree : grass ratio within the limit set by water availability also in savannas with an annual rainfall less than about 700 mm. Intense and
frequent fires generally reduce the woody component, at least over longer time frames
(Bond *et al.*, 2003). Herbivory by browsers and mixed feeders, particularly large-bodied
species such as the African elephant (*Loxodonta africana*) often but not always reduces
woody vegetation (Guldemond and van Aarde, 2008). In contrast, grazers at low densities can reduce woody growth but at high densities promote it (Eckhardt *et al.*, 2000;
Sankaran *et al.*, 2008). Often fire and herbivory interact in driving vegetation changes
(Chafota and Owen-Smith, 2009; Mapaure and Moe, 2009). The structure of savanna
vegetation is further influenced by soil nutrient availability and soil texture, with coarse
sandy soils, characterised by deep water infiltration, favouring woody growth (Walker
and Noy-Meir, 1982; Sankaran *et al.*, 2008). Soil water availability, soil nutrient availability, fire and herbivory are generally understood to be the main drivers influencing
savanna vegetation structure (Frost *et al.*, 1986; Walker, 1987). These drivers operate at
different spatial and temporal scales, with changes in water availability (climate) and
in soil properties usually being slow and covering large geographic areas, whereas fire
regime and herbivory can change rapidly, affecting vegetation structure and composition within one to a few years or decades, the time scale of interest for managers. In our
study in Chobe National Park in northern Botswana we were primarily interested in
the relationship between large herbivores and vegetation structure and composition.

In terms of the model proposed by Pickett *et al.* (2003; Chapter 1) climate (rainfall),
soil nutrients, herbivores and fire constitute agents and controllers of ecosystem change
at different spatial and temporal scales. At the scale of our interest, as well as for management decisions, elephants are the main agent shaping the vegetation (substrate),
whereas soil properties and, at a larger scale, climate constitute major controllers of
elephants' activities and of their effects on vegetation (Pretorius *et al.*, 2011). Nevertheless, elephants have not been the only agents of change in the Chobe ecosystem and its
vegetation during the 150 or more turbulent years covered by this chapter. There have
been others. We discuss the vegetation dynamics that took place concurrently with the
fall and rise of the elephant population following the ivory hunt in the end of the 20th
century (Chapter 6), and consider the relative importance of elephants, smaller herbivores and direct human impact through logging, burning and livestock grazing in causing these changes. Our study primarily concerns the Botswana-Norway Institutional
Co-operation and Capacity Building (BONIC) Project area, covering about 400 km^2
between the Kasane-Ngoma main road and the Chobe River (Preface; Figure 2.2). We
base our discussion on anecdotal information in travel and hunting accounts from the
19th and early 20th century, on reports and scientific publications about the area from
the 1960s onwards, and on data collected by the BONIC project. The most recent vegetation data originate primarily from 70 vegetation plots, each 400 m^2, distributed in
different elephant habitat types (Chapter 5), and from a study carried out specifically
to compare present vegetation with earlier records, as described in this chapter.

Vegetation in the Chobe area before the decline in elephants

The earliest accounts of vegetation in the Chobe area are from about 1850 (Living-stone, 1857) and 1870 (Selous, 1881). Selous was hunting, primarily for elephants, and he mentioned that the areas close to the Chobe River, probably referring to the flood-plain, were covered by short grasslands. He specifically described the crescent of raised alluvium extending almost 10 km mainly west of Serondela (Figure 2.2) as an open flat with scattered patches of bush and with many *Acacia* trees near the river. He named the open area Pookoo (puku) Flats, and saw large groups of puku, *Kobus vardonii*, and other game animals there. In spite of later vegetation changes, the name Puku Flats remained on maps until recently, when it confusingly was transferred to the nearby flood plain. The fringe of large trees along the river was mentioned also by Living-stone (1857). Selous (1881) further described the slope of the sand ridge facing the alluvial flat as covered by a dense 'jungle' of trees and shrubs, and south of that a more open forest or woodland (Figure 4.1). East and west of the Puku Flats he found sim-ilarly dense vegetation going down almost to the river bank and the open floodplain. Shortly after Selous' visit many changes took place in the Chobe ecosystem, including the near-extinction and later recovery of the elephant population and a sharp decline in other large herbivores as well as direct human impacts, leading to dramatic and still ongoing changes in the vegetation.

Elephants, germs, livestock and logging

When Selous (1881) hunted in the Chobe area around 1874 the elephant population was, according to him, declining, and by the time for the visit by Schultz and Hammar (1897) in 1884 elephants were rare. In 1899 they were reported to be nearly extinct (Reid, 1901; Chapter 6). Around the same time other large herbivores were being reduced by the rinderpest panzootic that reached northern Botswana in 1896–1897, decimating populations of even-toed ungulates including any domestic cattle in the area (Spinage, 2012; Chapter 13). Thus, by the end of the 19th century the density of large herbivores in the Chobe area was at its nadir. The relaxation of grazing and browsing by elephants and ungulates must have led to profound vegetation changes, directly by reduced consumption of plant material and indirectly by altered fire regime, nutrient cycling and soil water conditions. The release from herbivory provided a window of opportunity for plant species that had been suppressed by grazing or browsing to regenerate and lead to a reduction of other species that were weak competitors in the absence of large herbivores. Most noticeable was the establishment of woodlands dominated by *Acacia* spp. and *Combretum* spp. in the formerly open flats on alluvial soil, including the Puku Flats and nearby Kalwizikalkanga and Sidudu valleys 1900 (Simpson, 1974). Similar increase in woodland vegetation following the

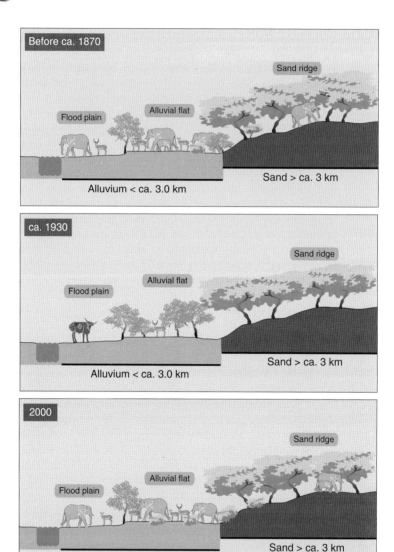

Figure 4.1   A simplified illustration of the ecosystem changes taking place around Seron-dela and southwards from Selous's visit in 1870 to our study around 2000. The scale is much compressed; in reality the sand ridge rises about 100 m above the river surface, and the area studied extends about 10 km south of the river. "Before ca. 1870" illustrates the situation around 1870 as described by Selous (1881) and as it had likely been for a long time; "ca. 1930" gives an idea of the situation after the ivory hunt had virtually eliminated the elephants, and ungulates had been reduced by rinderpest and competition with livestock, which was plenti-ful along the Chobe River. Logging of the woodlands on the sand had still not commenced. "2000" is the situation during our study with high densities of elephants and ungulates, fairly open shrub vegetation on the alluvial flat and woodlands on the sand ridge. See text for fur-ther explanation. (Source: Drawing by Marit Hjeljord.)

rinderpest outbreaks in the late 1800s has been recorded from places in East Africa (Prins and van der Jeugd, 1993).

Around 1870, when Selous travelled in what now is the Chobe National Park, the only people he mentioned living in the area were farming and livestock herding groups in Mababe and on Imparira Island (Figure 2.1) and San hunter-fisher-gatherers (Selous, 1881). West of the Park there were settled communities along the Linyanti. From about 1912 populations of humans and livestock increased along the Chobe and Linyanti rivers, and cattle posts were established at pans in the interior (Child, 1968). Heavy grazing by livestock, in combination with crop cultivation, particularly in the Kachikau area and at Serondela (Figures 2.1 and 4.1), and the virtually annual burning of the vegetation considered necessary to provide palatable grazing for the livestock, led to reduced cover of litter, grasses and forbs and to wind erosion of the sand (Child, 1968). The shortage of grazing, in combination with nagana sickness, caused by tsetse fly, and an outbreak of bovine streptotrichosis, led to abandonment of most of the cattle posts before 1955, although cattle still grazed along the Chobe River, including large numbers destined for export, waiting to be swum across the river from Kazungula to markets in Zambia. Anecdotal information from local residents (Simpson, 1974) suggests that the interruption of livestock grazing and of regular burning might be a reason for the spread of thicket vegetation along the river and in the riverine woodlands taking place around the 1950s and 1960s and reported by Child (1968).

Between 1935 and 1938 and from 1944 to 1955, logging took place in the northern part of the present Chobe National Park. At first, only the north-eastern corner was affected, but from 1944 onwards operations were based at sawmills at Serondela and Ihaha within our present core study area (Figure 2.2; Child, 1968). This logging first targeted all suitable trees within 20 miles (about 32 km) of the river, but was later restricted to *Pterocarpus angolensis*. Many of the large *Baikiaea plurijuga* trees found south of Serondela today originated as coppice shoots, as seen from now half overgrown but still visible sawn stumps. Frequent burning was practiced by the timber extractors as well as by the pastoralists (Child, 1968), with implications for both woody and herbaceous vegetation structure and plant recruitment.

When Chobe National Park was established in 1967, the remaining logging operations inside the park were discontinued and a strict non-burning policy adopted (Child, 1968), although wildfires frequently spread into the park from outside, and still do (Chapter 9). The last livestock was removed from Serondela in the late 1950s. From the 1960s the elephant population increased rapidly in the area (Chapter 6), whereas development of other mammal populations varied, some increasing and some decreasing (Chapter 13). Rinderpest, the reason for decline in Artiodactyla in the end of the 1800s, had its last outbreak in the area 1905 (Spinage, 2012). The vegetation in the previously logged areas on Kalahari sand was described in the 1960s as being characterised by dense shrubs, a low cover of perennial grasses – and correspondingly much bare ground – and a general absence of big trees (Child, 1968; Blair Rains and McKay, 1968).

Vegetation changes on the alluvium and on the sand

Simpson's vegetation types

Simpson (1974) collected information on the vegetation in the Chobe area in the early 1900s by talking to old residents of the area, and got the impression that around the 1940s five main vegetation types could be discerned along the river: (i) the seasonally inundated grassy floodplain; (ii) a riparian forest fringe on the river bank; (iii) riverine (*Acacia*) woodland on the raised alluvial flat; (iv) dense shrub- and woodland on the lower slope of the sand ridge and (v) typical Kalahari sand *Baikiaea*-dominated woodland on the upper slopes and further south (Simpson, 1974). He then compared this with the vegetation that he found during his own studies in 1969–1971, mainly targeting the area on alluvial soil between Ihaha and Kasane but surveying a larger area (Simpson, 1974, 1975).

Simpson (1974, 1975) defined a number of vegetation types, he called them habitat types, based on his own studies and related them to what he had learned about the older vegetation patterns and related these to the variations in soil and topography from the river to the crest of the sand ridge and beyond. The floodplain grassland encompassed the low-lying seasonally inundated and treeless floodplain along the river. The riparian forest fringe up to 70 yards (about 64 m) wide covered the river bank between the floodplain and the raised, alluvial flat. Riverine *Acacia* tree savanna grew on the raised alluvial flat between the river bank and the sand ridge, alongside *Dichrostachys* thicket, which was interpreted as a degraded phase of the riverine *Acacia* tree savanna. Regrowth *Combretum–Baphia* scrub had developed in some of the previously logged woodland areas on the sand, while relic patches of degraded Kalahari sand- or *Baikiaea* woodland were found further south on the sand ridge and beyond. *Colophospermum* tree/bush savanna, dominated by mopane, *Colophospermum mopane*, occurred in an area more than 7 miles (about 11 km) west of Serondela. The mixed tree-bush ecotone complex was confined to the main drainage lines, the Kalwizikalkanga and Sidudu valleys.

Vegetation transects in 1970, 1991 and 2003

Apart from the wider vegetation survey, Simpson (1974, 1975) used a small intensive study area east of Serondela at the easternmost edge of the raised flat with alluvial soil. There he surveyed four belt transects, 6 feet (1.8 m) wide and running about 500 m south from the water's edge into the zone with mixed sandy and alluvial soil. The transects were marked on a map, and approximately the same transects were analysed in 1991 by Addy (1993) and by us in 2003. Simpson (1974) included all trees with a basal diameter of more than 1 inch (about 2.5 cm) in his transect data, whereas Addy (1993) included trees more than 1 m tall. Our measurements of both height and basal diameters show that Addy's (1993) criteria included considerably more trees than did Simpson's (1975). We chose here to include trees more than 1 m tall, and used ranking as the measure of comparison to reduce the effect of the different inclusion criteria used.

Addy (1993) ranked the 15 most common woody species in the four belt transects in her and in Simpson's (1975) study, and we compared this with data from 2003 (Figure 4.2). As Simpson (1974, 1975) and Addy (1993) ranked all *Combretum* except *Combretum mossambicense* and *Combretum elaeagnoides* as one taxon, we used only 14 species from their studies. The rankings showed a succession of species that either decreased or increased over the 33-year period spanned by the three studies. Species that were abundant in 1970, but which had decreased or disappeared in the transects by 1991–2003, include *Commiphora pyracanthoides*, *Ximenia americana* and *Acacia schweinfurthii*, whereas *Flueggea virosa*, *Markhamia zanzibarica*, *Capparis tomentosa* and *Friesodielsia obovata* were among those that increased (nomenclature following Coates-Palgrave, 2002). *Combretum* spp. were common in 1970 and 1991 (not shown in Figure 4.2), but no *Combretum* other than *Combretum elaeagnoides* and *Combretum mossambicense* were ranked among the 15 most common species in 2003. *Baphia massaiensis* and *Dichrostachys cinerea* were among the most frequently recorded species in 1970 and 1991, and were also abundant in 2003, but as their average height then was less than 0.3 m, most individuals were not included in the ranking of plants more than 1 m tall.

Changes in Simpson's vegetation types from 1970 to 2000

We used Simpson's (1975) vegetation types from his large scale survey as a basis for comparing the vegetation around 2000, described from the 70 vegetation plots (Chapter 5), with the vegetation recorded in older literature and reports.

Floodplain

In the large scale vegetation survey, Simpson (1975) defined the Chobe floodplain as the seasonally inundated area between the river and the riverbank. Although there are more sections of floodplain of different width within the BONIC area we here mainly refer to the largest one between Serondela and Ihaha (Figure 2.2), that was described by Simpson (1975) and Child (1968; Plate 1). The widest floodplain is situated north of the Chobe River in Namibia. Selous (1881) mentioned that the flats adjacent to the river, referring to the floodplain, were covered by short grass. Child (1968) and Blair Rains and McKay (1968) listed more than 30 species of sedges, grasses and forbs occurring on the floodplains in about 1965, including *Cyperus articulatus*, *Cyperus esculentus*, *Echinochloa stagnina*, *Setaria sphacelata*, *Panicum maximum*,

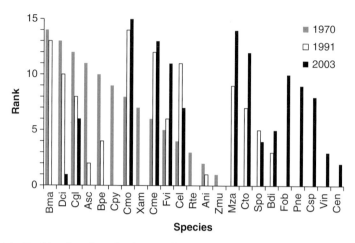

Figure 4.2   Ranking, based on the density of the 14 most abundant species in the four vegetation belt transects in 1970 and 1991 and the 15 most common species in 2003. Detailed information is given in the text. One is least and 14 (15) most abundant. Species are Bma, *Baphia massaiensis*; Dci, *Dichrostachys cinerea*; Cgl, *Canthium glaucum*; Asc, *Acacia schweinfurthii*; Bpe, *Bauhinia petersiana*; Cpy, *Commiphora pyracanthoides*; Cmo, *Combretum mossambicense*; Xam, *Ximenia americana*; Cme, *Croton megalobotrys*; Fvi, *Flueggea virosa*; Cel, *Combretum elaeagnoides*; Rte, *Rhus tenuinervis*; Ani, *Acacia nigrescens*; Zmu, *Ziziphus mucronata*; Mza, *Markhamia zanzibarica*; Cto, *Capparis tomentosa*; Spo, *Strychnos potatorum*; Bdi, *Berchemia discolor*; Fob, *Friesodielsia obovata*; Pne, *Philenoptera nelsii*; Csp, *Canthium* sp.; Vin, *Vangueria infausta*; Cen, *Combretum engleri*. In this figure, we have hypothesised that *Guibourtia coleosperma*, being important in Addy's (1993) sample but otherwise not recorded from the transects, is identical with *Bauhinia petersiana*, which was absent from her records but common in the transects in 1970 and 2003.

Brachiaria brizantha, *Hemarthria altissima* and *Digitaria eriantha*, many of which provide high-quality grazing (van Oudtshoorn, 1999). Of the grass species that were common in the 1960s many are tall (>1 m), tussock forming and dependent on sexual reproduction, traits not generally indicative of high tolerance of grazing (Díaz *et al.*, 2007). These species might have increased following the discontinuation of livestock grazing in the 1950s. Child (1968) described the floodplain vegetation as degrading with a large proportion of bare ground which might be the result of increasing grazing by wildlife on this vegetation. Both Child (1968) and Blair-Rains and McKay (1968) recorded an increase in *Cynodon dactylon*, a highly grazing-tolerant grass, and attributed its increase to grazing. However, about 5 years later, Simpson (1974, 1975) reported mostly the same genera of grasses and sedges as Child (1968) and Blair Rains and McKay (1968), and, surprisingly, did not mention *Cynodon dactylon*.

When our study was carried out around 2000, most of the species recorded by the old sources were still present in the floodplain, but *Cynodon dactylon* had become the dominant species with an average cover of around 33%, constituting 75% of the total plant cover (Aarrestad *et al.*, 2011; Chapter 5). *Cynodon dactylon* is a creeping, strongly clonal species, spreading both through stolons and by rhizomes. It is nutrient-rich and palatable to herbivores, tolerant of grazing, and is able to take up nitrogen directly from urea deposited in animal urine (van Oudtshoorn, 1999; Mathisen, 2005). Because of its low stature it is sensitive of competition for light in sparsely grazed vegetation, but it benefits from and expands with grazing mammals. On the Chobe floodplain it stays green and productive throughout the year.

Another grass that has increased on the floodplain, is the tall (in Chobe about 1 m), grazing-resistant, sharp and tough *Vetiveria nigritana*. This contains a vetiver-like oil (Kahlil and Ayoub, 2011), believed to act as a deterrent to herbivores. The species was recorded from the floodplains along the Chobe River by Blair Rains and McKay (1968), but neither Child (1968) nor Simpson (1974) mentioned it. At the time for our study, the *Vetiveria nigritana* was common (Aarrestad *et al.*, 2011, Chapter 5). The forb *Heliotropium ovalifolium*, not mentioned by the old sources, is now abundant on the floodplain (Chapter 5). It contains a potent alkaloid (Creeper *et al.*, 1999), and is totally avoided by herbivores. Thus, although the general structure of the floodplain vegetation, dictated by the seasonal inundations, has remained the same over the time for which records are available, there has been a considerable species turnover from the 1960s obviously caused by changing grazing and fire regimes. In recent decades grazing by wild animals has resulted in the dominance of short grasses, as was also recorded by Selous (1881). The high grazing pressure has caused a polarisation in the frequency of herbivory-related plant traits, with an increase in highly grazing-tolerant, palatable species at one end and in strongly defended unpalatable species at the other (Chapter 12).

Riparian forest fringe

The fringe of riparian forest between the floodplain and the raised alluvium was not mentioned by Selous (1881), who wrote only that ' … near the river grow many fine wide-branching camel-thorn trees'. Livingstone (1857) also mentioned that the steep banks of the Chobe were covered by magnificent trees, among which he recorded the genera *Ficus* and *Acacia*. The riparian forest fringe can have increased in density and width with the decline in large browsing animals around the year 1900. Child (1968), Simpson (1974) and Addy (1993) all described the riparian vegetation as a fringe, up to 70 yards (about 64 m) wide, along the riverbank north of the raised alluvial flat. Child (1968) and Simpson (1974) described it as varying from a closed canopy forest with, among other species (with modern nomenclature), *Acacia nigrescens*, *Acacia tortilis*, *Faidherbia albida*, *Garcinia livingstonei*, *Croton megalobotrys*, *Philenoptera violacea* (formerly *Lonchocarpus capassa*) and *Trichilia emetica*, through open canopy woodland to relic trees in dense shrub. All the mentioned species occurred along the river also during our study, but as scattered trees, forming a discontinuous line along the riverbank (Plate 6). Simpson (1974) further mentioned an undergrowth of several species of small trees, shrubs and climbers most of which, such as *Capparis tomentosa* and *Combretum mossambicense*, were also observed during our study, although some, such as *Rhus tenuinervis*, were rare. In 1991, the riparian fringe was dominated by *Dichrostachys cinerea*, *Croton megalobotrys*, *Combretum mossambicense* and *Canthium glaucum* (Addy, 1993). At the time of our study these species still occurred, and most of the former riparian vegetation, including scattered large trees, was floristically classified as *Capparis tomentosa – Flueggea virosa* shrubland vegetation (Chapter 5; Plates 7 and 8).

Riverine vegetation

Selous described the riverine vegetation on the raised alluvial flat between the riverbank and the sand ridge as ' ... a large flat piece of ground, in some parts over half a mile broad, lying between the steep forest-covered, jungly sand-belt and the bank of the river. This flat might be from six to eight miles long, and lay in the form of a semicircle, in a bend of the sand-belt, that rose abruptly behind it, and ran down to the water at each extremity. The greater part of this extensive tract – once, no doubt the ancient bed of the river – was open, though here and there patches of bush were scattered over its surface ... ' (Selous, 1881).

This open flat became invaded by woody vegetation – described as riverine woodland or riverine *Acacia* tree savanna (Simpson, 1974) – concurrently with the near-extinction of the elephants and the reduction of many ungulate populations by rinderpest. Child (1968) and Simpson (1974) described remnants of this vegetation as being dominated by large *Acacia nigrescens*, *Acacia tortilis*, *Berchemia discolor*, *Philenoptera violacea* and *Combretum imberbe* with a dense undergrowth of shrubs including *Dichrostachys cinerea*, *Combretum mossambicense* and *Combretum elaeagnoides*. Doubtless, the establishment of large trees in the area was caused by the release from browsing and debarking by elephants and from seedling predation by smaller browsing herbivores. Child (1968), and information from old residents (Simpson, 1974) indicated that the spread of thicket vegetation in these tree savannas mainly took place in the 1950s and 1960s, which suggests that it might have been related to the interruption of cattle grazing and of frequent burning. In the mid-1960s, open parts of the *Acacia* tree savanna had a grassy field layer with *Aristida* spp., *Panicum* spp. and *Eragrostis* spp. and patches of *Cynodon dactylon*. Child (1968) suggested that the latter signified direct human impact in the form of old cattle kraals and cultivated areas, as may also be the case with small patches of *Cenchrus ciliaris* found in our study (Denbow, 1979).

As the populations of elephants and other large herbivores built up again, particularly from the 1960s onwards, the relatively nutrient-rich vegetation on alluvial soils in the riverine *Acacia* tree savanna and the riparian fringe was preferentially used by grazing and browsing herbivores. Trampling by large-bodied species – elephant, buffalo and hippopotamus (*Hippopotamus amphibius*) – opened up riverine and riparian thickets, while browsing and debarking of trees by elephants reduced the tree layer (Child, 1968). The conversion of woodland and thicket to more open shrublands probably favoured impala, causing its population to increase (Sheppe and Haas, 1976; Rutina *et al.*, 2005; Chapter 13). Moe *et al.* (2009; Chapter 10; Plate 9) showed in an experiment with differential exclosures that impala, by browsing on seedlings and saplings, probably is a main controller in the system, preventing regeneration of the riparian and riverine woodlands after elephants have killed the big trees. Thus, impala could have contributed both to an increasingly open structure of the vegetation and to a shift in species composition from tree species such as *Faidherbia albida* and *Garcinia livingstonei*, the seedlings of which suffered heavy mortality from browsing, to shrubs such as *Combretum mossambicense*, which suffered less from browsing of seedlings (Moe *et al.*, 2009).

Remnants of the riverine woodland survived through the 1970s, but by the late 1990s, only few isolated trees, mainly *Berchemia discolor*, *Strychnos madagascariensis* and *Croton megalobotrys* remained alive. Instead smaller trees such as *Flueggea virosa* and *Erythroxylum zambesiacum* increased, together with shrubs from the former woodland undergrowth: *Combretum mossambicense*, *Combretum elaeagnoides* and *Capparis tomentosa*. Decaying stumps of large trees are still found scattered in the area, often hidden in scrambling *Capparis tomentosa* shrubs. A similar, recent, elephant-induced change of riverfront *Acacia*-dominated woodland into *Combretum mossambicense* shrub vegetation has been described from nearby Linyanti (Teren *et al.*, 2011).

In Chobe some of the increasing species, particularly *Capparis tomentosa*, *Combretum mossambicense*, *Flueggea virosa* and *Markhamia zanzibarica* are not heavily used by elephant (Stokke, 1999; Makhabu, 2005; Owen-Smith and Chafota, 2012; Chapter 12). This suggests that browsing by elephant might have been important in shifting species composition towards dominance by species not prone to elephant browsing (cf. Jachmann and Bell, 1985). However, these species constitute the main food resource for ruminant browsers in the area, impala, kudu (*Tragelaphus strepsiceros*) and giraffe (*Giraffa camelopardalis*) (Makhabu, 2005). Also the field layer vegetation has changed from the 1960s towards more grazing-tolerant grasses such as *Cynodon dactylon*, *Cenchrus ciliaris*, *Chloris virgata* and *Brachiaria deflexa*, and unpalatable forbs such as *Tribulus terrestris* (Aarrestad *et al.*, 2011; Chapter 5). In our study, the riverine vegetation was classified floristically as *Capparis tomentosa – Flueggea virosa* shrubland and *Combretum mossambicense – Friesodielsia obovata* wooded shrubland communities (Chapter 5; Plates 2 and 3).

Dichrostachys thicket

Simpson (1974) described *Dichrostachys* thicket as a separate vegetation type, characterising previously cultivated areas on alluvium, primarily close to Serondela. It was dominated by *Dichrostachys cinerea*, *Combretum mossambicense* and *Combretum elaeagnoides*, and seems to be the same type that was described by Addy (1993) as *Combretum–Dichrostachys* thicket. None of these types are distinguishable today from other parts of the former riverine *Acacia* tree savanna, currently a fairly open shrub vegetation (15–30% woody cover; Chapter 5). As said above *Dichrostachys cinerea* is still a common species in this vegetation, but mostly less than 0.5 m tall. In our floristic categorisation the former *Dichrostachys* thicket was a constituent of both *Capparis tomentosa–Flueggea virosa* shrubland and *Combretum mossambicense–Friesodielsia obovata* wooded shrubland communities (Chapter 5).

Mixed tree–bush ecotone complex

Travelling westward along the river from Imparira Island, Selous (1881) came across an open valley between two densely wooded sand ridges, running down towards the Chobe River. This was most likely the Sidudu valley. He did not describe the vegetation, other than saying that the valley was open and that he could find no shelter there when hunting buffalo. Simpson (1974, 1975) described the vegetation of the Sidudu and Kalwizikalkanga valleys and other drainage lines as a mixed tree-bush ecotone complex, containing mature trees, principally *Acacia tortilis*, *Philenoptera violacea*, *Guibourtia coleosperma* and *Piliostigma thonningii*, with an undergrowth of low shrubs and coppicing thicket, including *Ziziphus mucronata*, *Capparis tomentosa* and *Dichrostachys cinerea*. Many of the trees showed signs of elephant damage in 1970, and by the time of our study most were dead and those that were still alive were severely damaged by elephants. We found the valleys mostly open, but with patches of thicket, primarily of *Combretum elaeagnoides* (Plate 10).

Colophospermum tree-bush savanna

Simpson (1974, 1975) briefly mentioned *Colophospermum* tree-bush savanna as a vegetation type confined to the area west of Ihaha, with *Colophospermum mopane* as the dominant or co-dominant tree on heavily eroded and rocky soils. We did not record *Colophospermum mopane* in any of our plots, although some trees were seen in the area, and the species was common further south on more compact sand. Mopane is palatable to elephants (Pretorius *et al.*, 2011), and it is likely that the trees recorded by Simpson (1974, 1975) had established when elephants were virtually absent from the system, and that when elephant numbers later increased most of the trees were killed and disappeared. Also the few large baobab trees, *Adansonia digitata*, found in this area, primarily in rocky rugged terrain, likely established when elephant density was low. Like the mopane the baobab is palatable to elephants.

Kalahari sand vegetation or Baikiaea woodland

More than 90% of the area studied is covered by Kalahari sand, much of which supports *Baikiaea plurijuga* dominated woodlands. Selous (1881) described the northern slope of the sand ridge facing the alluvial flat as covered by dense vegetation of trees and shrubs, further south changing into more open forest. The logging operations 1935–1938 and 1944–1955 primarily targeted *Pterocarpus angolensis* and *Baikiaea plurijuga*, suggesting that these species were common in the area at that time. In 1965 and 1969–1971, respectively, Child (1968) and Simpson (1974) found only relic patches of the *Baikiaea* woodlands remaining after timber extraction, and that a 5–8 foot (about 1.5–2.5 m) high shrub vegetation, described by Simpson (1974) as regrowth *Combretum–Baphia* scrub, had developed from understory species such as *Baphia massaiensis*, *Bauhinia petersiana*, *Combretum* spp., *Terminalia sericea* and *Friesodielsia obovata*. Some coppicing of *Baikiaea plurijuga*, *Guibourtia coleosperma* and *Pterocarpus angolensis* was recorded, but was said to result in a bush-rather than a tree savanna (Simpson, 1974).

Most observers agree that from the 1960s, the woody vegetation on Kalahari sand south of the raised alluvium was regrowing from coppicing stumps, while being subjected to frequent fires and increasing impact by elephants. Around 1990, however, much of the bush savanna seems to have developed into woodland (Lindsay, 1990). Ben-Shahar (1993) described the area as covered by *Baikiaea plurijuga* woodland with a *Baikiaea plurijuga–Burkea africana* association. In our study the vegetation on the sandy areas south of the alluvial flat was divided into two structural habitat types (Skarpe *et al.*, 2004; Chapter 5). The lower and middle slopes of the sand ridge were covered by woody vegetation with scattered large *Baikiaea plurijuga* trees and many smaller trees, for example, *Croton gratissimus*, *Combretum apiculatum* and *Markhamia zanzibarica* and a dense shrub layer including *Combretum* spp., *Friesodielsia obovata*, *Baphia massaiensis* and *Bauhinia petersiana*. This type was referred to as mixed *Baikiaea* woodland (Skarpe *et al.*, 2004; Chapter 5). Floristically this habitat type was merged with the *Combretum* shrubland habitat into the *Combretum mossambicense–Friesodielsia obovata* wooded shrubland vegetation type (Chapter 5; Plate 3). On the higher slopes and crest of the sand ridge and beyond, the vegetation was, both floristically and structurally, classified as *Baikiaea plurijuga* woodland (Chapter 5) with the tree layer strongly dominated by *Baikiaea plurijuga*. The two types of woodland can be compared with Selous's (1881) distinction between the dense 'jungle' on the lower slope of the sand ridge and the open woodlands further south.

Elephants and the Chobe woodlands

By the 1990s, concern about elephant damage to woody vegetation in northern Botswana had been growing for some time (Child, 1968; Sommerlatte, 1976; Simpson, 1978; Moroka, 1984). This led to a number of studies in the Chobe woodlands and regionally of the impact of elephants on woodland structure and composition. The studies demonstrate that elephants use habitats, tree species and trees selectively, and that their impact on woodlands varies with tree species composition and site conditions.

Elephant impact on trees is species-specific, and in spite of their large size, elephants are selective browsers (Owen-Smith and Chafota, 2012). Relatively nutrient rich, weakly defended tree species often suffer severe browsing while heavily defended species such as *Baikiaea plurijuga* and *Erythrophleum africanum* are avoided (Holdo, 2003, 2006; Owen-Smith and Chafota, 2012; Chapter 12). Coppice regrowth of species not much browsed by elephants are more likely to reach maturity than coppice growth of palatable species. These are kept short or are killed by repeated browsing (Jachmann and Bell, 1985). This seems to be a reason for the strong dominance of *Baikiaea plurijuga*, which is not eaten by elephants, in the canopy tree layer in the Chobe woodlands on sand and the relative sparseness of species eaten by elephants such as *Pterocarpus angolensis* and *Guibourtia coleosperma*, although these species were present in the coppice growth after the logging (Simpson, 1974). Ben-Shahar (1996, 1997) found in northern Botswana that while *Colophospermum mopane* woodlands exhibited a reduction in tree density with increasing elephant density, this was not the case with *Baikiaea plurijuga* woodlands.

Woodlands in close proximity to permanent or temporary water sources often suffer heavy impact by elephants, and regeneration also of species not eaten can be impaired by trampling (Ben-Shahar, 1998). This might be the case in the mixed *Baikiaea* woodlands on the lower slopes of the sand ridge in our area, less than 3–4 km from the river, where density of *Baikiaea plurijuga* is reduced and density of smaller fast-growing species enhanced compared to woodlands further from the river. Elephant impact on nutrient cycling and availability close to the river (Chapter 9) might be another reason contributing to an increase in species of relatively fast growing small trees.

Elephant browsing in woodlands is spatially and temporally heterogeneous in response to, for example, terrain characteristics, fire, frost and exceptional rainfall (Ben-Shahar, 1998; Nellemann *et al.*, 2002; Baxter and Getz, 2005; Holdo, 2006, 2007; Chafota and Owen-Smith, 2009). We observed a patchy distribution of small trees and shrubs, *Combretum* spp., *Baphia massaiensis*, *Bauhinia petersiana* and others, heavily browsed by elephants within the *Baikiaea* woodland (Stokke *et al.*, 2003; Chapters 5 and 12). Chafota and Owen-Smith (2009) reported instances of patchy, intense damage by elephants to trees, particularly *Terminalia sericea* and *Burkea africana*, in the Chobe woodlands where elephants interacted with fire and intense frost. In the *Baikiaea* woodlands, south of Serondela and Ihaha lingering effects of the logging operations half a century ago can potentially influence spatial structure of the vegetation. The nature of such patches is not well known. Chafota and Owen-Smith (2009) and Gilson (2004) suggested that they might form a patch dynamic system with patches in different stages of regrowth, potentially with important consequences for ecosystem diversity and function (Lewin, 1986). If regeneration of canopy forming *Baikiaea plurijuga* is hampered by elephants, for example, through trampling, the intensely browsed patches might form nuclei of expanding shrub vegetation, whereas if fire is a main limitation, as suggested by Ben-Shahar (1997), the intensely utilised patches where trampling and grazing supress fires would revert to woodland. There are different predictions of the future of the Chobe *Baikiaea* woodlands (Ben-Shahar, 1997; Mosugelo *et al.*, 2002; Owen-Smith and Chafota, 2012), but compared to woodlands dominated by more palatable species, the *Baikiaea* woodlands are little affected by elephant browsing.

Thus, the fall and rise of the Chobe elephant population during the last 150 or so years, has affected the vegetation on the relatively nutrient rich alluvium differently from that on the nutrient-deficient sand. On the raised alluvium, the vegetation has shifted from open flat at unknown but likely high elephant density before the late 19th century to *Acacia* woodland when elephants were close to extinct in the early 20th century and again to fairly open shrub vegetation and high elephant density at present (Figure 4.1). The vegetation of the alluvium is of sufficiently high quality to allow smaller browsers, such as impala, to interact with the dynamic vegetation and prevent the regeneration of the woodland. Contrary, there is little evidence that the dynamics of the elephant population had any major effects on the *Baikiaea* woodland on Kalahari sand, although elephants in interaction with abiotic factors such as fire, frost and logging can modify species composition and spatial structure. Thus, elephants seem unlikely to cause a major state transition of the nutrient poor *Baikiaea plurijuga* woodland vegetation, as they have done with the vegetation on the more nutrient rich alluvium.

References

Aarrestad, P.A., Masunga, G.S., Hytteborn, H., Pitlagano, M.L., Marokane, W. & Skarpe, C. (2011) Influence of soil, tree cover, and large herbivores on field layer vegetation along a savanna landscape gradient in northern Botswana. *Journal of Arid Environments* 75, 290–297.

Addy, J.E. (1993) Impact on elephant induced vegetation change on the status of the bushbuck (*Tragelaphus scriptus ornatus*) along the Chobe River in northern Botswana. MSc Thesis, University of the Witwatersrand, Johannesburg, South Africa.

Baxter, P.W.J. & Getz, W.M. (2005) A model-framed evaluation of elephant effects on tree and fire dynamics in African savannas. *Ecological Applications* 15, 1331–1341.

Ben-Shahar, R. (1993) Patterns of elephant damage to vegetation in Northern Botswana. *Biological Conservation* 65, 249–256.

Ben-Shahar, R. (1996) Woodland dynamics under the influence of elephants and fire in Northern Botswana, *Vegetatio* 123, 153–163.

Ben-Shahar, R. (1997) Elephants and woodlands in northern Botswana: how many elephants should be there? *Pachyderm* 23, 41–43.

Ben-Shahar, R. (1998) Elephant density and impact on Kalahari woodland habitats. *Transactions of the Royal Society of South Africa* 53, 149–155.

Blair Rains, A. & McKay, A.D. (1968) *The Northern State Lands, Botswana.* Land Resource Study No 5. Land Resource Division, Directorate of Overseas Surveys, Tolworth, Surrey, England.

Bond, W.J., Midgley, G.F. & Woodward, F.I. (2003) What controls South African vegeta-tion – climate or fire? *South African Journal of Botany* 69, 79–91.

Chafota, J. & Owen-Smith, N. (2009) Episodic severe damage to canopy trees by elephants: inter-actions with fire, frost and rain. *Journal of Tropical Ecology* 25, 341–345.

Child, G. (1968) *An Ecological Survey of North-eastern Botswana*. Food and Agriculture Organi-zation of the United Nations, Rome.

Coates-Palgrave, K. (2002) *Trees of Southern Africa*. Struik Publishers, Cape Town, South Africa.

Creeper, J.H., Mitchell, A.A., Jubb, T.F. & Colegate, S.M. (1999) Pyrrolizidine alkaloid poison-ing of horses grazing a native heliotrope (*Heliotropium ovalifolium*). *Australian Veterinary Journal* 77, 401–402.

Denbow, J.R. (1979) *Cenchrus ciliaris*: an ecological indicator of iron age middens using aerial photography in eastern Botswana. *South African Journal of Science* 75, 405–408.

Díaz, S., Lavorel, S., McIntyre, S., Falczuk, V., Casanoves, F., Milchunas, D.G., Skarpe, C., Rusch, G., Sternberg, M., Noy-Meir, I., Landsberg, J., Zhang, W., Clark, H & Campbell, B.D. (2007) Plant trait responses to grazing – a global synthesis. *Global Change Biology* 13, 313–341.

Dublin, H.T., Sinclair, A.R.E. & McGlade, J. (1990) Elephants and fire as causes of multiple stable states in the Serengeti-Mara woodlands. *Journal of Animal Ecology* 59, 1147–1164.

Eagleson, P.S. & Segarra, R.I. (1985) Water-limited equilibrium of savanna vegetation systems. *Water Resources Research* 21, 1483–1493.

Eckhardt, H., van Wilgen, B. & Biggs, H. (2000) Trends in woody vegetation cover in the Kruger National Park, South Africa, between 1940 and 1998. *African Journal of Ecology* 38, 108–115.

Ellis, J.E. & Swift, D.M. (1988) Stability of African pastoral ecosystems: alternate paradigms and implications for development. *Journal of Range Management* 41, 450–459.

Frost, P., Medina, E., Menaut, J.-C., Solbrig, O., Swift, M. & Walker, B. (1986) Responses of savan-nas to stress and disturbance. *Biology International* 10, 1–82.

Gilson, L. (2004) Evidence of hierarchical patch dynamics in an east African savanna? *Landscape Ecology* 19, 883–894.

Guldemond, R. & van Aarde, R. (2008) A meta-analysis of the impact of African elephants on savanna vegetation. *Journal of Wildlife Management* 72, 892–899.

Holdo, R.M. (2003) Woody plant damage by African elephants in relation to leaf nutrients in western Zimbabwe. *Journal of Tropical Ecology* 19, 189–196.

Holdo, R.M. (2006) Elephant herbivory, frost damage and topkill in Kalahari sand woodland savanna trees. *Journal of Vegetation Science* 17, 509–518.

Holdo, R.M. (2007) Elephants, fire and frost can determine community structure and composi-tion in Kalahari woodlands. *Ecological Applications* 17, 558–568.

Jachmann, H. & Bell, R.H.V. (1985) Utilisation by elephants of the *Brachystegia* woodlands of the Kasungu National Park, Malawi. *African Journal of Ecology* 23, 245–258.

Kahlil, M.A.L. & Ayoub, S.M.H. (2011) Analysis of the essential oil of *Vertivaria nigritana* (Benth.) Stapf root growing in Sudan. *Journal of Medicinal Plants Research* 5, 7006–7010.

Lewin, R. (1986) In ecology, change brings stability. *Science* 234, 1071–1073.

Lindsay, K.W. (1990) Elephant/habitat interactions. In: Hancock, P., Cantrell, M. & Hughes, S. (eds.) *The Future of Botswana's Elephants. Proceedings of the Kalahari Conservation Society Symposium, Gaborone, Botswana*, pp. 19–23.

Livingstone, D. (1857) *Missionary Travels and Researches in South Africa*. John Murray, London, UK.

Makhabu, S.W. (2005) Resource partitioning within a browsing guild in a key habitat, the Chobe Riverfront, Botswana. *Journal of Tropical Ecology* 21, 641–649.

Mapaure, I. & Moe, S.R. (2009) Changes in the structure and composition of miombo woodlands mediated by elephants (*Loxodonta africana*) and fire over a 26-year period in north-western Zambia. *African Journal of Ecology* 47, 175–183.

Mathisen, I.E. (2005) Effects of clipping and nitrogen fertilization on a grazing tolerant grass in Chobe National Park, Botswana. MSc Thesis, Norwegian University of Science and Technology, Trondheim, Norway.

Moe, S.R., Rutina, L.P., Hytteborn, H. & du Toit, J.T. (2009), What controls woodland regeneration after elephants have killed the big trees? *Journal of Applied Ecology* 46, 223–230.

Moroka, D.N. (1984) *Elephant-Habitat Relationships in Northern Botswana*. Department of Wildlife and National Parks, Gaborone, Botswana.

Mosugelo, D.K., Moe, S.R., Ringrose, S. & Nellemann, C. (2002) Vegetation changes in a 36-year period in northern Chobe National Park, Botswana. *African Journal of Ecology* 40, 232–240.

Nellemann, C., Moe, S.R. & Rutina, L.P. (2002) Links between terrain characteristics and forage patterns of elephants (*Loxodonta africana*) in northern Botswana. *Journal of Tropical Ecology* 18, 835–844.

Owen-Smith, N. & Chafota, J. (2012) Selective feeding by a megaherbivore, the African elephant (*Loxodonta africana*). *Journal of Mammalogy* 93, 698–705.

Pickett, S.T.A., Cadenasso, M.L. & Benning, T.L. (2003) Biotic and abiotic variability as key determinants of savanna heterogeneity at multiple spatiotemporal scales. In: du Toit, J.T., Rogers, K.H. & Biggs, H.C. (eds.) *The Kruger Experience. Ecology and Management of Savanna Heterogeneity*. Island Press, Washington, DC, pp. 22–40.

Pretorius, Y., de Boer, W.F., van der Waal, C., de Knegt, H.J., Grant, R.C., Knox, N.M., Kohi, E.M., Mwakiwa, E., Page, B.R., Peel, M.J.S., Skidmore, A.K., Slowtow, R., van Wieren, S.E. & Prins, H.H.T. (2011) Soil nutrient status determines how elephants utilize trees and shape environments. *Journal of Animal Ecology* 80, 875–883.

Prins, H.H.T. & van der Jeugd, H. (1993) Herbivore population crashes and woodland structure in East Africa. *Journal of Ecology* 81, 305–314.

Reid, P.C. (1901) Journeys in the Linyanti Region. *Geographical Journal* 17, 573–588.

Rutina, L.P., Moe, S.R. & Swenson, J.E. (2005) Elephant *Loxodonta africana* driven woodland conversion to shrubland improves dry-season browse availability for impalas *Aepyceros melampus*. *Wildlife Biology* 11, 207–213.

Sankaran, M., Ratnam, J. & Hanan, N. (2008) Woody cover in African savannas: the role of resources, fire and herbivory. *Global Ecology and Biogeography* 17, 236–245.

Schultz, A.M.D. & Hammar, A.C.E. (1897) *The New Africa – A Journey up the Chobe and down the Okavango Rivers*. William Heinemann, London, UK.

Selous, F.C. (1881) *A Hunter's Wanderings in Africa*. Richard Bentley & Son, London, UK.

Sheppe, W. & Haas, P. (1976) Large mammal populations of the lower Chobe River, Botswana. *Mammalia* 2, 223–243.

Simpson, C.D. (1974) Ecology of the Zambezi Valley bushbuck *Tragelaphus scriptus ornatus* Pocock. PhD Thesis, Texas A&M University, College Station, Texas.

Simpson, C.D. (1975) A detailed vegetation study on the Chobe river in north-east Botswana. *Kirkia* 10, 185–227.

Simpson, C.D. (1978) Effects of elephant and other wildlife on vegetation along the Chobe River, Botswana. *Occasional Papers, The Museum Texas Technical University* 48, 1–15.

Skarpe, C. (1992) Dynamics of savanna ecosystems. *Journal of Vegetation Science* 3, 293–300.

Skarpe, C., Aarrestad, P.A., Andreassen, H.P., Dhillion, S., Dimakatso, T., du Toit, J.T., Halley, D.J., Hytteborn, H., Makhabu, S., Mari, M., Marokane, W., Masunga, G., Modise, D., Moe, S.R., Mojaphoko, R., Mosugelo, D., Motsumi, S., Neo-Mahupeleng, G., Ramotadima, M., Rutina, L., Sechele, L., Sejoe, T.B., Stokke, S., Swenson, J.E., Taolo, C., Vandewalle, M., & Wegge, P. (2004) The return of the giants: ecological effects of an increasing elephant population. *Ambio* 33, 276–282.

Sommerlatte, M.W. (1976) *A Survey of Elephant Populations in North-eastern Botswana.* Department of Wildlife and National Parks, Gaborone, Botswana.

Spinage, C.A. (2012) *African Ecology – Benchmarks and Historical Perspectives.* Springer-Verlag, Berlin.

Stokke, S. (1999) Sex differences in feeding-patch choice in a megaherbivore: elephants in Chobe National Park, Botswana. *Canadian Journal of Zoology* 77, 1723–1732.

Stokke, S., Motsumi, S., Skarpe, C. & Swenson, J.E. (2003) Ungulate population densities and composition in relation to habitat types and browsing pressure from elephant and other browsers in the Chobe River area, northern Botswana. In: Vandewalle, M. (ed.) *Effects of Fire, Elephants and Other Herbivores on the Chobe Riverfront Ecosystem.* Government Printer, Gaborone, Botswana, pp. 57–58.

Teren, G., Owen-Smith, N. & Erasmus, B.F.N. (2011) Structural and compositional riparian woodland change caused by extreme elephant impact in Northern Botswana. *South African Journal of Botany* 77, 560–561.

van Oudtshoorn, F. (1999) *Guide to Grasses of Southern Africa.* Briza Publications, Pretoria, South Africa.

Walker, B.H. & Noy-Meir, I. (1982) Aspects of the stability and resilience of savanna ecosystems. In: Huntley, B.J. & Walker, B.H. (eds.), *Ecology of Tropical Savannas.* Springer-Verlag, Berlin, pp. 556–590.

Walker, B.H. (1987) A general model of savanna structure and function. In: Walker, B.H. (ed.) *Determinants of Tropical Savannas.* IRL Press, Oxford, UK, pp. 1–12.

Walter, H. (1954) Die Verbuschung, eine Erscheinung der subtropischen Savannengebiete, und ihre Ökologischen Ursachen. *Vegetatio* 5/6, 6–10.

5

Vegetation: Between Soils and Herbivores

Per Arild Aarrestad[1], Håkan Hytteborn[2,3], Gaseitsiwe Masunga[4] and Christina Skarpe[5]

[1]Norwegian Institute for Nature Research, Norway
[2]Department of Plant Ecology and Evolution, Evolutionary Biology Centre, Uppsala University, Sweden
[3]Department of Biology, Norwegian University of Science and Technology, Norway
[4]Okavango Research Institute, University of Botswana, Botswana
[5]Faculty of Applied Ecology and Agricultural Sciences, Hedmark University College, Norway

The vegetation of the study area in Chobe National Park is influenced by a range of factors, including inundation by the Chobe River, soil moisture and fertility, and the impacts of different-size grazers and browsers. The vegetation constitutes a substrate (*sensu* Pickett *et al.*, 2003; Chapter 1) on which elephants and other large herbivores act. As such, vegetation change during the last century partly reflects fluctuations in elephant densities (Chapters 4 and 6). There is reason to believe that variations in elephant activity and the changes these induce in the vegetation, are also controlled by factors such as distance from the river, soil characteristics and the activities of meso-herbivores – medium-sized herbivores with a 4–450 kg body mass (Fritz *et al.*, 2002; Pickett *et al.*, 2003; Chapters 2, 8, 9 and 13).

Soil resource availability can control the effects of activities by megaherbivores, herbivorous animals typically weighing more than 1000 kg (Owen-Smith, 1988), on the vegetation by influencing the distribution of megaherbivores, the role of mesoherbivores as controllers and how different plant species and plant communities respond to herbivory (du Toit, 2003; Naiman *et al.*, 2003; Rutina *et al.*, 2005; Pretorius *et al.*,

Elephants and Savanna Woodland Ecosystems: A Study from Chobe National Park, Botswana,
First Edition. Edited by Christina Skarpe, Johan T. du Toit and Stein R. Moe.
© 2014 John Wiley & Sons, Ltd. Published 2014 by John Wiley & Sons, Ltd.

2011; Chapter 12). Large herbivores can also influence soil nutrient dynamics directly by depositing faeces and urine, and indirectly by influencing the quality and quantity of plant litter (McNaughton, 1979; Botkin et al., 1981; Hobbs, 1996; Ritchie et al., 1998; Persson et al., 2005; Chapter 9). In addition, herbivores can be linked to vegetation as responders, without necessarily having major influences on it (Chapter 12). Within the framework set by the evolutionary history of the flora and the environmental history of the region, including human land use, the species composition and structure of vegetation in general is a result of the availability of plant resources, such as light, water and soil nutrients, and of the loss of biomass, for example, by herbivory (Pickett and White, 1985; Huntly, 1991; Grime et al., 1997). In some grazing ecosystems the functional attributes and species composition of vegetation are explained more by the intensity of recent herbivory than by resource availability (McNaughton, 1979; Norton-Griffiths, 1979; Cumming, 1982; Skarpe, 1991; Augustine and McNaughton, 1998), whereas in other grazing systems, the effects of herbivory are not so apparent, and local vegetation variation is explained more by resource availability (O'Connor, 1991; Milchunas and Lauenroth, 1993).

In this chapter, we focus on how the structure and species composition of the present vegetation in northern Chobe National Park is related to recent herbivory by elephants, as agents shaping the vegetation, and by mesoherbivores acting as controllers or responders (Chapters 3, 4, 10, 12 and 13), along with abiotic controllers such as soil type and distance to the river (Chapter 9).

Habitat types

Using aerial photographs from 1998, together with field knowledge of the structure and composition of, primarily, the woody vegetation, we initially stratified the study area of the Botswana-Norway Institutional Co-operation and Capacity Building (BONIC) project into five landscape types. Each of these appeared to be used differently by elephants (Loxodonta africana; Chapter 12) and hence are referred to here as elephant habitat types: (i) floodplain, (ii) Capparis shrubland, (iii) Combretum shrubland, (iv) mixed woodland and (v) Baikiaea woodland (Skarpe et al., 2004). These habitat types were generally situated sequentially at increasing distance from the Chobe River. The floodplain is located on alluvial soils, mostly calcic gleysols, calcic luvisols and fluvisols (FAO, 1990; Farrar et al., 1994). The floodplain is inundated by the river from March to June and appears as vigorous grassland in the low-water season with almost no woody species. Away from the river, the vegetation becomes increasingly dependent on rainfall. Capparis shrubland is situated on flat areas close to the riverbanks on raised alluvium, mainly calcic luvisols. It is dominated by the shrub Capparis tomentosa, often acting as a scrambler and Flueggea virosa, which can be both a shrub and a tree. Combretum shrubland is situated on Kalahari sand, or a mixture of sand and alluvial deposits, and is dominated by the shrub or scrambler *Combretum mossambicense*. Mixed wood-land with several short-stature tree species and scattered large *Baikiaea plurijuga* forms a transitional zone between the shrublands and the taller *Baikiaea* woodland to the south, where *Baikiaea plurijuga* is the dominant tree species on deep, porous Kalahari sand (Skarpe *et al.*, 2004).

Plant communities, species diversity and structure of vegetation

To describe the flora and the species composition of the vegetation (as well as for other sampling within the BONIC project; Chapters 4, 9 and 13) we established five tran-sects along existing firebreaks, aligned north-south, perpendicular to the river, and running through the different habitat types (Skarpe *et al.*, 2004; Aarrestad *et al.*, 2011). The mean length of the transects was 10.3 km (range 9.5 – 12.0 km). Sampling of the vegetation and associated environmental variables was stratified according to the five habitat types. In each habitat along all five transects, we aimed to establish three per-manent sampling plots, each 20 m × 20 m, placed equidistant from one another at 50 m perpendicular distance from the firebreaks. Not all transects included all habitat types, however. We managed to establish 12 plots on the floodplain, 13 in *Capparis* shrubland, and 15 in each of the *Combretum* shrubland, mixed woodland and *Baikiaea* woodland habitats, a total of 70 plots overall.

For the present study each plot was surveyed once in the mid-wet season, January – March, during the period 1999 – 2002. We measured the percentage cover of each species in the field layer in five randomly placed 1 m × 1 m quadrats, together with the canopy cover and height of different layers within each quadrat, which we used for calculating mean cover and mean height for the sample plot. Cover measurements were done by visually projecting the canopies of each species on to the ground area of the sample plot. The field layer consisted of all herbs (forbs, grasses and sedges) and shrubs and young trees less than 0.5 m tall. The shrub layer was defined as woody plants 0.5 – 3.0 m tall, and the tree layer as all woody plants taller than 3 m.

Nomenclature for woody species follows Coates-Palgrave (2002) and that for forbs and graminoids, Barnes *et al.*, (1994), except the genus *Chamaecrista* Moench, and the species *Eclipta alba* (L.) Hassk. The species *Ambrosia artemisiifolia* L., *Sporobolus africanus* (Poir.) Robyns and Tournay and *Thunbergia reticulata* A. Rich., were not listed in Barnes *et al.*, (1994). Species that could not be identified were either classified by their genus name or growth form (shrub, grass or forb). No distinction was made between the grasses *Digitaria milanjiana* and *Digitaria eriantha*, because of their high variation in morphology and similarity in some inflorescence patterns (Theunissen, 1997); between the forbs *Sida alba* and *Sida cordifolia*; and between *Harpagophytum procumbens* and *Harpagophytum zeyheri*. Thus, these 'taxa' might consist of more than one species.

A two-way indicator species analysis (TWINSPAN version 2.3: Hill, 1979) classified the vegetation data from tree, shrub and field layers into four more or less distinct plant community groups (i) *Baikiaea plurijuga – Combretum apiculatum* woodland, (ii) *Combretum mossambicense – Friesodielsia obovata* wooded shrubland, (iii) *Capparis tomentosa – Flueggea virosa* shrubland and (iv) *Cynodon dactylon – Heliotropium ovalifolium* floodplain, named after the TWINSPAN indicator or preferential species with high cover, and the relative amount of shrubs and trees (Table 5.1). The indicators, the strongest preferential species in each of the communities, were all woody species in the woodland and shrublands, reflecting the prominence there of woody plants. This classification of the plant communities differs somewhat from that of the same sample plots reported by Aarrestad *et al.*, (2011), where only the field layer vegetation was analysed.

Detrended correspondence analysis (DCA: Hill and Gauch, 1980; ter Braak and Smilauer, 2002) of the sample plot data showed that the main variation in species composition was concentrated along the first DCA axis (eigenvalue, 0.81; sum of total eigenvalues, 5.17; standard deviation, 5.4: Figure 5.1). The TWINSPAN plant communities were fairly well separated along this axis, although there was some overlap of sample plots between the *Baikiaea plurijuga – Combretum apiculatum* woodland and the *Combretum mossambicense – Friesodielsia obovata* wooded shrubland community (Figure 5.1), indicating that some of these sample plots had several species in common.

Baikiaea plurijuga – Combretum apiculatum woodland

The *Baikiaea plurijuga – Combretum apiculatum* woodland community (Plate 4) was restricted to the *Baikiaea* woodland and mixed woodland habitat types. This community was species-rich and diverse, with 131 taxa recorded in 21 plots (a mean of 40 taxa per 0.04 ha plot) and a mean Shannon diversity index of 3.00, based on mean species-abundance values for each plant community (Table 5.1).The woodland was characterised by a sparse tree layer with a mean height of 7.0 m and a mean canopy cover of 10%, based on the five randomly selected measurements within the plot. Several woody species had their optimum distribution (occurring most frequently with high cover values) within this community (Table 5.1). Trees of *Baikiaea plurijuga* often reached a height of more than 10.0 m. In contrast, the mean height of the shrub layer, dominated by *Baphia massaiensis*, *Bauhinia petersiana*, *Combretum apiculatum* and *Croton gratissimus*, was 1.6 m. Shrub canopy cover averaged 9%. Some woody species such as *Acacia fleckii*, *Combretum molle*, *Diplorhynchus condylocarpon*, *Erythrophleum africanum*, *Ochna pulchra* and *Vitex mombassae* were only found in this community, all in low numbers.

Table 5.1 TWINSPAN derived plant communities of 70 (20 m × 20 m) sample plots along the Chobe River of Chobe National Park, Botswana with species showing an optimal distribution within the plant communities. Mean values and standard deviation (s.d.) of environmental variables from the same sample plots, found statistically significant to the overall species variation within the plant communities. Number of sample plots (N), total number of taxa per community (Tot sp.), mean number of taxa per sample plot (Mean sp.), Shannon mean diversity index (S-mean), herbivory (index) expressed in percentage (%). Faeces, alluvial soil and Kalahari sand as average of present/absent data. C (organic carbon) in wt%. All values shown are per community.

	Baikiaea plurijuga – Combretum apiculatum woodland	Combretum mossambicense – Friesodielsia obovata woodland/shrubland	Capparis tomentosa – Flueggea virosa shrubland	Cynodon dactylon – Heliotropium ovalifolium floodplain
Indicator species				
Trees and shrubs	Baikiaea plurijuga Combretum apiculatum	Combretum mossambicense	Capparis tomentosa Flueggea virosa	
Forbs				Heliotropium ovalifolium
Grasses				Cynodon dactylon
Preferential species				
Trees and shrubs	Baphia massaiensis Bauhinia petersiana	Acacia nigrescens Canthium glaucum ssp. frangula	Croton megalobotrys Philenoptera violacea	

(continued overleaf)

Table 5.1 (*Continued*)

Baikiaea plurijuga – Combretum apiculatum woodland	*Combretum mossambicense – Friesodielsia obovata* woodland/shrubland	*Capparis tomentosa – Flueggea virosa* shrubland	*Cynodon dactylon – Heliotropium ovalifolium* floodplain
Combretum molle	*Clerodendrum ternatum*		
Combretum collinum	*Combretum elaeagnoides*		
Croton gratissimus	*Friesodielsia obovata*		
Diplorhynchus condylocarpon	*Philenoptera nelsii*		
Erythrophleum africanum			
Markhamia obtusifolia			
Ochna pulchra			
Jasminium stenolobum			
Pavetta gardeniifolia			
Vitex mombassae			

Forbs	Commelina africana Corchorus olitorius Euphorbia crotonoides Evolvolus alsinoides var. linifolius Hemizygia bracteosa Hibiscus vitifolius Merremia tridentata ssp. angustifolia var. angustifolia Rhynchosia totta var. totta Spermacoce senensis	Ipomoea plebeia ssp. africana	Acanthospermum hispidum Asparagus sp. Bidens schimperi Cardiospermum halicacabum Corchorus tridens Duasperma quadrangulare Indigophera spp.	Acalypha sp. Ambrosia artemisiifolia Eclipta alba Hibiscus trionum Monsonia glauca
Grasses and sedges	Brachiaria nigropedata Cyperus margaritaceus Dactyloctenium giganteum		Brachiaria deflexa Cenchrus ciliaris Chloris virgata	Brachiaria eruciformis Digitaria maniculata Sporobulus africanus

(continued overleaf)

Table 5.1 (Continued)

	Baikiaea plurijuga–Combretum apiculatum woodland	Combretum mossambicense–Friesodielsia obovata woodland/shrubland	Capparis tomentosa–Flueggea virosa shrubland	Cynodon dactylon–Heliotropium ovalifolium floodplain
	Panicum maximum		Eragrostis cilianensis	Vetiveria nigritana
	Schmidtia pappophoroides		Eragrostis cylindriflora	
			Eragrostis superba	
			Eragrostis trichophora	
			Setaria sagittifolia	
			Sporobulus festivus	
			Tragus berteronianus	
			Urochloa spp.	
N	21	19	18	12
Tot sp.	131	113	129	48
Mean sp.	40.4 (8.9)	33.0 (8.4)	32.1 (5.5)	12.6 (3.2)
S-mean	3.00	3.43	3.18	1.31
S-mean rarified	2.95	3.36	3.12	–

Environmental variables

Light availability-herbivore impact

Woody cover (%)	41.7 (19,6)	32.1 (23.0)	15.5 (11.9)	0
Herbivory (%)	2.9 (3.4)	3.9 (3.8)	13.6 (10.0)	4.1 (10.5)
Twigs browsed (%)	21.9 (17.3)	26.3 (14.9)	15.3 (23.4)	–
Elephant bites	46.3 (39.5)	101.6 (54.9)	40.1 (62.8)	–
Faeces (+/−)	0.05 (0.14)	0.17 (0.22)	0.21 (0.21)	0.38 (0.32)
Soil resources				
Alluvial soil	0 (0)	0 (0)	0.7 (0.4)	1 (0)
Kalahari sand	1 (0)	0.9 (0.2)	0.2 (0.4)	0 (0)
pH	5.0 (0.35)	5.4 (0.5)	6.2 (0.7)	4.8 (0.3)
Ca (mg/100g)	24.7 (10.7)	36.1 (29.9)	131.8 (92.1)	420.2 (230.7)
Mg (mg/100g)	3.8 (1.5)	4.2 (2.1)	7.6 (2.7)	39.9 (35.0)
Na (mg/100g)	0.7 (0.5)	0.7 (0.4)	3.1 (2.4)	69.9 (120.2)
CEC (meq/100g)	2.3 (0.8)	2.8 (1.3)	5.8 (2.4)	32.6 (15.8)
C (%)	0.4 (0.1)	0.5 (0.2)	0.7 (0.4)	4.6 (3.0)
P (mg/100g)	11.9 (11.2)	20.5 (15.9)	65.5 (93.1)	43.3 (23.4)

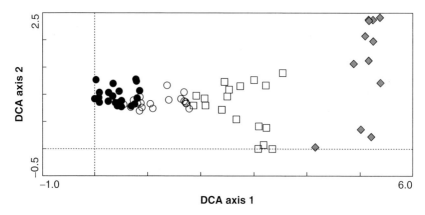

Figure 5.1 **Detrended correspondence analysis (DCA) diagram of the field, shrub and woody layer vegetation from 70 (20 m × 20 m) sample plots in Chobe National Park Botswana, with their TWINSPAN derived plant community membership. Axes 1 and 2 are in standard deviations. Black circles** *Baikiaea plurijuga – Combretum apiculatum* **wood-land, open circles** *Combretum mossambisence – Friesodielsia obovata* **wooded shrubland, open squares** *Capparis tomentosa – Flueggea virosa* **shrubland, grey dimonds** *Cynodon dactylon – Heliotropium ovalifolium* **floodplain.**

The field layer had a mean height of 0.65 m, with a mean canopy cover of 34%, mainly as a result of the dominance of tall grasses such as *Panicum maximum, Dactylocte-nium giganteum* and *Digitaria eriantha/milanjiana.* The grass *Brachiaria nigropedata* was restricted to this community, although it was a rather rare species that reached its optimum distribution close to the southern boundary of the study area along the Kasane-Ngoma road. Common sedges were *Abildgaardia hispidula, Cyperus margari-taceus* and *Mariscus dubius.* Many forbs reached their optimum in this community. The most common were *Commelina* spp., *Rynchosia totta* var. *totta* and *Spermacoce senen-sis,* with *Commelina africana, Euphorbia crotonoides* and *Hemizygia bracteosa* being found only in this community. The woodland community was also characterised by the presence of several uncommon species, such as the woody species *Kirkia acumi-nata, Ochna cinnabarina* and *Pseudolachnostylis maprouneifolia,* and the herbs *Gloriosa superba, Crabbea velutina* and *Hibiscus lobarius.*

This woodland community resembles 'Kalahari sand woodland', a herbivory-influenced, dry, deciduous woodland type in Hwange National Park in Zimbabwe (Mitchell, 1961), and '*Baikiaea plurijuga* woodland types' in the classification of Zimbabwean vegetation (Rattray, 1961). The Zimbabwean woodlands have a similar species composition, including the trees *Baikiaea plurijuga, Erythrophleum africanum, Ochna pulchra* and *Kirkia acuminata*; the shrubs *Baphia massaiensis, Combretum* spp., *Commiphora* spp. and *Diplorhynchus condylocarpon* and the grasses *Aristida* spp. and

Digitaria spp. With its comparatively high grass cover and patches of open woodland, this community also has much in common with open tree-savannas described by Rattray (1961) as mixed tree-bush-grass formations. A similar vegetation type, described from Pandamatenga, south-east of our study area in Botswana, has been called 'Mixed *Baikiaea* woodland' by Ringrose *et al.* (2003) and 'broad-leafed semi-deciduous savanna' by Scholes *et al.* (2004). *Commiphora mossambicensis*, *Baphia massaiensis*, *Bauhinia petersiana*, *Diplorhynchus condylocarpon* and *Terminalia* spp. were the most abundant woody species at Pandamatenga (Ringrose *et al.*, 2003; herbaceous species were not reported). With the exception of *Combretum mossambicense*, not present at Pandamatenga, the species composition is broadly similar to that of our study, even though the relative abundance of each species is somewhat different.

Combretum mossambicense – Friesodielsia obovata wooded shrubland

The *Combretum mossambicense – Friesodielsia obovata* wooded shrubland community (Plate 3) was restricted to the habitat types mixed woodland and *Combretum* shrubland. It had fewer species than the *Baikiaea plurijuga – Combretum apiculatum* woodland with 113 taxa occurring in 19 plots, a mean of 33 taxa per 0.04 ha plot. The mean Shannon diversity index of 3.43, however, was the highest of the four communities (Table 5.1). Felling of trees by elephants and regeneration of shrubs in open patches has created a mosaic of patches with tall trees alternating with open shrublands. With a mean tree layer height of 7.0 m and mean tree layer cover of 7%, the tree layer was less dense than that of *Baikiaea plurijuga – Combretum apiculatum* woodland. The dominant tree was *Baikiaea plurijuga*, although *Combretum mossambicense* often reached a height of more than 3.0 m. Mean height of the shrub layer was 1.8 m with a mean cover of 9%, reflecting a slightly more open community than *Baikiaea plurijuga – Combretum apiculatum* woodland. In addition to the dominant shrubs, *Combretum mossambicense* and *Friesodielsia obovata*, a few other woody species reached their optimum within this community, including *Acacia nigrescens*, *Combretum elaeagnoides* and *Philenoptera nelsii* (Table 5.1). Several common shrubs in the *Baikiaea plurijuga – Combretum apiculatum* woodland, such as *Combretum celastroides*, *Combretum engleri* and *Vangueria infausta*, were also found in this community, together with trees and shrubs of species more common in the *Capparis tomentosa – Flueggea virosa* shrubland closer to the river, such as *Berchemia discolor*, *Boscia albitrunca*, *Dichrostachys cinerea* and *Strychnos potatorum*.

The mean height of the field layer was 0.5 m with a mean cover of 20%, lower and less dense than that of *Baikiaea plurijuga – Combretum apiculatum* woodland. Only one forb, *Ipomoea plebeia* ssp. *africana*, reached its optimum here, although several field-layer species such as the forbs *Gisekia africana*, *Harpagophytum* spp., *Jacquemontia tamnifolia*, *Momordica kirkii*, *Thunbergia reticulata*, *Vigna* cf. *unguiculata*, and the graminoids *Mariscus laxiflorus*, *Digitaria eriantha/milanjiana*, *Panicum maximum* and *Pogonarthria squarrosa* were common both to this community and to *Baikiaea plurijuga – Combretum apiculatum* woodland. Some species, such as *Pupalia lappacea* var. *velutina*, *Tribulus terrestris* and *Urochloa* spp., were also common in the *Capparis tomentosa – Flueggea virosa* shrubland. This lack of strongly preferential forb and grass species suggests that the *Combretum mossambicense – Friesodielsia obovata* community is less clearly defined than the others. Nevertheless, it is reasonably distinct from the *Baikiaea plurijuga – Combretum apiculatum* woodland community because several species characteristic of that woodland were absent or less abundant. Species preferring disturbed areas were common, such as *Cleome hirta*, *Pogonarthria squarrosa*, *Tribulus terrestris* and *Urochloa* spp. (Blomberg-Ermatinger and Turton, 1988; van Oudtshoorn, 1999). This strongly suggests that species composition is affected by elephants opening up the woodland and creating a patchy structure.

Capparis tomentosa – Flueggea virosa shrubland

The *Capparis tomentosa – Flueggea virosa* shrubland community (Plate 2), occurring mainly in the *Capparis* habitat type, was relatively diverse, with 129 taxa recorded in 18 plots, a mean of 32 taxa per plot. The mean Shannon diversity index for plots of 3.18 was slightly higher than in the *Baikiaea* woodland, but lower than in the *Combretum mossambicense – Friesodielsia obovata* wooded shrubland (Table 5.1). The woody component was dominated by the shrubs *Capparis tomentosa* and *Flueggea virosa*, but *Capparis tomentosa* and *Croton megalobotrys*, another preferential species, sometimes grew taller than 3 m, thereby forming a sparse tree layer. This had a mean height of 4.0 m and a mean cover of 3%, whereas mean shrub height was 2.4 m with a mean cover of 7%. This low woody structure and sparse cover implies high light availability in the field layer.

The field layer had a mean height of 0.5 m and a mean cover of 28%. It was characterised by a high species richness of grasses. *Brachiaria deflexa*, *Cenchrus ciliaris*, *Chloris virgata*, several *Eragrostis* species, *Setaria sagittifolia*, *Sporobulus festivus* and *Tragus berteronianus*, were all preferential species for this community and, in general, apparently nutrient-demanding (Ernst and Tolsma, 1992). *Urochloa* spp. was abundant in open patches; this species was also prominent in *Combretum mossambicense – Friesodielsia obovata* wooded shrubland. Other preferential species were the forbs *Acanthospermum hispidum*, *Asparagus* sp., *Bidens schimperi*, *Cardiospermum halicacabum*, *Corchorus tridens*, *Duosperma quadrangulare* and *Indigofera* spp. The grass *Cynodon dactylon* occurred frequently in the sample plots close to the floodplain, although not abundantly so. The number of species preferring disturbed soil was even higher than in the *Combretum mossambicense – Friesodielsia obovata* wooded shrubland, reflecting pervasive animal impact.

Cynodon dactylon – Heliotropium ovalifolium floodplain

The *Cynodon dactylon – Heliotropium ovalifolium* floodplain community was markedly different from the woodland and shrubland communities, because of the absence of woody species (Plate 1). It was characterised by a low-growing field layer (mean height 0.25 m; mean cover 44%), influenced by grazing, heavy trampling by animals and flooding. It had the lowest and densest field layer of all four plant communities, dominated by a dense carpet of grazed *Cynodon dactylon* and the forb *Heliotropium ovalifolium*, both indicator species for this community. Species richness and diversity was also low, with only 48 taxa recorded in 12 sample plots, a mean of 12.6 taxa per plot and a mean Shannon diversity index of 1.31 (Table 5.1). Greatest diversity occurred among the grasses. Apart from the two dominant species, other preferential species were the forbs *Acalypha* sp., *Ambrosia artemisiifolia*, *Eclipta alba*, *Hibiscus trionum*, *Monsonia glauca* and the grasses *Brachiaria eruciformis*, *Digitaria maniculata*, *Sporobulus africanus* and *Vetiveria nigritana*, the last of which occurred mainly in fairly dense stands close to the river. Rarer species found only on the floodplain were the forb *Nidorella resedifolia* ssp. *resedifolia*, the sedges *Kyllinga alba*, *Scirpus microcephalus*, *Pycreus flavescens*, *Schoenoplectus erectus* and the grasses *Dactyloctenium aegyptium*, *Echinochloa colona*, *Echinochloa stagnina*, *Panicum repens* and *Sporobolus pyramidalis*. Despite the high number of species confined to this floodplain community, several species also occurred in the shrubland communities, although much less frequently. These included the grasses *Chloris virgata*, *Digitaria eriantha/milanjiana* and *Tragus berteronianus*, and the forbs *Chamaecrista* spp., *Sida* spp. and *Tribulus terrestris*. This similarity in species composition is probably related to the short dispersal distance between floodplain and shrubland plots (a few 100 m) and the fact that the plots were located on the middle and upper part of the floodplain, away from the main river channel and therefore only inundated for a short period between March and June. Flooding generally excludes woody plants, and different flooding regimes can greatly influence plant species composition, as described from the Okavango Delta in Botswana (Bonyongo *et al.*, 2000). The *Cynodon dactylon – Heliotropium ovalifolium* floodplain community in this study shared many species with the *Vetiveria nigritana – Setaria sphacelata* community and the *Sporobolus spicatus* community (*Cynodon dactylon* sub-community) of the Okavango floodplain, which occupied the highest elevations, least influenced by flooding.

Species richness and diversity

The length of the first DCA axis, 5.4 standard deviation units, reflects high compositional turnover of species and large between-community (β) diversity along our landscape gradient. The *Baikiaea plurijuga – Combretum apiculatum* woodland community was the richest community (Table 5.1). The mean number of species decreased successively from the *Combretum mossambicense – Friesodielsia obovata* wooded shrubland to the *Capparis tomentosa – Flueggea virosa* shrubland and the *Cynodon dactylon – Heliotropium ovalifolium* floodplain. This might support the intermediate disturbance hypothesis of Connell (1978) and Huston (1979) that moderate herbivory (as in the woodland) leads to high species richness, whereas substantial herbivory reduces species richness (e.g. the *Cynodon dactylon – Heliotropium ovalifolium* floodplain community). Nevertheless, many other factors, such as differences in soil conditions and in vegetation history (Chapters 4 and 9), can also influence species richness. On the floodplain, the low number of forbs might be the result of the dominance of, and competitive exclusion by, carpets of *Cynodon dactylon*, a species favoured by intense grazing. Species diversity was highest in *Combretum mossambicense – Friesodielsia obovata* wooded shrubland (Table 5.1), where elephants open up the woodland by browsing and pushing down trees. This disturbance can lead to a patchy landscape, less homogeneous vegetation and a diverse community with assorted species occurring in different plots (Table 5.1). The high diversity of these shrublands is probably related both to the nutrient-rich soil and to high light availability created by browsers opening up the woody vegetation. Because different numbers of plots were sampled in each community, these measures of diversity could be biased. Rarifying the mean Shannon indices to a base of 12 samples (Sanders, 1968), the lowest recorded number of sites in one community, the floodplain, did not significantly change the pattern of diversity across the plant communities however (Table 5.1).

Abiotic and biotic variables related to the present vegetation

To find the relationship between the vegetation and external factors (agents, controllers and responders), we recorded at each plot several environmental variables assumed both to affect and to respond to plant species composition. Light availability was measured as mean canopy cover of woody species above 0.5 m (**Woody-C**). Degree of animal impact was measured as (i) the occurrence of faeces as present/absent (**Faeces**); (ii) the proportion of biomass removed by grazing and browsing on plants in the field layer (**Herbivory**); (iii) browsing pressure exerted by elephants, measured as the number of twigs and branches bitten or broken by elephants, expressed as percentage of the total number of twigs with a diameter of 8 mm within 2.5 m of the ground, based on data from the whole 20 m × 20 m plot (**Twigs browsed**; 8 mm is the average diameter of twigs bitten off by elephants: Stokke and du Toit, 2000; Chapter 12) and (iv) the total number of twigs browsed or broken by elephants per plot (**Elephant bites**). Soil types were recorded as Kalahari sand (**K-sand**) and alluvial soil (**A-soil**), recorded as present or absent in each plot and expressed as an average for each community. Soil nutrient availability was assessed by chemical analyses of soil samples collected in the mid-wet season over the same 4-year period as the vegetation data.

The variables measured were pH, organic carbon (C), extractable phosphorous (P), exchangeable magnesium (Mg), calcium (Ca) and cation exchange capacity (CEC).

Direct gradient analysis in the form of canonical correspondence analysis (CCA) (ter Braak, 1987) was used to find the relationship between the variation in species composition and the observed environmental variables, including the distance to the Chobe River (**Dist. riv.**). Only those variables found to be significantly related to the variation in species composition at the 1% probability level, either in a CCA of each environmental variable alone (marginal effects) or in forward selection of the variables (conditional effects) in a Monte Carlo permutation test, were selected for analysis (ter Braak and Smilauer, 2002).

The CCA ordination revealed strong first and second axes with eigenvalues of 0.72 and 0.44, respectively; the sum of total eigenvalues was 5.17. The significance level of the environmental variables used in the CCA and their correlation with the species data showed that the most important variables were soil characteristics, followed by the distance to river and woody cover. Among the measured variables, signs of animal impact (**Herbivory** and **Faeces**) were less related to the variation in species composition. Elephant impact (**Twigs browsed** and **Elephant bites**) were significantly correlated with the species composition in the woodland and shrubland communities. Resource availability and primary soil resources have been shown to be key variables in other arid and semiarid regions (Milchunas and Lauenroth, 1993). Nevertheless, an analysis using variance partitioning, done by partial CCA (Borcard *et al.*, 1992), showed that some of the variation in species composition could be explained by covariation (interaction) between soil resources and herbivory.

The CCA biplot diagram (Figure 5.2), which illustrates the relationship between the statistically significant environmental variables and their associations with the sample plots, shows that the plant communities are distributed along the first CCA axis in the same sequence as they are along the landscape gradient. The *Cynodon dactylon – Heliotropium ovalifolium* floodplain community and the *Capparis tomentosa – Flueggea virosa* shrubland are well separated on both axes, whereas the woodland types show little separation along axis 1. On axis 2, there is a gradual change from the *Capparis tomentosa – Flueggea virosa* shrubland, at the negative end of the axis, to *Baikiaea plurijuga – Combretum apiculatum* woodland, with positive scores. The first axis is related mainly to differences in soil characteristics with Kalahari sand, consisting of nutrient-poor ferralic arenosols, at one end, and the nutrient-rich alluvial soil at the other, positive end (Figure 5.2). The distance to the river is negatively correlated with CCA axis 1, implying an increase in soil fertility along the landscape gradient from the woodlands to the floodplain. This is broadly consistent with positioning of the soil nutrient vectors, all of which are positively associated with axis 1, reflecting the predominance of calcic luvisols. These cover the higher elevated parts of the floodplain and adjacent riverbanks and flats behind, and are more clayey with a higher mineral nutrient content (FAO, 1990; Aarrestad *et al.*, 2011).

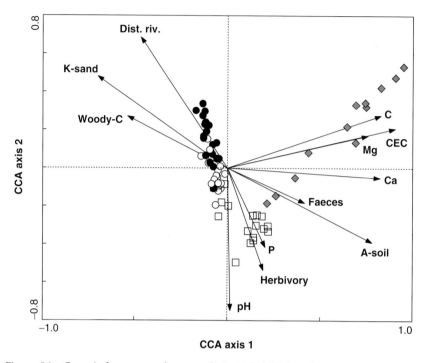

Figure 5.2   Canonical correspondence analysis (CCA) biplot diagram, axes 1 and 2, of environmental variables and vegetation data from the same 70 sample plots as in Figure 4.1 with their TWINSPAN derived plant community membership. Black circles *Baikiaea plurijuga – Combretum apiculatum* woodland, open circles *Combretum mossambisence – Friesodielsia obovata* wooded shrubland, open squares *Capparis tomentosa – Flueggea virosa* shrubland, grey dimonds *Cynodon dactylon – Heliotropium ovalifolium* floodplain.

The second CCA axis is negatively related to gradients in soil pH, phosphorous and herbivory, and positively correlated with distance to the river (Figure 5.2). This suggests that, in addition to an increase in soil fertility, the incidence of herbivory also increases along the gradient from woodland to shrubland. Woody cover, a measure of the amount of light reaching the ground, correlates negatively with axis 1 and positively with axis 2, with the highest values being in the woodland and decreasing towards the floodplain. At the same time, woody cover is negatively correlated with herbivory, which can be both a cause for, and an effect of, the open shrub-and-tree canopy.

Measures of herbivory on the floodplain were quite low, although the incidence of herbivore faeces, a proxy measure of the time spent by animals in the plots, were highest on this part of the gradient. Signs of herbivory were probably less noticeable in the low, mat-forming, *Cynodon dactylon*, as this species compensates quickly for grazed tissue (Mathisen, 2005). Consequently, unless whole tillers are eaten, evidence of herbivory will be largely hidden. Therefore, despite little direct evidence of intensive herbivory on the floodplain, the indirect evidence, most noticeably the clear pattern of variation in animal faeces, suggests that the frequency and intensity of herbivory also increases from the woodlands to the floodplain, and that herbivores are both affecting and responding to the distribution of the plant communities. **Elephant browsing** was not included in the CCA as it was not recorded in the floodplain, which lacked woody plants. Beyond that, indicators of elephant browsing (**Elephant bites** and **Twigs browsed**) were highest in the *Combretum mossambicense – Friesodielsia obovata* wooded shrubland and lowest in *Capparis tomentosa – Flueggea virosa* shrubland.

Overall, the CCA shows that the distribution of plant species and plant communities used by herbivores correlates best with soil properties (Phillips, 1992; Scholes and Walker, 1993; Ben-Shahar, 1995; Mapaure, 2001; Fornara and du Toit, 2008; Tessema *et al.*, 2011).

Baikiaea plurijuga – Combretum apiculatum woodland environment

The CCA diagram and the average values of environmental variables from the plant communities (Table 5.1) show that the species composition of the *Baikiaea plurijuga – Combretum apiculatum* woodland community is associated with low nutrient status of Kalahari sand.

The *Baikiaea plurijuga – Combretum apiculatum* woodland on Kalahari sand has a high woody canopy cover (42%) that reduces the amount of solar radiation reaching the field layer. Together with high litter cover and low exposure of bare soil (Aarrestad *et al.*, 2011), this reduces evaporation from the soil and should enhance nutrient cycling by creating a more favourable environment for decomposers to function. This in turn favours the growth of shade tolerant tall grasses and forbs, especially the grasses *Panicum maximum* and *Dactyloctenium giganteum*, which show a preference for growing in dense stands in the shade under tree canopies.

The species composition of this woodland is probably less affected by animal activity than that of the shrublands, judging by the rather low values of those variables reflecting the incidence and intensity of herbivory (**Herbivory**, **Elephant bites**, **Twigs browsed** and **Faeces**: Table 5.1). Nevertheless, elephants probably play an important indirect role in nutrient cycling in the woodlands, by turning large amounts of relatively nutrient-poor biomass into easily decomposable faeces and urine (Botkin *et al.*, 1981; Chapter 9). This should lead to faster and shallower cycling of nutrients compared with nutrients recycled through decomposing tree litter (Botkin *et al.*, 1981; Frost, 1985; Chapter 9), which in turn would favour shallow-rooted, nutrient-demanding species such as *Panicum maximum* and *Digitaria* spp. (Aarrestad *et al.*, 2011).

Combretum mossambicense – Friesodielsia obovata wooded shrubland environment

The *Combretum mossambicense – Friesodielsia obovata* wooded shrubland community was also highly associated with the nutrient-poor Kalahari sand, although the associated soils were marginally more fertile. Whether the dominant species in this community are more nutrient-demanding than those in *Baikiaea plurijuga – Combretum apiculatum* woodland is an open question. Woody cover (32%) was lower than in the latter woodland, indicating that more light reached the ground, thereby favouring light-tolerant species such as *Tribulus terrestris*. Given the higher indices of animal activity – **Herbivory**, **Twigs browsed**, **Faeces** and especially **Elephant bites** (Table 5.1) – we hypothesise that this community is a result of substantial animal disturbance, especially by elephants, of an earlier, more homogeneous *Baikiaea plurijuga*-dominated woodland.

Capparis tomentosa – Flueggea virosa shrubland environment

In contrast to the woodland communities, the *Capparis tomentosa – Flueggea virosa* shrubland community occurred mainly on fine textured alluvial soil with high clay content. The low percentage cover of trees and shrubs (16%) ensures high light availability in the field layer. The soils were distinctly different from those of the plots on Kalahari sand, and had the highest average values of soil pH and phosphorous and higher concentrations of basic cations favouring the more nutrient-demanding plant species.

The ground was substantially disturbed by animal trampling. The abundance of faeces on the ground (**Faeces**) and the percentage of grazed and browsed plants in the field layer (**Herbivory**) were the highest across all communities, although there were fewer signs of elephant impact on the woody vegetation (**Elephant bites** and **Twigs browsed**) than in the *Combretum mossambicense – Friesodielsia obovata* wooded shrubland (Table 5.1). Apart from this, however, the indicators all reflected heavy grazing and browsing pressure in this community. This has produced an open shrubland, favouring light-tolerant and pioneer species such as *Tribulus terrestris*, *Indigophera* spp., *Urochloa* spp. and several other forbs and grasses.

Cynodon dactylon – Heliotropium ovalifolium floodplain environment

The *Cynodon dactylon – Heliotropium ovalifolium* floodplain community was found only on alluvial soil. The community is inundated annually and thus functions hydrologically in a quite different way from the other communities. Many of the soil variables such as exchangeable cations, organic carbon and CEC were markedly higher on the floodplain than in the other communities.

We speculate that the mosaic of more-or-less open soil patches is caused by the impacts of substantial animal trampling and the removal of whole plants during grazing. The ordination diagrams (Figures 5.1 and 5.2) show considerable habitat diversity along the DCA axis 2, reflecting large variation in species composition between the sample plots (Shmida and Wilson, 1985). Even though the incidence of grazed and browsed plants was rather low, faeces were abundant. Given this and the likelihood that the impact of grazers on the floodplain was probably underestimated, and the considerable numbers of animals visiting the river daily to drink, the *Cynodon dactylon – Heliotropium ovalifolium* floodplain is probably the most animal-affected of the plant communities that we studied.

Life-form and species distribution

The CCA analysis suggests that species distribution along the gradient from sand to alluvium is largely related to variation in resource availability (indexed by the soil variables and vegetation structure), with herbivory, as indicated by the relevant variables, being less important. Conversely, variation in species traits along the gradient agrees in many cases more with general hypotheses related to herbivory than to resource availability (Crawley, 1983; Coley *et al.*, 1985; Grime *et al.*, 1997). Nevertheless, the contrasting characteristics of the sandy and alluvial soils strongly affect vegetation composition, thereby diminishing the evidence of herbivore impact.

Woody species

Natural regeneration of the overstorey species was most apparent in *Baikiaea plurijuga – Combretum apiculatum* woodland, where small plants (<0.5 m tall) of tree and shrub species were common, indicating their ability to regenerate in the shade on the nutrient-poor Kalahari sand. Large tree species, such as *Baikiaea plurijuga*, were under-represented as seedlings in the field layer, compared with their presence in the tree and shrub layers, whereas smaller tree and shrub species, such as *Combretum* spp., were proportionally more common. This might be interpreted as a sign of a regression from woodland to shrub vegetation (Rattray, 1961; Mosugelo *et al.*, 2002), but could equally well reflect differences in regeneration strategies between the long-lived overstorey trees and the smaller trees and shrubs with presumably shorter generation times (Harper, 1977).

At the other end of the gradient, the floodplain was almost completely devoid of woody species, perhaps a result of regular flooding and probably substantial animal impact as well. Above the river bank, on raised alluvium close to the river, trees were largely absent, despite the nutrient-rich soils. Given the compact, clayey nature of these soils, moisture availability could be limiting, potentially restricting the growth of woody plants (Walker and Noy-Meir, 1982). Nevertheless, these areas, which today support sparse *Capparis tomentosa – Flueggea virosa* shrubland, and which in 1870, before the disappearance of elephants, were described as open flats (Selous, 1881), were covered by tall *Acacia – Combretum* woodland during the first half of the 20th century, showing that they can support woody growth. These woodlands became established when herbivore populations were low following ivory hunting and the outbreak of rinderpest. When elephant and impala populations again increased, these woodlands were replaced by the present shrub vegetation (Simpson, 1975, Skarpe *et al.*, 2004, Rutina, 2004; Moe *et al.*, 2009; Chapters 4 and 10). Thus, both structure and species composition of the vegetation on the raised alluvium along the river seems to be largely a result of herbivore activities.

Most of the woody species occurring as overstorey trees and in the field layer vegetation are vigorous resprouters, an adaptation to fire or herbivory, or both (Bond and Midgley, 2001). *Capparis tomentosa*, *Combretum mossambicense* and *Dichrostachys cinerea*, occurring at the resource-rich end of the gradient, are spinescent; *Capparis tomentosa* is thought to be poisonous (Coates-Palgrave, 2002; Chapter 12). *Capparis tomentosa* and *Flueggea virosa* are nutrient-demanding species, and are said to grow preferentially on termite mounds (Coates-Palgrave, 2002). In our transects they grew on the resource-rich raised alluvial flats adjacent to the floodplain, an area of intense herbivory.

Forbs

The forb vegetation in the nutrient-poor Kalahari sand under the woodland and shrubland overstorey contained many erect, comparatively tall perennial and annual species such as *Commelina* spp., *Hibiscus vitifolius*, *Jacquemontia tamnifolia* and *Spermacoce senensis*, as well as many climbers such as *Ipomoea* spp., *Monsonia kirkii*, *Rhynchosia totta* var. *totta* and *Vigna*. cf. *unguiculata*. This could reflect competition for light as well as more favourable microclimatic conditions under the tree canopies, coupled with low levels of herbivory. In the middle of the gradient, where there was much bare soil, small or creeping annuals such as *Acanthospermum hispidum*, *Corchorus tridens*, *Gisekia africana* and *Tribulus terrestris* predominated. There were, however, some erect, comparatively tall annuals such as *Chamaecrista* spp. Both a creeping growth form and an ephemeral life history can be interpreted as strategies for escaping herbivory by large herbivores. In both South Africa and Botswana, small forbs of the genera *Cleome*, *Corchorus*, *Gisekia*, *Indigofera*, *Limeum*, *Tribulus* and *Tephrosia*, several of which are annuals, have been reported to be prominent species in sandveld areas under pressure from heavy grazing and trampling (Skarpe, 1986; van Rooyen *et al.*, 1991; O'Connor and

Pickett, 1992). In contrast, on the resource-rich floodplains where levels of herbivory are high, the dominant forbs were perennial and non-creeping, and apparently heavily chemically defended (Chapter 12).

Graminoids

Grasses in the resource-poor and little browsed and grazed end of the gradient tended to be tall, tufted and sexually reproducing (e.g. *Panicum maximum, Dactyloctenium giganteum* and *Brachiaria nigropedata*). These grasses were found almost exclusively in the woodlands, where the canopy cover of overstorey trees and shrubs reduces light availability, evapotranspiration and temperature extremes in the field layer. This could be a prerequisite for the persistence of tall growing, broad-leafed lush species such as *Dactyloctenium giganteum* and *Panicum maximum*, both of which typically grow under trees (Yeaton *et al.*, 1986; Ernst and Tolsma, 1989; van Oudtshoorn, 1999). Despite resource limitations in Kalahari sand, all these grasses are nutrient-rich and palatable to grazers, and are unlikely to persist under heavy grazing. *Brachiaria nigropedata, Digitaria eriantha* and *Panicum maximum* have all been listed as species that decrease with increased herbivory (Novellie, 1988; Wenzel *et al.*, 1991).

With increasing resource availability and herbivory, and reduced shade in the middle sections of the gradient, the grasses tended to be smaller. Tufted perennial species, such as *Eragrostis trichophora, Pogonarthria squarrosa* and many *Aristida* spp. which do not withstand grazing well are largely unpalatable. In contrast, more palatable species, such as the annuals *Chloris virgata* and *Urochloa* cf. *trichopus*, escape herbivory in time or, like *Schmidtia pappophoroides*, are perennial and stoloniferous with good resprouting ability (van Oudtshoorn, 1999). Several of these species have been listed in other studies as increasing in heavily grazed areas (Wenzel *et al.*, 1991; Dahlberg, 2000).

The resource-rich floodplains experience the heaviest grazing pressure, at least in the dry season (Taolo, 2003; Mathisen, 2005). They were dominated by *Cynodon dactylon*, a palatable, creeping and clonal grass with substantial resprouting ability (Mathisen, 2005). Although the creeping growth form can be an escape strategy, by maintaining a proportion of the biomass inaccessible to grazers, clonal growth and resprouting ability indicates high tolerance of grazing. *Cynodon dactylon* is probably also adapted to the high sodium concentrations in the soil (Table 5.1), although it does not grow in areas with highest sodium concentrations (Ellery *et al.*, 1993). The species also has the ability to use urea nitrogen, probably highly available on the floodplain soils (Mathisen, 2005; Chapter 9). In contrast to *Cynodon dactylon*, the second most common species, *Vetiveria nigritana*, has hard, sharp leaves and chemical defence in the form of a vetiver-like oil (Kahlil and Ayoub, 2011).

Seed-bank of woody species

The considerable vegetation changes occurring over time along the Chobe River (Chapter 4) might be mirrored in the soil seed-bank. If so, buried seeds, if viable, could be a reservoir for future vegetation changes. To estimate which species occurred in the seed bank and in what quantities, the seeds of woody species were sampled and identified in the upper 0.1 m of the soil in two soil types along the Chobe River – pure alluvial soil and in alluvium with a windblown layer of Kalahari sand on top – at both fairly undisturbed and disturbed sites. The distribution of the seeds was generally highly clumped. Seeds were tested for viability with triphenyl-tetrazolium chloride method (Mathumo, 2003). The number of viable seeds in fairly undisturbed sites was about 64 seeds m^{-2} to 0.1 m depth, and in disturbed sites about 12 seeds m^{-2}, which is somewhat higher than in the Nylsvley savanna (Scholes and Walker, 1993). The number was higher in loose sand overlying alluvial soil, about 58 m^{-2}, than in the more compacted alluvial soil, about 21 m^{-2} (Mathumo, 2003).

Nineteen woody species were found in the soil seed-bank. Fourteen of these were tree species; the other five, shrub species. The dominant species in the seed-bank were *Berchemia discolor*, *Diospyros mespiliformis* and *Flueggea virosa*, constituting 52.0%, 15.2% and 14.1%, respectively, of the total seed-bank, although few seeds of *Berchemia discolor* and *Flueggea virosa* were viable, just 2.2% and 0.8% respectively. *Adansonia digitata*, *Erythroxylum zambesiacum*, *Sterculia africana*, *Sclerocarya birrea* and *Vangueria infausta* were found in the seed-bank, but were not recorded in the aboveground vegetation at the seed-bank survey sites. Three species, *Markhamia zanzibarica*, *Trichilia emetica* and *Combretum mossambicense*, present in the vegetation at these sites, were not found in the seed-bank by Mathumo (2003), although both we and Mathumo (2003) found *Trichilia emetica* and *Combretum mossambicense* during a preliminary seed-bank survey. The number of seeds of *Croton megalobotrys* was significantly higher in disturbed sites than in moderately undisturbed ones (Mathumo, 2003). The similarity between the above-ground vegetation and the soil seed-bank was low, both with the present vegetation (Jaccard's similarity index: 39.9%), and with the woody vegetation recorded by Simpson, (1975) (Jaccard index: 22.6%). Thus the soil seed-banks in both undisturbed and disturbed sites, and on both soil types, were not quantitatively representative of either the present or the previously recorded aboveground vegetation.

Some species, such as *Adansonia digitata*, *Dichrostachys cinerea*, *Sterculia africana*, *Philenoptera violacea* and *Acacia* species, had high seed viability (>90%), but in others, such as *Berchemia discolor*, *Flueggea virosa*, *Garcinia livingstonei* and *Sclerocarya birrea*, viability was low (<10%). Several species had clearly damaged seeds (e.g. about 98% of the seeds of *Berchemia discolor*). The low abundance of seeds of most woody species suggests that they might have limited potential for regenerating from the soil seed-bank (Mathumo, 2003). For species with viable seeds, however, the seed-bank could still be important for their regeneration and for vegetation dynamics generally.

Concluding remarks

The species composition of the vegetation changes significantly from the woodlands in the south to the Chobe River in the north, a change strongly related to soil characteristics and the structure of the shrub and tree cover. Despite the historic importance of elephants and other large herbivores in the dynamics of the vegetation in the Chobe area (Chapter 4), we found only a weak correlation between plant species composition and signs of the presence of, or foraging by, mesoherbivores and elephants. This could reflect a mismatch in scales. Herbivores influence the vegetation at many different spatial and temporal scales, whereas vegetation surveys are typically limited to just one or a few years and mostly to areas less than $100 \, km^2$. In such studies, the importance of biotic and abiotic environmental factors in structuring the plant communities is generally deduced by correlating spatial variation in plant species composition with a range of environmental variables, as was done in this study. This implicitly assumes that the scales of variation are the same. If the agent shaping the vegetation operates at a much larger (or smaller) spatial or temporal scale than that sampled, its effects will not be readily detected. Macroclimate is typically one such factor, and also herbivores, particularly large-bodied habitat generalists such as elephant, can influence much larger areas than those sampled in most vegetation studies. In our study, therefore, the factors accounting for most of the variation in the vegetation were soil type and distance from the river, both varying within the roughly 10 km sampled south of the river. These are controllers, both of the vegetation and its responses to elephant browsing, and of the elephants themselves. There was, however, a significant correlation between vegetation composition and signs of herbivory in the field layer along the vegetation gradient from the woodlands to the heavily grazed shrublands and floodplain. The mesoherbivores responsible for this grazing and browsing have more specific habitat requirements than elephants, and are primarily responders to the vegetation (Chapter 13) or, as with impala, *Aepyceros melampus*, and African buffalo, *Syncerus caffer*, controllers (Chapters 10 and 11) rather than as agents causing the variation. The mosaic structure of the reasonably dense *Baikiaea plurijuga – Combretum apiculatum* woodland and the more open *Combretum mossambicense – Friesodielsia obovata* wooded shrubland, however, might yet be a result of elephant activity, perhaps in interaction with fire, frost and direct human impact (Chapter 3).

References

Aarrestad, P.A., Masunga, G.S., Hytteborn, H., Pitlagano, M.L., Marokane, W. & Skarpe, C. (2011) Influence of soil, tree cover and large herbivores on field layer vegetation along a savanna landscape gradient in northern Botswana. *Journal of Arid Environment* 75, 290–297.

Augustine, D.J. & McNaughton, S.J. (1998) Ungulate effects on the functional species composition of plant communities: herbivore selectivity and plant tolerance. *Journal of Wildlife Management* 62, 1165–1183.

Barnes, J.E., Turton, L.M. & Kalake, E. (1994) *A List of the Flowering Plants of Botswana*. The Botswana Society and the National Museum, Monuments and Art Gallery, Gaborone, Botswana.

Ben-Shahar, R. (1995) Habitat classification in relation to movements and densities of ungulates in a semi-arid savanna. *African Journal of Ecology* 33, 50–63.

Blomberg-Ermatinger, V. & Turton, L. (1988) *Some Flowering Plants of South-Eastern Botswana*. Botswana Society, Gaborone, Botswana.

Bond, W.J. & Midgley, J.J. (2001) Ecology of sprouting in woody plants: the persistence niche. *Trends in Ecology and Evolution* 16, 45–51.

Bonyongo, M.C., Bredenkamp, G.J. & Veenendaal, E. (2000) Floodplain vegetation in the Nxaraga Lagoon area, Okavango Delta, Botswana. *South African Journal of Botany* 66, 15–21.

Borcard, D., Legendre, P. & Drapeau, P. (1992) Partialling out the spatial component of ecological variation. *Ecology* 73, 1045–1055.

Botkin, D.B., Mellilo, J.M. & Wu, L.S.-Y. (1981) How ecosystem processes are linked to large mammal population dynamics. In: Fowler, C.W. & Smith, T.D. (eds.) *Dynamics of Large Mammal Populations*. John Wiley & Sons, New York, pp. 373–387.

Coates-Palgrave, K. (2002) *Trees of Southern Africa*. Struik Publishers, Cape Town, South Africa.

Coley, P.D., Bryant, J.P. & Chapin, F.S. III., (1985) Resource availability and plant antiherbivore defense. *Science* 230, 895–899.

Connell, J.H. (1978) Diversity in tropical rainforests and coral reefs. *Science* 199, 1302–1310.

Crawley, M.J. (1983) *Herbivory: The Dynamics of Animal–Plant Interactions*. Blackwell, Oxford, UK.

Cumming, D.H.M. (1982) The influence of large herbivores on savanna structure in Africa. In: Huntley, B.J. & Walker, B.H. (eds.) *Ecology of Tropical Savannas*. Ecological studies 42. Springer Verlag, Berlin, pp. 217–245.

Dahlberg, A.C. (2000) Vegetation diversity and change in relation to land use, soil and rainfall – a case study from North-East District, Botswana. *Journal of Arid Environments* 44, 19–40.

du Toit, J.T. (2003) Large herbivores and savanna heterogeneity. In: du Toit, J.T., Rogers, K.H. & Biggs, H.C. (eds.) *The Kruger Experience – Ecology and Management of Savanna Heterogeneity*. Island Press, Washington, DC, pp. 292–309.

Ellery, W.N., Ellery, K. & McCarthy, T.S. (1993) Plant distribution in islands of the Okavango Delta, Botswana: determinants and feedback interactions. *African Journal of Ecology* 31, 118–134.

Ernst, W.H.O. & Tolsma, D.J. (1989) Mineral nutrients in some Botswana savanna types. In: Proctor, J. (ed.) *Mineral Nutrients in Tropical Forest and Savanna Ecosystems*. Special Publication of the British Ecological Society, Vol. 9. Blackwell, Oxford, UK, pp. 97–120.

Ernst, W.H.O. & Tolsma, D.J. (1992) Growth of annual and perennial grasses in a savanna of Botswana under experimental conditions. *Flora* 186, 287–300.

FAO (1990) *Soil Map of the Republic of Botswana*. Soil Mapping and Advisory Services Project FAO/BOT/85/011. FAO and Government of Botswana, Gaborone, Botswana.

Farrar, T.J., Nicholson, S.E. & Lare, A.R. (1994) The influence of soil type on the relationships between NDVI, rainfall, and soil moisture in semiarid Botswana. II. NDVI response to soil moisture. *Remote Sensing of Environment* 50, 121–133.

Fornara, D.A. & du Toit, J.T. (2008) Community-level interactions between ungulate browsers and woody plants in an African savanna dominated by palatable-spinescent *Acacia* trees. *Journal of Arid Environments* 72, 534–545.

Fritz, H., Duncan, P., Gordon, I.J. & Illius A.W. (2002) Megaherbivores influence trophic guilds structure in African ungulate communities. *Oecologia* 131, 620–625.

Frost, P.G.H. (1985) Organic matter and nutrient dynamics in a broadleafed African savanna. In: Tothill, J.C. & Mott, J.J. (eds.) *Ecology and Management of the World's Savannas*. Australian Academy of Sciences, Canberra, pp. 200–206.

Grime, J.P., Thompson, K., Hunt, R., Hodgson, J.G., Cornelissen, J.H.C., Rorison, I.H., Hendry, G.A.F., Ashenden, T.W., Askew, A.P., Band, S.R., Booth, R.E., Bossard, C.C., Campbell, B.D., Cooper, J.E.L., Davison, A.W., Gupta, P.L., Hall, W., Hand, D.W., Hannah, M.A., Hillier, S.H., Hodkinson, D.J., Jalili, A., Liu, Z., Mackey, J.M.L., Matthews, N., Mowforth, M.A., Neal, A.M., Reader, R.J., Reiling, K., Ross-Fraser, W., Spencer, R.E., Sutton, F., Tasker, D.E., Thorpe, P.C. & Whitehouse, J. (1997) Integrated screening validates primary axes of specialization in plants. *Oikos* 79, 259–281.

Harper, J.L. (1977) *Population Biology of Plants*. Academic Press, London, UK.

Hill, M.O. & Gauch, H.G. (1980) Detrended correspondence analysis: an improved ordination technique. *Vegetatio* 42, 47–58.

Hill, M.O. (1979) *TWINSPAN – A FORTRAN Program for Arranging Multivariate Data in an Ordered Two-way Table by Classification of Individuals and Attributes*. Cornell University, Ithaca, New York.

Hobbs, N.T. (1996) Modification of ecosystems by ungulates. *Journal of Wildlife Management* 60, 695–713.

Huntly, N. (1991) Herbivores and the dynamics of communities and ecosystems. *Annual Review of Ecology and Systematics* 22, 477–503.

Huston, M. (1979) A general hypothesis of species diversity. *American Naturalist* 113, 81–101.

Kahlil, M.A.L. & Ayoub, S.M.H. (2011) Analysis of the essential oil of *Vertivaria nigritana* (Benth.) Stapf root growing in Sudan. *Journal of Medicinal Plants Research* 5, 7006–7010.

Mapaure, I. (2001) Small-scale variations in species composition of miombo woodland in Sengwa, Zimbabwe: the influence of edaphic factors, fire and elephant herbivory. *Systematics and Geography of Plants* 71, 935–947.

Mathisen, I.E. (2005) Effects of clipping and nitrogen fertilization on a grazing tolerant grass in Chobe National Park, Botswana. MSc Thesis, Norwegian University of Science and Technology, Trondheim, Norway.

Mathumo, I. (2003) The potential of the soil seed bank in regenerating riparian woodland along Chobe River, Botswana. MSc Thesis, Agricultural University of Norway, Ås, Norway.

McNaughton, S.J. (1979) Grazing as an optimization process: grass-ungulate relationships in the Serengeti. *The American Naturalist* 113, 691–703.

Mitchell, B.L. (1961) Some notes on the vegetation of a portion of the Wankie National Park. *Kirkia* 2, 200–209.

Milchunas, D.G. & Lauenroth, W.K. (1993) Quantitative effects of grazing on vegetation and soils over a global range of environments. *Ecological Monographs* 63, 327–366.

Moe, S.R., Rutina, L.P., Hytteborn, H. & du Toit, J.T. (2009) What controls woodland regeneration after elephants have killed the big trees? *Journal of Applied Ecology* 46, 223–230.

Mosugelo, D.K., Moe, S.R., Ringrose, S. & Nellemann, C. (2002) Vegetation changes during a 36-year period in northern Chobe National Park, Botswana. *African Journal of Ecology* 40, 232–240.

Naiman, R.J., Braak, L., Grant, R., Kemp, A.C., du Toit, J.T. & Venter, F.J. (2003) Interactions between species and ecosystem characteristics. In: du Toit, J.T., Rogers, K.H. & Biggs, H.C (eds.) The *Kruger Experience – Ecology and Management of Savanna Heterogeneity*. Island Press, Washington, DC, pp. 221–241.

Norton-Griffiths, M. (1979) The influence of grazing, browsing and fire on the vegetation dynamics of the Serengeti. In: Sinclair, A.R.E. & Norton-Griffiths, M. (eds.) *Serengeti – Dynamics of an Ecosystem*. The Chicago University Press, Chicago, pp. 310–352.

Novellie, P. (1988) The impact of large herbivores on the grassveld in the Addo Elephant National Park. *South African Journal of Wildlife Research* 18, 6–10.

O'Connor, T.G. (1991) Influence of rainfall and grazing on the compositional change of the herbaceous layer of a sandveld savanna. *Journal of the Grassland Society of Southern Africa* 8, 103–109.

O'Connor, T.G. & Pickett, G.A. (1992) The influence of grazing on seed production and seed banks of some African savanna grasslands. *Journal of Applied Ecology* 29, 247–260.

Owen-Smith, R.N. (1988) *Megaherbivores: The Influence of Very Large Body Size on Ecology*. Cambridge University Press, Cambridge, UK.

Persson, I.-L., Pastor, J., Danell, K. & Bergström, R. (2005) Impact of moose population density on the production and composition of litter in boreal forests. *Oikos* 108, 297–306.

Phillips, M.C. (1992) A survey of the arable weeds of Botswana. *Tropical Pest Management* 38, 13–21.

Pickett, S.T.A. & White, P.S. (1985) *The Ecology of Natural Disturbance and Patch Dynamics*. Academic Press, Orlando, Florida.

Pickett, S.T.A., Cadenasso, M.L. & Benning, T.L. (2003) Biotic and abiotic variability as key determinants of savanna heterogeneity at multiple spatiotemporal scales. In: du Toit, J.T., Rogers, K.H. & Biggs, H.C. (eds.) *The Kruger Experience – Ecology and Management of Savanna Heterogeneity*. Island Press, Washington, DC, pp. 22–40.

Pretorius, Y., de Boer, W.F., van der Waal, C., de Knegt, H.J., Grant, R.C., Knox, N.M., Kohi, E.M., Mwakiwa, E., Page, B.R., Peel, M.J.S., Skidmore, A.K., Slowtow, R., van Wieren, S.E. & Prins, H.H.T. (2011) Soil nutrient status determines how elephants utilize trees and shape environments. *Journal of Animal Ecology* 80, 875–883.

Rattray, J.M. (1961) Vegetation types of Southern Rhodesia. *Kirkia* 2, 68–93.

Ringrose, S., Matheson, W., Wolski, P. & Huntsman-Mapila, P. (2003) Vegetation cover trends along the Botswana Kalahari transect. *Journal of Arid Environments* 54, 297–317.

Ritchie, M.E., Tilman, D. & Knops, J.M.H. (1998) Herbivore effects on plant and nitrogen dynamics in oak savanna. *Ecology* 79, 165–177.

Rutina, L.P., Moe, S.R. & Swenson, J.E. (2005) Elephant *Loxondonta africana* driven woodland conversion improves dry-season browse availability for impala *Aepyceros melampus*. *Wildlife Biology* 11, 207–213.

Rutina, L.P. (2004) Impalas in an elephant-impacted woodland: browser-driven dynamics of the Chobe riparian zone, Northern Botswana. PhD Thesis, Agricultural University of Norway, Ås, Norway.

Sanders, H.L. (1968) Marine benthic diversity: a comparative study. *American Naturalist* 102, 243–282.

Scholes, R.J. & Walker, B.H. (1993) *An African Savanna: Synthesis of the Nylsvley Study*. Cambridge University Press, Cambridge, UK.

Scholes, R.J., Frost, P.G.H. & Tian, Y. (2004) Canopy structure in savannas along a moisture gradient on Kalahari sands. *Global Change Biology* 10, 292–302.

Selous, F.K. (1881) *A Hunter's Wanderings in Africa*. Richard Bentley & Son, London, UK.

Shmida, A. & Wilson, M.V. (1985) Biological determinants of species diversity. *Journal of Biogeography* 12, 1–20.

Simpson, C.D. (1975) A detailed vegetation study on the Chobe river in north-east Botswana. *Kirkia* 10, 185–227.

Skarpe, C. (1986) Plant community structure in relation to grazing and environmental changes along a north–south transect in the western Kalahari. *Vegetatio* 68, 3–18.

Skarpe, C. (1991) Impact of grazing in savanna ecosystems. *Ambio* 20, 351–356.

Skarpe, C., Aarrestad, P.A., Andreassen, H.P., Dhillion, S., Dimakatso, T., du Toit, J.T., Halley, D.J., Hytteborn, H., Makhabu, S., Mari, M., Marokane, W., Masunga, G., Modise, D., Moe, S.R., Mojaphoko, R., Mosugelo, D., Motsumi, S., Neo-Mahupeleng, G., Ramotadima, M., Rutina, L., Sechele, L., Sejoe, T.B., Stokke, S., Swenson, J.E., Taolo, C., Vandewalle, M. & Wegge, P. (2004) The return of the giants: ecological effects of an increasing elephant population. *Ambio* 33, 276–282.

Stokke, S. & du Toit, J.T. (2000) Sex and size related differences in the dry season feeding patterns of elephants in Chobe National Park, Botswana. *Ecography* 23, 70–80.

Taolo, C.L. (2003) Population ecology, seasonal movement and habitat use of African buffalo (*Syncerus caffer*) in Chobe National Park, Botswana. PhD Thesis, Norwegian University of Science and Technology, Trondheim, Norway.

ter Braak, C.J.F. & Smilauer, P. (2002) *CANOCO Reference Manual and CanoDraw for Windows User's Guide: Software for Canonical Community Ordination (Version 4.5)*. Microcomputer Power, Ithaca, New York.

ter Braak, C.J.F. (1987) The analysis of vegetation-environment relationships by canonical correspondence analysis. *Vegetatio* 69, 67–77.

Tessema, Z.K., de Boer, W.F., Baars, R.M.T. & Prins, H.H.T. (2011) Changes in soil nutrients, vegetation structure and herbaceous biomass in response to grazing in a semi-arid savanna of Ethiopia. *Journal of Arid Environments* 75, 662–670.

Theunissen, J.D. (1997) Selection of suitable ecotypes within *Digitaria eriantha* for reclamation and restoration of disturbed areas in southern Africa. *Journal of Arid Environments* 35, 429–439.

van Oudtshoorn, F. (1999) *Guide to Grasses of Southern Africa*. Briza Publications, Pretoria, South Africa.

van Rooyen, N., Bredenkamp, G.J. & Theron, G.K. (1991) Kalahari vegetation: veld condition trends and ecological status of species. *Koedoe* 34, 61–72.

Walker, B.H. & Noy-Meir, I. (1982) Aspects of the stability and resilience of savanna ecosystems. In: Huntley, B.J. & Walker, B.H. (eds.) *Ecology of Tropical Savannas*. Springer-Verlag, Berlin, pp. 556–590.

Wenzel, J.J., Bothma, J.du.P. & van Rooyen, N. (1991) Characteristics of the herbaceous layer in preferred grazing areas of six herbivore species in the south-eastern Kruger National Park. *Koedoe* 34, 51–59.

Yeaton, R.I., Frost, S. & Frost, P.G.H. (1986) Direct gradient analysis of grasses in a savanna. *South African Journal of Science* 82, 482–487.

Part III
The Agent

In the model by Pickett *et al.* (2003), which is described in Chapter 1 and around which this book is structured, the agent is the driver that creates, transforms or maintains structural and functional heterogeneity in the substrate and, subsequently, in responders that react to heterogeneity in the substrate. Agents can be abiotic or biotic. Rivers can be agents eroding riverbanks and depositing sediments on floodplains, and isolated savanna trees can be agents transforming soil conditions and microclimates under their canopies. The distribution and actions of agents, and the scale and type of heterogeneity they create, can be modified by controllers. In this section of the book we see the elephant population, *Loxodonta africana*, as the main agent driving temporal and spatial heterogeneity in the Chobe ecosystem. We describe how historical variations in overall elephant density, as well as present variations in distribution and activities of elephants in the area, account for heterogeneity in the vegetation (substrate) and in organisms and processes responding to that variation (Drawing by Marit Hjeljord).

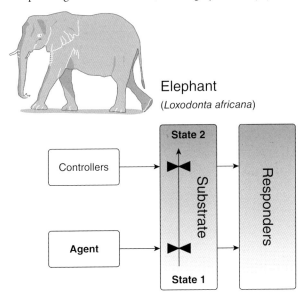

Elephant
(*Loxodonta africana*)

Elephants and Savanna Woodland Ecosystems: A Study from Chobe National Park, Botswana,
First Edition. Edited by Christina Skarpe, Johan T. du Toit and Stein R. Moe.
© 2014 John Wiley & Sons, Ltd. Published 2014 by John Wiley & Sons, Ltd.

References

Pickett, S.T.A., Cadenasso, M.L. & Benning, T.L. (2003) Biotic and abiotic variability as key determinants of savanna heterogeneity at multiple spatiotemporal scales. In: du Toit, J.T., Rogers, K.H. & Biggs, H.C. (eds.) *The Kruger Experience. Ecology and Management of Savanna Heterogeneity*. Island Press, Washington, Covelo, London, pp. 22–39.

6

Guns, Ivory and Disease: Past Influences on the Present Status of Botswana's Elephants and their Habitats

Mark. E. Vandewalle[1] *and Kathy. A. Alexander*[2]

[1]CARACAL, Botswana
[2]Department of Fisheries and Wildlife Conservation, Virginia Tech, USA

Introduction

Today there are more elephants *Loxodonta africana* in Botswana than in any other African country (Blanc *et al.*, 2007) due largely to the successful protection efforts of its government. The situation has been quite different historically, however. The 20th century was a period of recovery for the elephant population following its decimation in the 19th century by European and indigenous ivory hunters. Being landlocked, Botswana was somewhat isolated from colonising settlers arriving on the continent by sea. This had an important bearing on the history of Botswana's elephant population, starting with protection through isolation, then over-exploitation through uncontrolled ivory hunting and then political change that brought protection through legislation and its enforcement. Conversely, there were no barriers to exotic cattle-borne diseases that spread across the African continent, profoundly affecting livestock and various wildlife populations, with ripple-effects that are still influencing the dynamics of Botswana's ecosystems today.

Pre- and post-colonial hunting of elephants in southern Africa

Before the arrival of the European explorers and settlers, indigenous people hunted elephants, rhinoceros and other large animals and, although the methods used were

Elephants and Savanna Woodland Ecosystems: A Study from Chobe National Park, Botswana,
First Edition. Edited by Christina Skarpe, Johan T. du Toit and Stein R. Moe.
© 2014 John Wiley & Sons, Ltd. Published 2014 by John Wiley & Sons, Ltd.

primitive and dangerous for the hunters, they were very effective (Campbell, 1990). Elephants were killed for ivory, which was used for domestic carving, tributes and personal decoration, as well as for their meat and skin. There are also reports of people (seemingly Tswana-speaking groups) living in the interior of the sub-continent who traded ivory as far back as the mid-1660s when the early Dutch East India Company colonised the Cape. Trade occurred mainly along the coastal regions where ivory and other wares could be easily transported to the coast and shipped to Europe and the East. Ivory was derived from hunting elephants, as well as collecting tusks from elephants that had died from natural causes (Spinage, 1973). As the demand for ivory increased and foreign markets opened up, however, tribal groups in southern Africa intensified their hunting efforts to obtain tradable unweathered tusks. Hunting methods were diverse, including the use of pit traps, entrapment by fire or in water bodies, spearing, heavily weighted harpoons dropped from above by a trigger mechanism, and barbed spikes in shallow holes that would pierce an elephant's foot, pinning it to the spot (Klein, 1977; Campbell, 1990).

Europeans visited and established settlements along the navigable regions of the African coast from the 15th century onwards; by the 17th century even the distant southern tip of the continent was a frequent stopover point. When the first Dutch settlers arrived at the Cape of Good Hope in the mid-1600s, elephants occurred in suitable habitats throughout the southern African sub-continent (Hall-Martin, 1992). Although numbers were low in the more arid regions such as the Karroo, the Kalahari and the higher altitudes of the Highveld interior, the early settlers encountered herds of elephants as far south as present-day Cape Town. As European settlement expanded over the next few centuries, big game hunting, trophy collection and the export of animal products, especially ivory, became favoured activities. The resulting decimation of local elephant populations proceeded northwards through present-day South Africa so that by the early 1800s both their density and distribution had been substantially reduced throughout the region (Campbell, 1990; Hall-Martin, 1992), largely because of the ivory trade (Spinage, 1973). This led to big game hunters and explorers travelling into and exploring the African interior, including areas now within Zimbabwe, Zambia, Angola and Botswana – then called Bechuanaland.

The elephants of Bechuanaland

There are no estimates of the numbers of elephants in Bechuanaland before or during the period of intense hunting for ivory in the 19th century, although books and journals of early hunters, traders, explorers and entrepreneurs describe the presence of large elephant herds, high numbers being killed, and many wagonloads of ivory being exported from various parts of the country. Based on elephant population surveys elsewhere in Africa, Parker and Graham (1989) modelled density estimates of elephants using annual rainfall, fertility class (based on soil types) and human population densities. With their model, Campbell (1990) estimated the Bechuanaland population at 200,000–400,000 elephants around the beginning of the 19th century. An assumption of the model is that mortality by humans is absent or low, which might be invalid in this case given that hunting excursions into southern Bechuanaland had already occurred by 1800. Nevertheless, an independent estimate of over 200,000 elephants for the country's northern range only, based on population trends (Spinage, 1990), indicate that the countrywide population could reasonably have been in the range of Campbell's estimate. Systematic surveys of Botswana's current elephant population have produced estimates of 155,000 individuals in 2006 (unpublished Government of Botswana report, 2006) and 133,000 in 2010 (Chase, 2011), much of which is presently limited to northern Botswana, although a small population of less than 1500 occurs in the Tuli Block near Francistown. The historic range of elephants was more widespread across regions of the country from where they are now excluded by human settlement and lack of permanent surface water. Therefore, at a conservative minimum, the population in the late 1700s, prior to the period of heavy exploitation by humans, could quite feasibly have been twice that of today's estimate given that elephants then had access to more than twice the range presently available.

Exploitation of Bechuanaland's elephants

European hunters, explorers, missionaries and local traders conducted major excursions into Bechuanaland from south and east of the country from about 1810 (Campbell, 1990). Elephants were seen and shot, often in great numbers, around the present-day locations of Gaborone (now the capital), Mochudi, Shoshong and Ghanzi and along the Nossob and Molopo rivers in the south, as well as in the east around Francistown and along the Shashe and Boteti rivers. Only a few decades later, explorers and hunters were conducting expeditions as far north as Lake Ngami in central Bechuanaland, and along the Linyanti, Chobe and Zambezi rivers in the north of the country. Despite the objections of the *diKgosi* (Tswana chieftains) to the exploitation of their elephants by foreigners (Campbell, 1990), there are references to large numbers of animals being killed for their ivory during this period (Andersson, 1856; Livingstone, 1857; Baldwin, 1863; Baines, 1864; Chapman, 1868; Selous, 1893; Oswell, 1900). There were expeditions along the Limpopo River as early as 1820 where some 600 elephants were shot by one Boer hunting group alone. Over the next few decades the killing rate increased, with known kills of 900 elephants on the Boteti River and Lake Ngami in 1849 and one Boer was known to have killed over 1000 elephants on his own in 1860. In the mid-19th century 2000–3000 elephants were killed annually in the southern and central regions resulting in local extermination and causing the hunting effort to move northwards. On his southward return from a hunting excursion on the Thamalakane River near Maun, Baldwin (1863) was somewhat taken aback by the large quantities of ivory in the wagons of Boer hunters returning from the Mababe region while he alone carried more than 2 tons in his own wagons. In 1872, a hunter and his sons shot 104 elephants in 1 day near Makakung in the north-west. By the end of the 19th century, the vast herds of elephants that ranged throughout most of Bechuanaland were reduced to only a few thousand individuals. Later travellers to Lake Ngami, Chobe, Linyanti, Savuti and Mababe seldom encountered elephants. Most of the people in present-day Botswana now live in areas where several generations of them have never seen elephants.

Disease and ecological transformation: the rinderpest panzootic arrives in 1896

The power of infectious disease as a transformative agent is clearly evident in human history from outbreaks of bubonic plague, cholera, malaria and other pathogens that have influenced societies across the globe (McNeill, 1976). Animal and plant communities are equally as vulnerable, as is seen in the history of rinderpest, an exotic virus introduced into Africa that directly affected populations of many ungulate species and indirectly changed savanna plant communities across the continent (Barrett and Rossiter, 1999; Spinage, 2003).

Rinderpest, now eradicated worldwide (Anderson et al., 2011), is a negative stranded RNA virus that caused severe disease in cattle and other artiodactyls. This multi-host pathogen is considered to have been one of the most contagious and lethal pathogens infecting these species (Spinage, 2003) with mortality rates over 90% (Barrett and Rossiter, 1999). Rinderpest is spread by direct contact between infected and susceptible hosts through respiratory and ocular secretions, faeces or urine. Environmental contamination and persistence is limited as the virus is rapidly destroyed outside the host (Rossiter, 1994). There was considerable variability in the effects of the disease among wild ungulate hosts with the disease being more severe in African buffalo, *Syncerus caffer*, giraffe, *Giraffa camelopardalis*, common eland, *Taurotragus oryx* and warthog, *Phacochoerus aethiopicus* (Rossiter, 1994). Impala, *Aepyceros melampus*, showed only mild or atypical disease (Scott *et al.*, 1960), whereas in hippos, *Hippopotamus amphibius*, the symptoms were persistently sub-clinical (Plowright and Laws, 1964). Cattle breeds also varied in their susceptibly to the pathogen and the severity with which the disease manifested itself (Spinage, 2003, 2012).

Rinderpest is reported to have entered Botswana in March 1896 through infected oxen travelling on the Bulawayo-Mafeking road. Cattle populations were decimated with mortality rates upwards of 90%. Historical records also show that rinderpest had a large-scale impact on cloven-hoofed wildlife populations throughout the country (Spinage, 2003).

Across Africa's savannas, the effects of the great rinderpest panzootic (1889–1897) reverberated through food webs and modified ecological systems through changes in herbivory and fire (McNaughton, 1992). The vegetation communities in turn responded in ways that persist today (Plowright, 1982; Spinage, 2003, 2012). The almost concurrent decimation of populations of various artiodactyl species (by rinderpest) and elephants (by over-hunting), during a period of relatively high rainfall, appears to have provided unique conditions for a pulse of woodland regeneration in Botswana and elsewhere on the continent. This particularly affected the acacia, mopane and riverine-woodland communities, as well as populations of 'soft-barked' trees such as baobab, *Adansonia digitata*, and star chestnut, *Sterculia africana*. Even-aged stands of such tree species are now found across eastern and southern Africa (Lewin, 1986; Prins and van der Jeugd, 1993). The development of woodland along the Chobe riverfront is an example of this regeneration pulse, which later became transformed by the recovering elephant population (Chapter 1).

Recovery of Botswana's elephant population in the 20th century

Surveys and population estimates

Following its decimation by ivory hunters, the elephant population in Botswana at the turn of the 20th century was extremely low in numbers. In 1933 only a single herd of about 25 elephants was known to live on the Chobe River between Kasane and Kazungula; otherwise, the species was scarce to the west between Kasane and Ngoma (Child, 1968). San people living in Ngwezumba in the centre of Chobe National Park had not seen elephants in the area prior to 1945. Ideas of protecting wildlife in certain areas in Botswana started around 1930 and many of the tribal chieftains declared areas under their authority as non-hunting areas (Tlou and Campbell, 1997). Over time, the administration of some of these areas was taken over by the Botswana government. The present-day Moremi Game Reserve was gazetted in 1963. Chobe Game Reserve was gazetted in 1960 and became a national park in 1967. From about 1950 onwards, elephant numbers started to increase in northern and eastern Botswana and herds moved back into areas from where they had been eradicated. The herds along the Chobe River increased steadily and in 1963 about 500 were estimated there, based on spoor counts (Child, 1968). In northern Botswana today, some of the highest densities of elephants on the continent occur along the Chobe River in the dry season. In the east, growth in the resident population has been boosted by immigration from Zimbabwe. In the southern parts of the country, however, where surface water is scarce, the elephant population has not recovered at all.

The first scientific estimates of the elephant population in Botswana were carried out in the mid-1970s. Ground and aerial surveys were conducted in the Chobe District in north-eastern Botswana and a population of 12,000 elephants was estimated to occur in the region with approximately half of these in the Chobe National Park (W. von Richter *et al.*, unpublished Government of Botswana report, 1974; Sommerlatte, 1975). K. Tinley (unpublished report, 1973) mentions elephants in the Okavango Delta and Moremi Game Reserve, but no estimate of their numbers was made. L. Patterson (unpublished Government of Botswana report, 1980) conducted an aerial survey of the Kwaai, Savuti and Kwando drainages and estimated about 8800 elephants there. The first aerial survey of the total elephant population of northern Botswana was carried out in 1978 (L. Patterson, unpublished Government of Botswana report, 1980). To evaluate elephant population trends, systematic aerial surveys were then repeated every second year, occasionally annually, by the Kalahari Conservation Society (unpublished reports to the Government of Botswana, 1984, 1985) until 1987. After that, the Botswana Department of Wildlife and National Parks (DWNP) conducted country-wide aerial surveys of all wildlife but with a focus on elephants. The surveys were carried out in conjunction with consultants (unpublished reports to the Government of Botswana by Bonifica, 1992 and ULG, 1993, 1994), then later under internal control. Logistics and finances permitting, surveys were carried out in some years in both the wet and the dry seasons. The last DWNP count occurred in 2006, then all surveys were discontinued until the Kasane-based non-governmental organisation, Elephants Without Borders, carried out a survey in 2010 (Chase, 2011).

The survey results show an impressive increase in population across Botswana's elephant range from an estimated 39,500 in 1981 to over 155,000 in 2006 (Figure 6.1). The trajectory of growth is not smooth, however. Several sharp increases occurred in the late 1990s and early 2000s, possibly reflecting immigration from Zimbabwe to the east and Namibia to the north. Those borders are unfenced, other than one disease fence between Botswana and the western Caprivi Strip of Namibia, which has not been maintained and is largely ineffective as a barrier to elephants. Extensive cross-border movements of elephants have been observed (G. C. Calef, unpublished report to the Government of Botswana, 1991; Verlinden and Gavor, 1998; M. E. Vandewalle, unpublished report to the Government of Botswana, 2003; Chase, 2007, 2011), with some marked animals from the Okavango recently having moved through the Caprivi to Angola (Chase and Griffin, 2011).

The Government of Botswana banned trophy-hunting of elephants between 1983 and 1994 because of declines in trophy size, but no measurable increase in population size resulted. Nevertheless, anecdotal evidence from hunting concessionaires in northern Botswana suggests that during the ban elephants dispersed from protected areas into previously hunted concession areas. This exemplifies the 'landscape of fear' concept, one that underlies the speculation that elephants are concentrating in northern Botswana because it is a safe haven, instead of dispersing into habitats in neighbouring

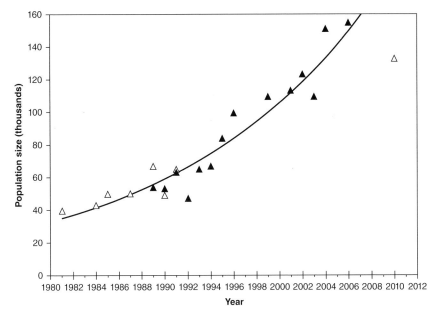

Figure 6.1   **Estimates of elephant numbers in northern Botswana from aerial surveys conducted by the Botswana Department of Wildlife and National Parks (solid triangles) and various non-governmental organisations (NGOs) (open triangles). The best-fit exponential curve ($R^2 = 0.925$) was calculated using estimates from all surveys and is defined by $N_t = N_{(t-1)} \cdot e^{r \cdot t}$ where growth rate (r) = 0.058 and N is the population size at time t. References are provided in the text.**

countries where human–elephant conflict is more common. With recent surveys indicating that population density is not increasing (Chase, 2011), however, it seems that, if compression is occurring, it must be localised. The design of future surveys will need to allow specifically for tests of whether the population is growing but expanding its range through density-dependent dispersal, or stabilising through density-dependent resource limitation (Junker *et al.*, 2008).

Population growth rate

Annual growth rates of 9.6% and 11.2% have been calculated using linear regression models for data from the aerial surveys of northern Botswana in wet and dry seasons, respectively, from 1973 to 1993 (Junker *et al.*, 2008). Those rates are, however, higher than the 7% maximum potential population growth rate estimated for elephants

(Calef, 1988), so immigration has to be considered a factor during that period. By excluding surveys in the 1970s, because they did not include the entire elephant range (Sommerlatte, 1975), we find the best-fit exponential equation indicates an overall increase in the northern Botswana elephant population of 5.8% per annum since the 1980s (Figure 6.1).

To assess the reproductive growth rate of Botswana's elephant population, M. Vandewalle (unpublished report to the Government of Botswana, 2003) conducted ground surveys to estimate the ages of elephants in herds in Chobe and Ngamiland Districts using established criteria from Amboseli and Botswana (C. J. Moss, unpublished reports, 1990, 1991). The estimated proportion of calves (<1 year old) indicate that the intrinsic growth rate of the population approximates 6.7% overall, with a small difference between herds in the Chobe and Ngamiland Districts. Age-specific mortality rates are unknown but overall mortality resulting from shooting (problem animal control and the annual hunting quota), natural mortality and diseases such as anthrax, is estimated at less than 1% per annum (R. B. Martin, unpublished report to Kalahari Conservation Society, 1990). Subtracting this from the calf recruitment of 6.7%, which matches previous estimates from Botswana and elsewhere (G. W. Calef and C. J. Moss, unpublished report to Government of Botswana, 1991), leaves a net annual population recruitment rate of about 5.7%. This conforms closely with the overall growth rate of 5.8% per annum established from the annual aerial survey data since the 1980s (Figure 6.1). Furthermore, the age pyramid indicates that the growth rate in the northern Botswana elephant population has remained stable over the last few decades (Figure 6.2). In Chobe, however, population density has also remained fairly stable over the past decade (Chase, 2011) suggesting that, while growing, the population is expanding its range through density-dependent dispersal (Junker et al., 2008).

Range and seasonal distribution patterns

The distribution of elephants across the landscape is strongly influenced by the distribution of water, a resource on which they depend (Laws, 1970; Western and Lindsay, 1984; Viljoen, 1989). The density of elephants in Botswana, a semi-arid country with high inter-seasonal and inter-annual spatial variability in surface water availability, is equally variably distributed in space and time.

Much of southern Africa experienced an extended period of drought in the early 1800s that reduced the availability of perennial surface water despite some unusually wet episodes involving flooding (Nash and Endfield, 2002). Those anomalies were associated with volcanic activity elsewhere that influenced climates globally (Ballard, 1986), as well as moderate to strong El Niño–southern oscillation events (Nash and Endfield, 2002). Prior to the 19th century, surface water availability in Botswana appears to have

Figure 6.2   **Age class distribution (%) of the Botswana elephant population in 2000 and 2001. (Source: M. E. Vandewalle, unpublished report to the Government of Botswana, 2003.)**

been more widespread (Campbell, 1990), with permanent water in the Makgadikgadi Pans, Lake Ngami and around Ghanzi. Now-ephemeral rivers like the Molopo, Nossob, Shashe, Limpopo, Boteti, Nata and Nogotwane were more permanent then than now, and elephants occurred throughout most of the country except for the arid interior of the Kalahari.

Following the decimation of the elephant population by ivory hunters, and despite subsequent recovery in the north, it is unlikely that elephants will ever re-establish in southern Botswana because of limited perennial surface water and expanding human settlement there. The difference is that, in the north, trypanosomiasis, carried by tsetse fly, prevented cattle production and kept human densities low while the elephant population was recovering in those areas with permanent water. This return of elephants to their former habitats in northern Botswana is illustrated by the expansion of their dry season range during 1989–2006 (Figure 6.3). Most of the expansion occurred in the north-west towards the Caprivi Strip, southward into the Nxai Pan – Makgadikgadi area, south-east along the Zimbabwe border towards Nata and throughout the Okavango Delta. Telemetry research has shown some Botswana elephants dispersing into neighbouring countries (M. E. Vandewalle, unpublished report to the Government of Botswana, 2003; Chase, 2007) or else, as indicated by some marked individuals, remaining resident but moving over smaller home range sizes than before. In a 10-year period,

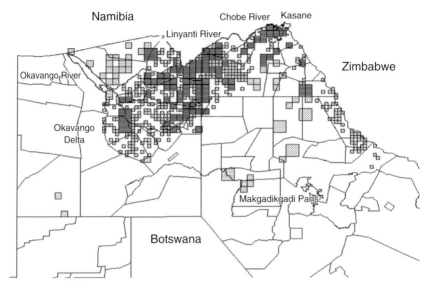

Figure 6.3 **Elephant distribution based on actual observations of herds from countrywide aerial surveys conducted in the dry seasons of 1989 (hatched) and 2006 (stippled) showing expansion in the population range.**

the mean home range size of radio-collared elephant cows decreased from 2338 km² (*n* = 18; G. W. Calef, unpublished report to the Government of Botswana, 1991; Verlinden and Gavor, 1998) to 1214 km² (*n* = 15; M. E. Vandewalle, unpublished report to the Government of Botswana, 2003).

The provision of water from boreholes for wildlife has increased across the dry interior of the Chobe and Ngamiland Districts and these attract some elephants during the dry season, mainly bulls. The important water sources for elephants in Botswana during the dry season (June–October) remain restricted to the north, where there are three concentrations: along the Chobe River, along the Linyanti–Kwando river system and in the Okavango Delta (Figure 6.3).

Overview

From an historical perspective, as outlined here, it is obvious that current issues associated with elephants and their habitats in Botswana are still strongly influenced by long-past disturbances imposed by people. The historical landmarks are summarised as follows.

- 1830–1890: elephant population decimated by uncontrolled hunting for ivory.
- 1896–1900: rinderpest panzootic causes major die-offs in artiodactyl wildlife and livestock populations, providing a 'window of opportunity' for the establishment of tree seedlings that leads to a pulse of woodland regeneration along the major river systems.
- 1930–present: protection of Botswana's wildlife resources by traditional and then later government leaderships, facilitating the recovery of most wildlife populations including elephant.
- 1978–present: regular aerial surveys monitor the total elephant population of northern Botswana.
- 1980–present: browsing and bark-stripping by elephants causes transition from woodland to shrubland along the major river systems in northern Botswana, resulting in stakeholder concern for tourist areas such as the Chobe riverfront.
- 2000–present: elephant density stabilises while the population continues to expand its range.

Over the past century and a half, the elephant population and its habitats in northern Botswana have been continually changing in response to two major anthropogenic disturbances: free-for-all ivory hunting and the exotic cattle-borne rinderpest virus. The system finally appears to be equilibrating, but what land managers and the custodians of Botswana's elephants do next could set off another chain of ecological responses. How historical insights might contribute to future decision making remains to be seen, but the aim of this chapter will be achieved if it provides a vision of the linkages that connect humans with wildlife and their habitats, and the long-term feedbacks resulting from disturbances within social-ecological systems.

References

Anderson, J., Baron, M., Cameron, A. & Kock, R. (2011) Rinderpest eradicated; what next? *Veterinary Record* 169, 10–11.

Andersson, C.J. (1856) *Lake Ngami: Explorations and Discoveries during Four Years Wandering in the Wilds of Southwestern Africa*. Hurst and Blackett, London, UK (facsimile edition, reprinted 1987, C. Struik, Cape Town).

Baines, T. (1864) *Explorations in South-West Africa*. Longman, Roberts and Green, London, UK.

Baldwin, W.C. (1863) *African Hunting from Natal to the Zambezi including Lake Ngami, the Kalahari Desert, etc. from 1852 to 1860*. Richard Bentley, London, UK.

Ballard, C. (1986) Drought and economic distress: South Africa in the 1800s. *The Journal of Interdisciplinary History* 17, 359–378.

Barrett, T. & Rossiter, P. (1999) Rinderpest: the disease and its impact on humans and animals. *Advances in Virus Research* 53, 89–110.

Blanc J.J., Barnes R.F.W., Craig G.C., Dublin, H.T., Thouless, C.R., Douglas-Hamilton, I. & Hart, J.A. (2007) *African Elephant Status Report 2007. An Update from the African Elephant Database*. Occasional Paper of the IUCN Species Survival Commission No. 33, IUCN, Gland, Switzerland.

Calef, G.W. (1988) Maximum rate of increase in the African elephant. *African Journal of Ecology* 26, 323–327.

Campbell, A.C. (1990) History of elephants in Botswana. In: Hancock, P., Cantrell, M. & Hughes, S. (eds.) *The Future of Botswana's Elephants. Kalahari Conservation Society Symposium, Gaborone, Botswana*, pp. 5–15.

Chapman, J. (1868) *Travels in the Interior of South Africa, 1849–1863: Hunting and Trading Journeys from Natal to Walvis Bay and Visits to Lake Ngami and Victoria Falls*. Bell and Daldy, London, UK.

Chase, M.J. (2007) *Home Ranges, Transboundary Movements and Harvest of Elephants in Northern Botswana and Factors Affecting Elephant Distribution and Abundance in the Lower Kwando River Basin*. ProQuest Information and Learning Company, Ann Arbor, Michigan, USA.

Chase, M. (2011) *Dry Season Fixed-wing Aerial Survey of Elephants and Wildlife in Northern Botswana*. Elephants Without Borders, Kasane, Botswana; Department of Wildlife and National Parks, Botswana; and Zoological Society of San Diego, USA. [online] http://www.elephantdatabase.org/population_submission_attachments/102.

Chase, M.J. & Griffin, C.R. (2011) Elephants of south and east Angola in war and peace: their decline, recolonization and recent status. *African Journal of Ecology* 49, 353–361.

Child, G. (1968) *An Ecological Survey of North-eastern Botswana*. Food and Agriculture Organization of the United Nations, Rome.

Hall-Martin, A. (1992) Distribution and status of the African elephant, *Loxodonta africana*, in South Africa, 1652–1992. *Koedoe* 35, 65–88.

Junker, J., van Aarde, R.J. & Ferreira, S.M. (2008) Temporal trends in elephant *Loxodonta africana* numbers and densities in northern Botswana: is the population really increasing? *Oryx* 42, 58–65.

Klein, R.G. (1977) The ecology of early man in southern Africa. *Science* 197, 115–126.

Laws, R.M. (1970) Elephants as agents of habitat and landscape change in East Africa. *Oikos* 21, 1–15.

Lewin, R. (1986) In ecology, change brings stability. *Science* 234, 1071–1073.

Livingstone, D. (1857) *Missionary Travels and Researches in South Africa*. John Murray, London, UK.

McNaughton, S. (1992) The propagation of disturbance in savannas through food webs. *Journal of Vegetation Science* 3, 301–314.

McNeill, W.H. (1976) *Plagues and Peoples*. Anchor Books, New York.

Nash, D.J. & Endfield, G.H. (2002) A 19th century climate chronology for the Kalahari region of central southern Africa derived from missionary correspondence. *International Journal of Climatology* 22, 821–841.

Oswell, W.E. (1900) *William Cotton Oswell, Hunter and Explorer*. William Heinemann, London, UK.

Parker, I.S.C. & Graham, A.D. (1989) Elephant decline: downward trends in African elephant distribution and numbers (Part I). *International Journal of Environmental Studies* 34, 287–305.

Plowright, W. (1982) The effects of rinderpest and rinderpest control on wildlife in Africa. *Symposium of the Zoological Society of London* 50, 1–28.

Plowright, W. & Laws, R.M. (1964) Serological evidence for the susceptibility of the hippopotamus (*Hippopotamus amphibius* Linnaeus) to natural infection with rinderpest virus. *Journal of Hygiene* 62, 329–336.

Prins, H.H.T. & van der Jeugd, H.P. (1993) Herbivore population crash and woodland structure in East Africa. *Journal of Ecology* 81, 305–314.

Rossiter, P.B. (1994) Rinderpest. In: Coetzer, J.A.W., Thompson, G.R., Tustin, R.C. & Kriek, N.P. (eds.) *Infectious Diseases of Livestock with Special Reference to South Africa*, Vol. 2. Oxford University Press, Cape Town, pp. 735–757.

Scott, G.R., Cowan, K.M. & Elliott, R.T. (1960) Rinderpest in impala. *Veterinary Record* 72, 787–788.

Selous, F.C. (1893) *Travel and Adventure in South-East Africa*. Rowland Ward & Co., London, UK.

Sommerlatte, M.W.L. (1975) A preliminary report on the number, distribution and movement of elephants in the Chobe National Park with notes on browse utilization. *Botswana Notes and Records* 7, 121–129.

Spinage, C.A. (1973) A review of ivory exploitation and elephant population trends in Africa. *African Journal of Ecology* 11, 281–289.

Spinage, C.A. (1990) Botswana's problem elephants. *Pachyderm* 13, 15–19.

Spinage, C.A. (2003) *Cattle Plague: A History*. Kluwer Academic/Plenum Publishers, New York.

Spinage, C.A. (2012) *African Ecology – Benchmarks and Historical Perspectives*. Springer-Verlag, Berlin.

Tlou, T. & Campbell, A. (1997) *History of Botswana*. Macmillan Boleswa, Gaborone, Botswana.

Verlinden, A. & Gavor, I.K.N. (1998) Satellite tracking of elephants in northern Botswana. *African Journal of Ecology* 36, 105–116.

Viljoen, P. (1989) Spatial distribution and movements of elephants (*Loxodonta africana*) in the northern Namib Desert region of the Kaokoveld, South West Africa/Namibia. *Journal of Zoology* 219, 1–19.

Western, D. & Lindsay, W. (1984) Seasonal herd dynamics of a savanna elephant population. *African Journal of Ecology* 22, 229–244.

7

The Chobe Elephants: One Species, Two Niches

Sigbjørn Stokke[1] and Johan T. du Toit[2]

[1]Norwegian Institute for Nature Research, Norway
[2]Department of Wildland Resources, Utah State University, USA

Botswana is a stronghold of the African savanna elephant, *Loxodonta africana africana*. There are more elephants in Botswana – about 130,000 of them – than any other country in Africa (Blanc *et al.*, 2007; Chase, 2011). They are all in the northern part of the country, with about 30,000 occurring in Chobe National Park (Chase, 2011) where they have never been culled, are very seldom poached, and have continuous free range over vast tracts of wilderness extending beyond the unfenced park boundaries. As a result, elephant family units in Chobe are more stable than elsewhere (Stokke, 2000) and their behaviour is little influenced by human disturbance. These were the conditions we needed to test an unsupported observation that patterns of ranging and feeding behaviour of elephant bulls differ from those of cows and their offspring (Guy, 1976). Our study, the first to focus on such differences in elephants, was conducted in Chobe in 1995–1996 (Stokke, 1999; Stokke and du Toit, 2000, 2002). Since then it has become recognised that the substantial ecological differences between adult elephant bulls and members of family units merit special consideration in management planning (Owen-Smith *et al.*, 2006; Kerley *et al.*, 2008). Nevertheless, elephant population surveys still produce sexually undifferentiated estimates of the total number or density of animals in a survey area and those data become incorporated into analyses aimed at investigating elephant effects on woody vegetation (e.g. Guldemond and van Aarde, 2008; Asner and Levick, 2012). Scientists obviously have to work with the best data they can get, but in this chapter, we show that total population estimates are only crude

Elephants and Savanna Woodland Ecosystems: A Study from Chobe National Park, Botswana,
First Edition. Edited by Christina Skarpe, Johan T. du Toit and Stein R. Moe.
© 2014 John Wiley & Sons, Ltd. Published 2014 by John Wiley & Sons, Ltd.

predictors of elephant effects on woody vegetation because of significant sex differences in elephant ecology, which occur at multiple scales.

Sexual size-dimorphism and social organization

African savanna elephants are the largest living terrestrial animals and the adult males can be twice as big as the females. Males reach a body mass of 5500–6000 kg and a shoulder height just over 3 m; females reach 2500–2800 kg in body mass and about 2.7 m in shoulder height (Owen-Smith, 1988). Although pronounced in elephants, sexual size-dimorphism is common among large mammals and probably evolved through competition among males for mating opportunities in polygynous breeding systems (Clutton-Brock and Harvey, 1978; Loison *et al.*, 1999). Elephants reach puberty at ages of 11–14 years in females and 10–15 years in males (Owen-Smith, 1988; Moss, 2001). Males undergo a growth surge from about 20 years of age until reaching full social maturity and competitive ability at about 30 years, whereas growth in females levels off after 10–15 years (Moss, 1983). The social system exhibited by African elephants is distinctive in that, an experienced female, the matriarch, aged 35 years or more, guides each 'core' social group as a family unit comprised of her young offspring, her mature daughters and their progeny (Archie *et al.*, 2006). During maturation the males gradually become expelled by their matriarchs from their family units, so at maturity they live solitarily or in temporary bachelor groups with only transient associations with other family units when scouting for receptive females (Spinage, 1994).

Chobe is renowned for its large aggregations of elephants on the floodplain, where 100 or more elephants can be seen together in the dry season. This typically happens in the late afternoon when they gather to drink and then wallow in mud along the shore or completely immerse themselves in the main river channel (Stokke, 2000). They then break away and move back into the savanna in their family units. The typical family unit size (mean = 12.2; s.e. = 0.9; median = 11.0; n = 251) recorded in our study in Chobe was large when compared with other elephant populations: Tsavo East, 9 (Poole, 1989); Masai Mara, 8 (Dublin, 1996); Tsavo West, 7 (Poole, 1989) and Mikumi, 6 (Poole, 1989). Although we focused on elephants moving to and from the Chobe River, it seems that large family units are typical of the Chobe population as a whole. Family units observed moving to and from pumped water points in the Savuti area of Chobe averaged 9 animals (Power and Compion, 2009). Large family units indicate an abundance of food resources and are thus observed mainly during wet seasons (Dublin, 1996). Our study period, however, was spread over the wet and dry seasons of 1995 and 1996, during which the rainfall pattern was not unusual. Moreover, among adults, only 4.4% of males and 4.8% of females were visually scored in the 'poor' body-condition class (Stokke and du Toit, 2000), so the population was evidently not limited nutritionally during that period.

Sex differences in the use of plant parts

As elsewhere in their range (Spinage, 1994) the Chobe elephants are primarily grazers in the wet season and browsers in the dry season (browsing recorded in 94% of all
dry season foraging bouts: Stokke, 2000). With their specialised feeding apparatus and
physical strength, they are adapted for the efficient harvesting of woody plants during feeding by plucking twigs and fruits, debarking stems and roots, stripping leaves,
uprooting saplings and felling trees. Because the biggest management concern associated with elephants is their disturbance of the tree layer of savanna vegetation (Scholes
and Mennell, 2008), we examined elephant browsing patterns with particular attention
to sex differences in their use of woody plant parts. We expected to find the same pattern as in other populations of sexually size-dimorphic herbivores, in which adult males
and females usually have contrasting diets because of their different dietary tolerances
resulting from dissimilar body sizes (reviewed by du Toit, 2005). In addition, pregnant
or lactating females might have different nutritional needs to those of males. These sex
differences underpin the forage selection hypothesis, or FSH, by which adult males and
females are expected to have qualitatively different diets (Main, 2008). According to the
FSH, elephant bulls are expected to feed less selectively on abundant forage of lower
quality (higher fibre, lower digestibility) when compared with cows and their offspring
in family units.

We located elephants by driving along 10 fixed routes, each of which traversed all the
main elephant habitat types in northern Chobe (Stokke, 1999). One browsing adult was
selected at random from the group – or the only animal, as was usually the case with
bulls – to be the focal animal. The tree being browsed at the time was used to define the
centre of a 5 m radius feeding plot, which we assumed to be representative of the food
patch selected by that animal. Every second feeding plot was paired with a control plot
of the same size, located 50 m away in a direction perpendicular to the foraging path of
the focal animal. We assumed that the control plots were representative of the matrix
between the food patches selected by the focal animal while moving along its foraging
path. Once a focal animal had voluntarily moved a safe distance along its foraging
path, the food plot was examined on foot. Fresh signs of elephant browsing are clearly
visible: records were made of all parts removed from all plants of all woody species
browsed within each food plot; all woody species present; and the number of individuals of each woody species present in the plot ($n = 267$ food plots). The same data were
recorded in each paired control plot, except data on browsing ($n = 113$ control plots).

In accordance with the FSH we were unsurprised to find that bulls browsed from
a lower diversity of woody plants but exploited a wider range of plant parts within
the species browsed (Figure 7.1). In contrast, cows and their young picked out the
'best' parts from many plants selected from a higher number of species. Furthermore,
bulls were more likely to consume larger branches (bigger bite diameter) and break
off thicker stems (Figure 7.2). In so doing, they spent more time at each browsing site

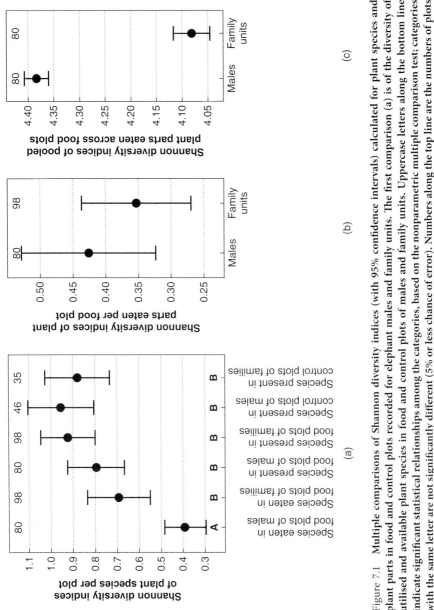

Figure 7.1 Multiple comparisons of Shannon diversity indices (with 95% confidence intervals) calculated for plant species and plant parts in food and control plots recorded for elephant males and family units. The first comparison (a) is of the diversity of utilised and available plant species in food and control plots of males and family units. Uppercase letters along the bottom line indicate significant statistical relationships among the categories, based on the nonparametric multiple comparison test; categories with the same letter are not significantly different (5% or less chance of error). Numbers along the top line are the numbers of plots surveyed in each case. The second comparison (b) is of the diversity of food items eaten by males and family units per food plot. The third comparison (c) is of the diversity of all plant parts eaten in 80 food plots for males and family units respectively, based on 100 bootstrap replications each. Each replication was a subset of records from 80 food plots randomly drawn from their respective data sets. (Source: Stokke and du Toit, 2000. Reproduced with permission of John Wiley & Sons.)

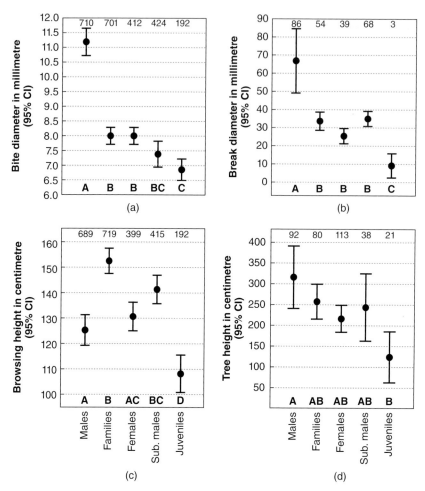

Figure 7.2 **Bite diameters (a), break diameters (b), browsing heights (c) and heights of browsed woody plants (d) recorded in food plots used by each elephant sex and age class. Uppercase letters along the bottom line indicate significant statistical relationships among the categories, based on the nonparametric multiple comparison test; categories with the same letter are not significantly different (5% or less chance of error). Numbers along the top line are the numbers of plots surveyed in each case. Error bars are 95% confidence intervals. (Source: Stokke and du Toit, 2000. Reproduced with permission of John Wiley & Sons.)**

and engaged in more uprooting, breaking branches and felling tree than did cows and their young, which fed more by stripping leaves, thereby ingesting less lignin and more digestible matter. By having a distinctly different browsing pattern to those of cows and their offspring in family units, bull elephants in Chobe exert a disproportionately large influence on the structure of the tree layer (Stokke and du Toit, 2000). Similar results have subsequently been reported from studies on elephants feeding in and around the Kruger, Phinda and Pongola wildlife areas in South Africa (Greyling, 2004; Shannon *et al.*, 2006).

Browsing height stratification

In various species of large herbivorous mammal, smaller-bodied females and their young are more efficient than adult males in extracting high quality plant parts from shared food plants when the abundance of such food is relatively low. This can force males to shift their feeding to other patches (in grazers) or higher levels in the canopy (in browsers) as an outcome of intraspecific scramble competition (reviewed by du Toit, 2005). By this scramble competition hypothesis, or SCH, we expected elephant bulls to browse higher in the canopy than females and their young offspring, resulting in browsing height stratification within the population. What we found, however, was that the browsing height of all elephant age and sex classes occurred in a narrow stratum mostly below 2 m on trees that were generally less than 4 m high. This could be because elephants tend to rebrowse coppice from stumps of previously felled or browsed trees, as has been found elsewhere (Guy, 1976; Jachmann and Bell, 1985; Owen-Smith, 1988; Kabigumila, 1993). Interestingly, when comparing cows and bulls we found no evidence of feeding-height stratification, which we expected to find in the late dry season when the taller bulls could escape intraspecific competition by browsing above the level used by cows and subadults. This is despite the bulls having the opportunity to browse above 2 m in the canopy, with the mean canopy height of trees browsed by bulls being more than 3 m (Figure 7.2d). This nonconformity with the SCH could be because the abundance of elephant food below 2 m was not limiting, at least not during the two consecutive years of our study. But a more compelling explanation, as we explain later, is that elephant bulls segregate themselves spatially from family units and so seldom feed together in the same patches, with the result that instantaneous scramble competition is avoided.

An especially interesting finding was that the cows fed significantly higher in the canopy when in a family unit than when alone (Figure 7.2c). That could be argued as evidence for the SCH, with the smaller-bodied young members of the family unit selecting high quality plant parts within easy reach, forcing the bigger cows to feed above them. Alternatively, because all members of family units are genetically related, an argument based on kin selection could be that the cows voluntarily avoid competing with their offspring and young relatives when feeding together as a family unit.

Sex differences in the use of food patches

Moving upscale from plant-part and feeding-height selection on individual plants, we investigated whether sex differences occur in the use of food patches by elephants. We think of a patch as a relatively homogeneous area that differs in some quantifiable way from its surroundings. Patches of trees, for example, can be visually identified and described compositionally by frequency distributions of trees, tree species, tree heights or other such measurements (Forman, 1995). With regard to food patches a consumer's decisions include where to search for patches, which food items to eat among those encountered in a patch, and how long to feed within a patch before searching for another patch or seeking refuge from predators (Brown, 1988). In the case of elephants we assume that patch selection is minimally influenced by predation risk and the use of food patches by each family unit is governed by the foraging decisions of its matriarch, an adult cow. We were interested in whether elephant cows and bulls differed in terms of the diversity of woody plant species included in their diets, and in their discrimination of food patches from the surrounding matrix of woody vegetation. The prediction of the FSH is that elephant cows should select food patches to derive relatively high-quality diets, whereas the much larger bulls should select patches that enable them to maximise ingestion rate.

To summarise our previously published results (Stokke, 1999; Stokke and du Toit, 2000), elephant cows and bulls both spread their feeding activity over woody plant species in general proportion to availability. Ranked selection indices (Table 7.1) showed that some common woody plants in the Kalahari-sand woodlands were used significantly less than expected, from availability, indicating that they are unpalatable to elephants: *Bauhinia petersiana, Baikiaea plurijuga, Combretum mossambicense* and *Ochna pulchra*. Family groups, dominated by females, fed on a higher diversity of woody plant species than bulls and selected patches with marginally higher species diversity than occurred in the surrounding matrix (Figure 7.1a). In contrast, plots in which bulls fed were less diverse than control plots, but not statistically significantly so, indicating no detectable patch selection based on species composition. Within food plots, however, bulls used a higher diversity of plant parts than did cow-dominated family units (Figure 7.1b), and spent more time feeding before moving on.

Our finding that family units, led by cows, are more selective than bulls at the food-patch scale is consistent with the FSH. Bulls eat more of what they find available to them at each feeding stop, thereby maximising ingestion rate, whereas cows select the best plant parts from a wider range of plant species and move between patches that offer a wider choice than that occurring in the surrounding matrix. This increased diversity of plant species in the cows' diets, and therefore in family units as a whole, deserves further research. Nevertheless, it is consistent with the suggestion made by Owen-Smith and Chafota (2012), also based on research in Chobe, that elephants are sensitive to toxic secondary compounds in plants. By spreading their diets over

Table 7.1 **Selection indices for woody plant species utilised by elephant family units and bulls respectively.**

Family units		Bulls	
Plant species	B_i	Plant species	B_i
Combretum engleri	0.062	Burkea africana	0.007
Acacia tortilis	0.062	Baikiaea plurijuga*	0.005
Commiphora mossambicensis	0.062	Combretum mossambicense*	0.002
Diospyros batocana	0.062	Acacia nigrescens	0.063
Peltophorum africanum	0.062	Garcinia livingstonei	0.063
Gardenia volkensii	0.052	Combretum molle	0.063
Acacia nigrescens	0.041	Manilkara mochisia	0.063
Combretum zeyheri	0.040	Sterculia africana	0.063
Pterocarpus angolensis	0.036	Guibourtia coleosperma	0.063
Croton megalobotrys	0.033	Terminalia prunioides	0.063
Guibourtia coleosperma	0.031	Commiphora mollis	0.063
Acacia sieberiana	0.031	Combretum apiculatum	0.059
Grewia spp.	0.031	Dichrostachys cinerea	0.044
Commiphora caerulea	0.031	Commiphora mossambicensis	0.044
Combretum adenogonium	0.031	Croton megalobotrys	0.032
Flueggea virosa	0.026	Acacia tortilis	0.032
Philenoptera nelsii	0.026	Albizia anthelmintica	0.032
Combretum apiculatum	0.026	Acacia fleckii	0.025
Terminalia sericea	0.025	Baphia massaiensis	0.025
Albizia harveyi	0.025	Philenoptera nelsii	0.021
Dichrostachys cinerea	0.024	Philenoptera violacea	0.021
Colophospermum mopane	0.022	Gardenia volkensii	0.021
Boscia albitrunca	0.021	Brachystegia boehmii	0.021
Combretum imberbe	0.021	Colophospermum mopane	0.019
Combretum elaeagnoides	0.020	Combretum elaeagnoides	0.017
Baphia massaiensis	0.019	Terminalia sericea	0.016
Capparis tomentosa	0.017	Friesodielsia obovata	0.011
Combretum molle	0.016	Burkea africana	0.011
Friesodielsia obovata	0.011	Bauhinia petersiana*	0.007
Brachystegia boehmii	0.009	Ochna pulchra*	0.005
Bauhinia petersiana*	0.008	Combretum mossambicense*	0.005

*The standardised selection index, B_i, is ranked from highest to lowest selection, with species used significantly less than expected indicated (*), where $B_i = \log(1 - f_i)/\Sigma \log(1 - f_i)$, where f_i is the proportion of items in the ith available food category that were browsed.

many plant species, each with different types and concentrations of toxins, generalist mammalian herbivores such as elephants are able to meet their dietary requirements without overloading any of the multiple detoxification pathways operating in their guts and livers (Freeland and Janzen, 1974). Elephant cows, carrying the added physiological loads of gestation (including teratogenic risks) and lactation, can be expected to be more sensitive than bulls to toxins in their diets.

Sexual segregation at the habitat scale

Moving even further upscale, we investigated whether the Chobe population was sexually segregated through differential habitat use. A common phenomenon among sexually size-dimorphic mammals, especially large herbivores, is for sexual segregation to occur outside the rutting period. Several hypotheses have been advanced to explain this: differences in body size and nutritional needs could cause feeding segregation; separate social hierarchies could cause social segregation; size-related differences in activity patterns could cause spatial and temporal segregation; differences in predation risk could cause adult females to keep their offspring in safer habitats than those used by adult males (reviewed by Main and du Toit, 2005). These hypotheses are not mutually exclusive and can be subsumed under the overriding hypothesis that, where males and females have different patterns of habitat use, it is because of the different constraints and requirements of their separate reproductive 'strategies' (Main, 2008).

Our research in northern Chobe involved driving a total distance of 570 km along 43 separate survey routes, each arranged to transect the 10 major elephant habitat types there. Observations were made while driving slowly during all daylight hours and survey effort was spread over the 1995 and 1996 wet and dry seasons. To avoid a bias against sighting bulls, which were often solitary, all sightings used for analyses of habitat occupancy had to be within 50 m of the observer. A sighting was recorded as a 'group' whether it included just one animal or several. Sightings of family units were recorded as such even if they included one or more bulls, resulting in more sightings of family units than bull groups. A total of 219 sightings were used (family units, $n = 121$; bull groups, $n = 98$).

Our results revealed statistically significant (<5% chance of error) sex differences in the use of habitats by Chobe elephants (Stokke and du Toit, 2002). In brief, family units restricted themselves to habitats near the Chobe and Zambezi rivers in the dry season, occupying a lower diversity of habitats than in the wet season, the reverse applied to bulls. The pattern for family units can be explained by their dependence on drinking water and the locomotory limitations of their young members. In the dry season, they have to remain close to the rivers, but can be widely distributed in the wet season, drinking from rain-filled pans dotted across the landscape. An explanation for the opposite pattern among bulls is less apparent but we suggest it is driven by the need to avoid male-male conflict.

When elephant bulls reach sexual maturity they begin displaying episodes of elevated aggression, roving behaviour and association with cows in family units. The onset of this phenomenon, or musth, occurs at about 29 years of age and is initially sporadic but stabilises into an annual occurrence lasting 2 – 5 months in bulls older than 35 years (Poole, 1987). During this time, bulls produce temporal gland secretions, dribble urine, walk in a distinctive posture, tusk the soil, vocalise with a distinctive rumble and generally make their dangerous status obvious. Non-musth bulls respond through avoidance or subordination (Slotow *et al.*, 2000). This behaviour minimises risky conflicts in a mating system characterised by extreme competition among males for mating opportunities, with females being receptive for only 2 – 10 days about once every 5 years (Moss, 1983). We suggest year-round avoidance of conflict among males – driven by a few aggressive males replacing one another in areas occupied by females, thereby causing what appears to be sexual segregation – is a phenomenon of exceptionally large-bodied species. It has been documented for the sperm whale (*Physeter macrocephalus*), which displays social behaviour similar to that of elephants (Best, 1979; Weilgart *et al.*, 1996). With such large-bodied animals the risk of physical injury from male–male conflict is extreme and females have short time-windows of fertility between lengthy periods of gestation and lactation. In such species an adaptive behaviour pattern for each big male is to search for mating opportunities in finite bouts of advertised aggression and then keep clear while the others have a chance.

In the dry season, male-male conflict avoidance can explain why most bulls, not being in musth, only briefly visit rivers to drink near where the cows congregate to meet the needs of their young offspring. This is where the dangerous musth bulls are. But why does the pattern of sexual segregation switch in the wet season? We suggest that elephant cows benefit from spacing themselves across the landscape to reduce male harassment, but they can only do so in the wet season when drinking water is widely available. Cows in oestrus frequently avoid mating attempts, especially by young bulls, by running away with one or more bulls in pursuit. Only one-third of those chases end in copulation, about two-thirds of which involve a large bull more than 35 years old (Moss, 1983). Chases are energetically costly and disruptive to family units, so oestrous cows solicit mate guarding by musth bulls (Poole, 1989). Social stability and feeding efficiency are thus maximised in the wet season by family units being spaced across the landscape, in home ranges perhaps determined by the dominance hierarchy among matriarchs (Wittemyer *et al.*, 2007). In this arrangement oestrous cows can be consorted by musth bulls. It is energetically expensive and dangerous for non-musth bulls to roam widely for opportunistic encounters with unguarded oestrous cows, so they remain close to the rivers. Furthermore, because elephants are grazers in the wet season, the non-musth bulls can benefit from the abundant green grass on the Chobe floodplain.

Implications for management and further research

Our research in Chobe has shown that, within the same population of elephants, bulls occupy a separate ecological niche from cows and their offspring in family units (Stokke, 1999; Stokke and du Toit, 2000, 2002). Our results have subsequently been replicated in research elsewhere in southern Africa (Greyling, 2004; Shannon *et al.*, 2006, 2008a; de Knegt *et al.*, 2011), confirming sex differences in the use of woody browse resources by elephants at plant, patch and habitat scales. That is not surprising, considering elephant bulls grow to twice the size of cows and sexual size-dimorphism is commonly associated with sexual segregation in vertebrates (Ruckstuhl and Neuhaus, 2005). Nevertheless, elephant population surveys still routinely lump elephant sex and age classes together, so managers and researchers have to work with crude total-population data when addressing issues of elephant-driven disturbance to vegetation (Guldemond and van Aarde, 2008; Scholes and Mennell, 2008). Such surveys are typically conducted from the air in the dry season and are biased towards drainage lines where densities are highest (Smit and Ferreira, 2010), resulting in the non-musth bulls being undercounted. From our research in Chobe, those undercounted bulls can be expected to have a disproportionately high impact on the woody vegetation and might be responsible for some seemingly anomalous patterns of disturbance to large trees in savanna landscapes. In southern Kruger, for example, the proportion of large trees (reaching 5 m or more in height) used by elephants, the intensity of use and the proportion pushed over, all increase with distance along 4 km transects starting at permanent water (Shannon *et al.*, 2008b). With the now compelling body of theoretical and empirical evidence for niche segregation between the sexes, future elephant management has to be guided by methods that accurately incorporate population structure into elephant surveys.

Elephant management is difficult enough as it is (Scholes and Mennell, 2008) without us complicating matters further. Nevertheless, a safe starting point is to accept that elephant bulls make a substantially greater per capita contribution to reducing tree cover than do members of family units. That fact is relevant to any area where there is concern about declining tree cover and which might lead to a management decision to reduce the elephant population. In such cases, the more effective strategy should be to remove only bulls rather than reduce numbers across the whole population. An ongoing process of research-driven adaptive management could then determine what adult sex ratio, age structure among bulls and density of bulls per habitat maintains both social stability and conserves tree cover within desired limits.

Finally, we must acknowledge that the central message of this book – elephant populations are agents of heterogeneity in African savannas – is a generalisation. In fact, elephant bulls influence savanna vegetation differently from the ways that elephants in family units do, and those influences vary seasonally in intensity and spatial distribution. An agent of heterogeneity can thus be heterogeneous in itself.

References

Archie, E.A., Moss, C.J. & Alberts, S.C. (2006) The ties that bind: genetic relatedness predicts the fission and fusion of social groups of wild African elephants. *Proceedings of the Royal Society B* 273, 513–522.

Asner, G.P. & Levick, S.R. (2012) Landscape-scale effects of herbivores on treefall in African savannas. *Ecology Letters* 15, 1211–1217.

Best, P. (1979) Social organisation in sperm whales, *Physeter macrocephalus*. In: Winn, H.E. & Olla, B.L. (eds.) *Behaviour of Marine Animals: Current Perspectives in Research, 3: Cetaceans.* Plenum Press, New York, USA, pp. 35–52.

Blanc J.J., Barnes R.F.W., Craig G.C., Dublin, H.T., Thouless, C.R., Douglas-Hamilton, I. & Hart, J.A. (2007) *African Elephant Status Report 2007. An Update from the African Elephant Database.* Occasional Paper of the IUCN Species Survival Commission No. 33, IUCN, Gland, Switzerland.

Brown, J.S. (1988) Patch use as an indicator of habitat preference, predation risk, and competition. *Behavioural Ecology and Sociobiology* 22, 37–47.

Chase, M. (2011) *Dry Season Fixed-wing Aerial Survey of Elephants and Wildlife in Northern Botswana.* Elephants Without Borders, Kasane, Botswana; Department of Wildlife and National Parks, Botswana; and Zoological Society of San Diego, USA. [online] http://www.elephantdatabase.org/population_submission_attachments/102.

Clutton-Brock, T.H. & Harvey, P.H. (1978) Mammals, resources and reproductive strategies. *Nature* 273, 191–195.

de Knegt, H.J., van Langevelde, F., Skidmore, A.K., Delsink, A., Slotow, R., Henley, S., Bucini, G., de Boer, W.F., Coughenour, M.B., Grant, C.C., Heitkönig, I.M.A., Henley, M., Knox, N.M., Kohi, E.M., Mwakiwa, E., Page, B.R., Peel, M., Pretorius, Y., van Wieren, S.E. & Prins, H.T.T. (2011) The spatial scaling of habitat selection by African elephants. *Journal of Animal Ecology* 80, 270–281.

Dublin H.T. (1996) Elephants of the Masai Mara, Kenya: seasonal habitat selection and group patterns. *Pachyderm* 22, 25–35.

du Toit, J.T. (2005) Sex differences in the foraging ecology of large herbivores. In: Ruckstuhl, K.E. & Neuhaus, P (eds.) *Sexual Segregation in Vertebrates: Ecology of the Two Sexes.* Cambridge University Press, Cambridge, UK, pp. 35–52.

Forman R.T.T. (1995) *Land Mosaics: The Ecology of Landscapes and Regions.* Cambridge University Press, Cambridge, UK.

Freeland, W.J. & Janzen, D.H. (1974) Strategies in herbivory by mammals: the role of plant secondary compounds. *American Naturalist* 108, 269–289.

Greyling, M.D. (2004) Sex and age related distinctions in the feeding ecology of the African elephant, *Loxodonta africana*. PhD Thesis, University of the Witwatersrand, Johannesburg, South Africa.

Guldemond R. & van Aarde R. (2008) A meta-analysis of the impact of African elephants on savanna vegetation. *Journal of Wildlife Management* 72, 892–899.

Guy P.R. (1976) The feeding behaviour in elephant *Loxodonta africana* in the Sengwa area, Rhodesia. *South African Journal of Wildlife Research* 61, 55–63.

Jachmann, H. & Bell, R.H.V. (1985) Utilisation by elephants of the *Brachystegia* woodlands of the Kasungu National Park, Malawi. *African Journal of Ecology* 23, 245–258.

Kabigumila, J. (1993) Feeding habits of elephants in Ngorongoro Crater, Tanzania. *African Journal of Ecology* 31, 156–164.

Kerley, G.I.H., Landman, M., Kruger, L. & Owen-Smith, N. (2008) Effects of elephants on ecosystems and biodiversity. In: Scholes, R.J. & Mennell, K.G. (eds.) *Elephant Management: A Scientific Assessment for South Africa*. Wits University Press, Johannesburg, South Africa, pp. 146–205.

Loison, A., Gaillard, J.-M., Pelabon, C. & Yoccoz, N.G. (1999) What factors shape sexual size dimorphism in ungulates? *Evolutionary Ecology Research* 1, 611–633.

Main, M.B. (2008) Reconciling competing ecological explanations for sexual segregation in ungulates. *Ecology* 89, 693–704.

Main, M.B. & du Toit, J.T. (2005) Sex differences in reproductive strategies affect habitat choice in ungulates. In: Ruckstuhl, K.E. & Neuhaus, P. (eds.) *Sexual Segregation in Vertebrates: Ecology of the Two Sexes*. Cambridge University Press, Cambridge, UK, pp. 148–161.

Moss, C.J. (1983) Oestrous behaviour and female choice in the African elephant. *Behaviour* 86, 167–196.

Moss, C.J. (2001) The demography of an African elephant (*Loxodonta africana*) population in Amboseli, Kenya. *Journal of Zoology* 255, 145–156.

Owen-Smith, R.N. (1988) *Megaherbivores: The Influence of Very Large Body Size on Ecology*. Cambridge University Press, Cambridge, UK.

Owen-Smith, N. & Chafota, J. (2012) Selective feeding by a megaherbivore, the African elephant (*Loxodonta africana*). *Journal of Mammalogy* 93, 698–705.

Owen-Smith N., Kerley G.I.H., Page B., Slotow R. & van Aarde R.J. (2006) A scientific perspective on the management of elephants in the Kruger National Park and elsewhere. *South African Journal of Science* 102, 389–394.

Poole, J.H. (1987) Rutting behaviour in African elephants: the phenomenon of musth. *Behaviour* 102, 283–316.

Poole, J.H. (1989) Mate guarding, reproductive success and female choice in African elephants. *Animal Behaviour* 37, 842–849.

Power, R.J. & Compion, R.X.S. (2009) Lion predation on elephants in the Savuti, Chobe National Park, Botswana. *African Zoology* 44, 36–44.

Ruckstuhl, K.E. & Neuhaus, P (eds.) (2005) *Sexual Segregation in Vertebrates: Ecology of the Two Sexes*. Cambridge University Press, Cambridge, UK.

Scholes, R.J. & Mennell, K.G. (eds.) (2008) *Elephant Management: A Scientific Assessment for South Africa*. Wits University Press, Johannesburg, South Africa.

Shannon, G., Page, B.R., Duffy, K.J. & Slotow, R. (2006) The role of foraging behaviour in the sexual segregation of the African elephant. *Oecologia* 150, 344–354.

Shannon G., Page B.R., Mackey R.L., Duffy K.J. & Slotow R. (2008a) Activity budgets and sexual segregation in African elephants (*Loxodonta africana*). *Journal of Mammalogy* 89, 467–476.

Shannon, G., Druce, D.J., Page, B.R., Eckhardt, H.C., Grant, R. & Slotow, R. (2008b) The utilization of large savanna trees by elephant in southern Kruger National Park. *Journal of Tropical Ecology* 24, 281–289.

Slotow, R., van Dyk, G., Poole, J., Page, B. & Klocke, A. (2000) Older bull elephants control young males. *Nature* 408, 425–426.

Smit, I.P.J. & Ferreira, S.M. (2010) Management intervention affects river-bound spatial dynamics of elephants. *Biological Conservation* 143, 2172–2181.

Spinage, C.A. (1994) *Elephants*. T. & A.D. Poyser Ltd., London, UK.

Stokke, S. (1999) Sex differences in feeding-patch choice in a megaherbivore: elephants in Chobe National Park, Botswana. *Canadian Journal of Zoology* 77, 1723–1732.

Stokke, S. (2000) Sexual segregation in the African elephant (*Loxodonta africana*). PhD Thesis, Norwegian University of Science and Technology, Trondheim, Norway.

Stokke, S. & du Toit, J.T. (2000) Sex and size related differences in the dry season feeding patterns of elephants in Chobe National Park, Botswana. *Ecography* 23, 70–80.

Stokke, S. & du Toit, J.T. (2002) Sexual segregation in habitat use by elephants in Chobe National Park, Botswana. *African Journal of Ecology* 40, 360–371.

Weilgart, L., Whitehead, H. & Payne, K. (1996) A colossal convergence. *American Scientist* 84, 278–287.

Wittemyer, G., Getz, W.M., Vollrath, F. & Douglas-Hamilton, I. (2007) Social dominance, seasonal movements, and spatial segregation in African elephants: a contribution to conservation behavior. *Behavioral Ecology and Sociobiology* 61, 1919–1931.

(8)

Surface Water and Elephant Ecology: Lessons from a Waterhole-Driven Ecosystem, Hwange National Park, Zimbabwe

Simon Chamaillé-Jammes[1], Marion Valeix[2],
Hillary Madzikanda[3] and Hervé Fritz[2]

[1]Centre d'Ecologie Fonctionnelle et Evolutive, France
[2]Laboratoire de Biométrie et Biologie Evolutive, France
[3]Scientific Services, Zimbabwe Parks and Wildlife Management
Authority, Zimbabwe

Elephants are heavy drinkers. Captive adult elephants can drink as much as 200 L day^{-1} (Olson, 2004), and although what happens in natural environments is likely to differ somewhat, water consumption by elephants surely remains huge when considered at the scale of whole populations. Nevertheless, there are reports of elephants wandering without water for days and living in some of the most arid places in Africa (Viljoen, 1989). Thus one is left wondering about the importance of surface water in the ecology of elephants, and how this often scarce resource in savannas might affect elephant distribution and dynamics.

We have studied these issues in Hwange National Park (hereafter Hwange NP), located in north-western Zimbabwe where it borders Botswana. This ecosystem, as well as the larger Kalahari sand region, offers a great opportunity to study the relationship between water and elephants: surface-water is often patchily distributed in a few ephemeral ponds and artificial waterpoints pumped pans, which make the spatial constraint on elephant populations easily detectable.

Elephants and Savanna Woodland Ecosystems: A Study from Chobe National Park, Botswana,
First Edition. Edited by Christina Skarpe, Johan T. du Toit and Stein R. Moe.
© 2014 John Wiley & Sons, Ltd. Published 2014 by John Wiley & Sons, Ltd.

A brief description of Hwange National Park

Hwange NP (14,651 km^2) has limited availability of natural surface water during the dry season. It has no permanent rivers, and only a few pans retain water in an average year (mean annual rainfall ~600 mm). Only 20% and 50% of the park are within 5 and 10 km, respectively, of natural water sources under these conditions (Chamaillé-Jammes *et al.*, 2007a). These percentages increase by approximately 10% when existing boreholes are in use, but the biggest effect of artificial water provision is to virtually eliminate the otherwise large variability in accessible surface water associated with fluctuating rainfall. Thus pumping principally prevents the occurrence of a complete lack of surface water during extreme droughts. No studies have been undertaken so far to assess the sustainability of the groundwater resource, however.

Two-thirds of the park lies on the eastern fringes of Kalahari sands that cover most of Botswana. The vegetation is characteristic of the wider region with its semi-arid climate and dystrophic soils, being a mix of bushland and patches of *Baikiaea plurijuga* woodland. Monospecific stands of mopane trees *Colophospermum mopane* cover much of the north of the park, where the soils are derived from basalt and the climate is more arid. Further details of the vegetation communities and their determinants can be found in Rogers (1993). Thus Hwange NP is a semi-arid savanna ecosystem superficially similar in habitats to other large NPs hosting large numbers of elephants, such as Kruger NP (South Africa) and Chobe NP (Botswana), but it differs in the way surface water is distributed and changes during the season, because in these other parks elephants can also access one or several permanent rivers.

Movement patterns reveal the dry-season trade-off between foraging and drinking

In Hwange NP, aerial censuses show that elephant distribution in the dry season varies between years and is strongly linked to surface-water availability from both natural and artificial pans (Chamaillé-Jammes *et al.*, 2007b). More elephants are present where waterhole density is highest. Few occur beyond 15 km from water, although bulls are more likely to be found far from water than are family herds (Conybeare, 1991). This pattern has been observed repeatedly in other protected areas and suggests a universal influence of surface-water availability on the dry-season distribution pattern of elephants (e.g. Stokke and du Toit, 2002; Smit *et al.*, 2007a; Loarie *et al.*, 2009).

Tracking of family herds in Hwange NP, using very high frequency (VHF) telemetry in the early 1980s (Conybeare, 1991) and by global positioning system (GPS) telemetry in 2009–2011 (S. Chamaillé-Jammes, unpublished data; Figure 8.1) has provided further details on how elephants use water sources during the dry season. Tracking

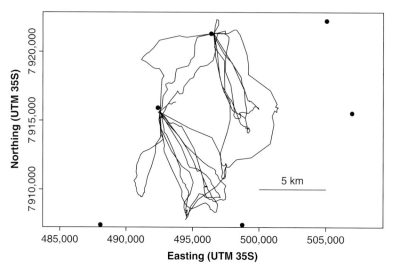

Figure 8.1   **Typical elephant movement during the dry season in Hwange NP. Data are from a 3-week tracking by GPS collar of a family herd. Solid circles show locations of waterholes.**

revealed that during the hot, dry season family herds most commonly visit waterholes every 24–48 h, and virtually never less frequently than every 72 h. The few other studies available have reported similar results (Young, 1970 in van Aarde et al., 2008). Even in the Namib desert, elephant family herds drink every 48–72 h (Leggett, 2006). Keepers of captive elephants however recommend providing water at least twice a day (Olson, 2004), which suggests that if offered the opportunity to drink without constraint elephants will do so more often. Between two drinking events, elephants travelled at an average speed of about 0.5 km h^{-1} (calculated over 1 h intervals), but moved noticeably faster, closer to 1.0 km h^{-1}, both when coming to and when leaving from waterholes, suggesting that they do not actively forage for several kilometres while moving to and from feeding areas. This might also suggest that commuting between drinking and foraging places acts as a significant time-constraint on elephants. Altogether these results highlight how elephants trade-off foraging for drinking during the dry-season, the most critical time for survival.

Evidence that water defines key-resource areas: population-level processes

When Hwange NP was gazetted in 1928 elephant numbers were low (about 1000: Davison, 1967), at least partly the result of the low surface-water availability, notwithstanding the widespread reduction in numbers caused by ivory hunting during the late 19th century (Chapter 1; Spinage, 1973). They might also have roamed this area only during the wet season. The creation of artificial waterholes, aimed at increasing the abundance of large herbivores, followed soon after (Davison, 1967). By the 1980s, over 60 had been built. Although some boreholes have been closed and new ones opened, the number is similar today, varying from 40 to 60, year to year, mostly because of logistical or financial constraints on pumping. Elephants responded rapidly to water provision so that by 1968 numbers were estimated to be almost 8000 (Williamson, 1975). A culling programme was initiated in 1966 and that policy was later adjusted to maintain a population of approximately 13,000 elephants (Cumming, 1981). Culling was abandoned in 1986 because of a conjunction of logistical and financial difficulties, international public pressure, and maybe other, non-ecological, reasons (Child, 2004). The population then doubled in a few years, indicating substantial immigration, although its sources remain unknown. Since 1992, however, elephant abundance has fluctuated around 30,000–35,000 individuals (Chamaillé-Jammes et al., 2008), with the last census, in 2007, producing an estimate of 34,321 individuals or $2.26\,\mathrm{km^{-2}}$ (Dunham et al., 2007: note that this area includes the neighbouring $510\,\mathrm{km^2}$ Deka Safari Area). This represents one of the highest elephant densities over a large area (Blanc et al., 2007), similar to that observed in Chobe NP, where local dry-season densities can be as high as $10–15$ elephants $\mathrm{km^{-2}}$, and which was one of the first populations to show density dependence (Junker et al., 2008).

The dynamics of the population since culling was abandoned demonstrates the importance of surface-water in limiting the population. The rapid increase in elephant abundance during the dry season was spatially patterned, with greater increases in numbers observed around waterholes where elephant numbers were initially lower (Chamaillé-Jammes et al., 2008). Elephant were also more evenly distributed among waterholes when the mean number of elephants at waterholes was high because of either high elephant density or low waterhole availability, or perhaps both (Chamaillé-Jammes et al., 2008). Such density-dependent use of space could be linked to interference competition for water. At the most crowded waterholes, we have regularly observed more than 1500 elephants coming to drink over a 24-h period. More than 300 individuals can be present at a waterhole concurrently, sometimes with only a 2-m radius trough from which to drink (see Figures 8.2 and 8.3). Aggressive interactions are then common, and time lost waiting to access water is evident, albeit not yet quantified. In dry years, when elephant numbers at waterholes are high, the temporal window during which elephants drink expands, with some groups coming in the afternoon, rather than after sunset, when most elephants drink (Valeix et al., 2007a). This suggests that some groups may avoid competition by drinking at times when the presence of other elephants at a waterhole is less likely. Indeed, even during years of greatest elephant abundance, water is still accessible at waterholes for much of the day. Exploitation competition for forage accessible from waterholes during the

Figure 8.2   **Elephants drinking at an artificial waterhole in Hwange NP during the dry season. Groundwater is pumped into the trough that will soon be the only source of water as the pan dries up. (Source: © S. Chamaillé-Jammes.)**

dry season may therefore be the critical factor driving density-dependent use of space. During severe droughts, carcasses of young and old individuals are commonly found in the vicinity of waterholes still containing water (Dudley *et al.*, 2001), supporting the view that lack of water is not the cause of death. The occurrence of starvation-induced mortality rather than thirst-induced mortality was also observed in Tsavo NP, Kenya, during the of 1970–1971 drought (Corfield, 1973).

Building on these observations and the movement data, the scenario we propose is that during the dry season elephants have to commute between two complementary resources, water and food. As the dry season progresses and food becomes depleted, the weakening animals are required to take longer trips. The observation that elephants were located further from waterholes in the late dry season than in the early dry season (Conybeare, 1991), supports this hypothesis. The weakest animals eventually die, either from starvation and physiological stress, or from increased lion predation, possibly facilitated by poor condition and stress on individuals (Loveridge *et al.*, 2006). We now need to test this hypothesis by studying the loss of body condition and increased physiological stress levels in this population during the course of the dry season.

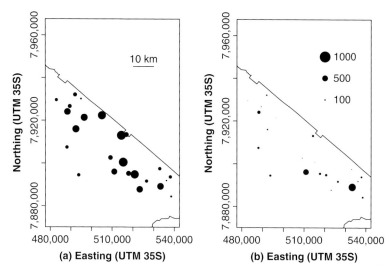

Figure 8.3 **Interannual variability in elephant numbers using waterholes in the dry season in Hwange NP. Data are elephant numbers counted over 24 h in a dry (a) and a wet (b) year.**

How commonly water defines the key-resource areas (*sensu* Illius and O'Connor, 2000) that drive elephant-population regulation in savannas remains to be ascertained. Because many protected elephant populations are still recovering from past harvesting, we have limited data with which to test this hypothesis more widely. Nevertheless, recent studies point in the same direction. In Kruger NP, even though elephants remain distributed close to water sources following the closure of about half the artificial waterholes and the cessation of culling, they still distribute themselves more evenly across the landscape as density increases (Young *et al.*, 2009). This suggests that some form of competition is occurring. In Chobe NP, herds appear to be foraging further from water than in earlier years, perhaps a result of decade-long forage depletion on the riverfront (Chapter 4). More generally, data from a number of southern African protected areas suggest that the survival of weaned elephant calves declines with increasing daily-displacement distances in the dry season, itself affected by elephant density and water provision (Young and van Aarde, 2010). Bridging the gap from use of space to demography is now an important objective, which might best be conducted by long-term monitoring of individually tracked herds.

Adaptive use of space is clearly an important strategy used by elephants to adjust to competitive pressures during the dry season, within the constraints imposed by surface-water availability. Understanding dry-season range selection is therefore important. The recent 3-year telemetry study in Hwange NP showed that herds return to the same dry-season ranges in consecutive years, a pattern also found elsewhere (e.g. Verlinden and Gavor, 1998; Loarie *et al.*, 2009; Young *et al.*, 2009). During the dry season, elephants seem to rely on previous knowledge of an area and maybe also the location of water sources. When and under what conditions then do changes in the dry-season distribution of populations occur? Population censuses in Hwange have shown that immigration occurred soon after culling ended, reflecting the attraction of some herds, previously living outside the park, to an area now little disturbed by humans. In Tarangire NP, Tanzania, elephants were forced to use different dry-season ranges when the park suffered a severe drought that disrupted the usual patterns of surface water availability (Foley *et al.*, 2008). Have some of these individuals experienced and responded to such conditions before, or are the observed patterns novel responses to previously unknown conditions? We are still far from having a predictive understanding of how elephant distributions will change when faced with unusual patterns of surface water availability, possibly associated with changes in intraspecific competition.

How does the absence of rivers, a striking characteristic of Hwange, affect the generality of our findings from this waterhole-driven ecosystem? First, interference competition for access to water is unlikely to occur when elephants drink at a river. Although elephants do not use the riverfront evenly, access to water is more readily available than at waterholes. Second, large permanent rivers are, by definition, a non-depletable resource in respect of water consumption by elephants. Indeed an important consequence of the large amount of water consumed by elephants is the depletion of waterholes during the course of the dry season. Although we have yet to analyse the sensitivity of waterholes to water consumption by elephants, relative to other factors affecting water levels, such as rainfall, evaporation and consumption by other animals, there seems little doubt that, given the huge number of elephants, their consumption represents a critical factor determining when pans dry up in the season. If so, this would be a major source of intraspecific resource-depletion and competition. Pumping groundwater reduces depletion, but at waterholes used intensively by elephants it is insufficient to offset consumption. Researchers at a workshop held in Hwange NP in 1995 calculated that 22 artificial waterholes would deliver only 20–40% of the water required by the approximately 25,000 elephants there (Owen-Smith, 1996). Such a waterhole-driven, depletion-prone system, makes Hwange peculiar compared with Kruger and Chobe, two other big NPs hosting large numbers of elephants. Both have a least one large permanent river, although areas of Chobe away from the riverfront may be similar to Hwange because, there too, artificial waterholes are the only source of water. In particular, water depletion may cause Hwange elephants to travel more as the dry season advances, moving from one water source to the next as each dries up. The demographic consequences of such regional differences are unknown and need investigating.

Beyond water, habitats and social interactions

A critical aspect of our proposed scenario of elephant population regulation is the importance of the distribution of dry-season food resources in relation to water, and how these food resources resist depletion and long-term destruction from elephant foraging. For instance, most Chobe elephants feed on floodplain grasses that stay green long into the dry season (Spinage, 1990). This would explain why massive die-off are not reported from this area despite high elephant density and intermittent severe droughts. In contrast, Hwange is dominated by woody vegetation communities, and those with many species palatable to elephants are heavily impacted. Seasonal depletion of forage (Figure 8.2) and elephant transformation of vegetation structure by breaking, pushing and toppling of trees is evident (Valeix *et al.*, 2011), particularly in the vicinity of waterholes when elephant densities increase sharply. Interestingly, despite this heavy use of vegetation around waterholes, long-term vegetation change does not seem to be detrimental to those elephants that survive the dry season: 7 out of the 10 herds tracked using GPS collars remained in the same area in the wet season. More generally, this suggests that the long-term elephant-induced vegetation change does not decrease the attractiveness of these sites to elephants. Why then do some elephants move away from their dry-season range in the wet season and undertake seasonal migration, as observed in Hwange where, at the onset of the rains, some elephants move towards the centre of the park? One answer could be social interactions. For instance in Samburu and Buffalo Springs National Reserves in Kenya, contests occur between groups even in the absence of immediate competition for resources, especially in the wet season (Wittemyer and Getz, 2007), and dominant groups remain closer to water sources than subordinates even during the wet season (Wittemyer *et al.*, 2007).

If surface water acts as a definitive constraint on elephant distribution during the dry season, and thus its availability across the landscape becomes a major driver of population dynamics, then a deeper understanding of its effect will emerge only by accounting for the configuration of both forage and water, and how elephants interact socially to use that landscape.

Surface-water driven management of elephants and savanna ecosystems

Our studies clearly show that surface-water availability is a key driver of elephant dynamics in semi-arid savannas. Artificial provision of surface water is now recognised as a cause of unnaturally high elephant densities (van Aarde and Jackson, 2007), and thus should be discussed where the necessity of controlling elephant numbers is considered. Closure of artificial waterholes may appear as an obvious solution to reducing elephant numbers. It is a realistic alternative to culling which is ethically debatable, particularly where management actions are the direct cause of over-abundance in the first place. Nevertheless, closing artificial waterholes raises many unanswered questions about the generality of this proposition (Chamaillé-Jammes et al., 2007c; Smit et al., 2007b).

First, overlap between accessible areas defined by artificial waterholes and other permanent water-sources is important. In ecosystems such as Kruger or Chobe, characterised by the presence of both perennial rivers and artificial waterholes, the effectiveness of waterhole closure close to and away from rivers could be contrasted. Closure could be effective in areas far from rivers where artificial waterholes would allow elephants to exploit areas during the dry season that they would otherwise be unable to access. Conversely, closure of waterholes located only a few kilometres from permanent rivers would likely be ineffective because these are probably unsuccessful at attracting elephants away from rivers, where the forage is often preferred (Smit and Ferreira, 2010). Areas close to rivers will also remain accessible to elephants in the dry season even without waterholes, and therefore may be used eventually by elephants when competition for riverine forage increases. Generally, at the scale of a park, restricting water availability by closing artificial waterholes will likely increase elephant pressure on riverine woodland (Smit and Ferreira, 2010).

Second, little is known about the impact of such closures on other water-dependent species. At waterholes, despite occasional aggressive interactions between elephants and other species, elephants rarely prevent other animals from drinking. Most elephants come to drink after sunset, and other herbivore species can shift their time of drinking to avoid interference (Valeix et al., 2007a). Abandoning water pumping could result in elephants at the remaining water sources, increasing water depletion and adversely affecting other water-dependent species. Pumping currently partially replenishes the waterholes during elephants' absences, thereby maintaining availability of water to other species. Waterhole closure will also lead to redistribution and possibly changes in abundance of other species, irrespectively of the elephants' responses, probably benefitting water-independent species or competitors displaced by species ranging close to water. Socio-economic factors also come into play (tourism, source-sink dynamics with neighbouring safari hunting areas). These, alongside any ecological effects of water provision, decrease the likelihood of finding a universally successful policy, in turn emphasising the need for well-defined, explicit, management objectives.

In spite of these potential drawbacks, managing surface water supplies holds promise for managing large populations of elephants and is amenable to modelling to improve decision-making (e.g. Martin et al., 2010). Nevertheless, in many places, debates on the actual need for elephant management and population reductions remain unresolved. Claims that elephants drastically reduces tree cover are numerous and have sometimes been demonstrated (Guldemond and van Aarde, 2008), but this does not always occur

(Valeix *et al.*, 2007b; Chamaillé-Jammes *et al.*, 2009), or is sometimes limited to tall trees with the density of woody plants increasing simultaneously (Kalwij *et al.*, 2010). Elephants can also influence the composition of woody plant communities (O'Connor *et al.*, 2007), although the ecological relevance of these changes can be hard to establish (but might endanger some species of conservation interest: Lombard *et al.*, 2001). The cascading effects on other fauna are also unclear: for instance negative correlations in abundance that suggest competition between elephants and medium-sized browsers (across-sites comparison: Fritz *et al.*, 2002; in Hwange: Valeix *et al.*, 2007b, 2008) are not supported by field studies. On the contrary, they suggest that elephant-induced vegetation change improves browse availability (Rutina *et al.*, 2005; Makhabu *et al.*, 2006, Kohi *et al.*, 2011) and, by increasing visibility, enhances safety from predation (Valeix *et al.*, 2011).

In Hwange NP the current management policy is to maintain pumping and therefore the current patterns of elephant distribution and dynamics. Recent tourism development within the park – required to bring in revenue – will lead to an increase in the number of waterholes, redistribution of the elephants and a likely increase in their numbers with associated changes in the local vegetation. Although no formal consultation has been undertaken, opinions among stakeholders and within institutions clearly differ on the necessity to revise the current water management policy to manipulate elephant numbers and impacts.

Research should inform policy makers on the potential match or mismatch between elephant densities, associated impacts and their acceptability and the level at which elephant population regulation will occur (Smit *et al.*, 2007b). As we have shown, these can be affected by the distribution of artificial waterpoints. Despite substantial knowledge gained over the years, we feel that we have just started paving the way to developing a definitive understanding of the effects of surface water on both elephant and savanna ecology. These effects may well be highly context-specific, making it unlikely that there will be a one-solution-fits-all for the many diverse objectives governing surface-water management.

Acknowledgements

The Director General of the Zimbabwe Parks and Wildlife Management Authority (ZPWMA) is acknowledged for having provided the opportunity to carry out this research and for permission to publish this manuscript. We sincerely thank the Wildlife Environment Zimbabwe association for having provided waterhole census data over the years. This work was mostly funded by the 'Centre National de la Recherche Scientifique', the 'Centre de coopération Internationale en Recherche Agronomique pour le Développement', the French Ministry of Foreign Affairs and grants from the 'Agence Nationale de la Recherche' (ANR-08-BLAN-0022; ANR-11-CEPL-003-06).

Finally, this work could not have been possible without the help of numerous students, field assistants (particularly Martin Muzamba) and staff from ZPWMA to whom we are indebted. The original manuscript was significantly improved by comments from Izak Smit.

References

Blanc J.J., Barnes R.F.W., Craig G.C., Dublin, H.T., Thouless, C.R., Douglas-Hamilton, I. & Hart, J.A. (2007) *African Elephant Status Report 2007. An Update from the African Elephant Database.* Occasional Paper of the IUCN Species Survival Commission No. 33, IUCN, Gland, Switzerland.

Chamaillé-Jammes, S., Fritz, H. & Murindagomo, F. (2007a) Climate-driven fluctuations in surface-water availability and the buffering role of artificial pumping in an African savanna: potential implication for herbivore dynamics. *Austral Ecology* 32, 740–748.

Chamaillé-Jammes, S., Valeix, M. & Fritz, H. (2007b) Managing heterogeneity in elephant distribution: interactions between elephant population density and surface-water availability. *Journal of Applied Ecology* 44, 625–633.

Chamaillé-Jammes, S., Valeix, M. & Fritz, H. (2007c) Elephant management: why can't we throw out the babies with the artificial bathwater? *Diversity and Distributions* 13, 663–665.

Chamaillé-Jammes, S., Fritz, H., Valeix, M., Murindagomo, F. & Clobert, J. (2008) Resource variability, aggregation and direct density dependence in an open context: the local regulation of an African elephant population. *Journal of Animal Ecology* 77, 135–144.

Chamaillé-Jammes, S., Fritz, H. & Madzikanda, H. (2009) Piosphere contribution to landscape heterogeneity: a case study of remote-sensed woody cover in a high elephant density landscape. *Ecography* 32, 871–880.

Child, G. (2004) Elephant culling in Zimbabwe. *ZimConservation Opinion* 1, 1–6.

Conybeare, A.M. (1991) Elephant occupancy and vegetation change in relation to artificial water points in a Kalahari sand area of Hwange National Park. PhD Thesis, University of Zimbabwe, Harare, Zimbabwe.

Corfield, T. (1973) Elephant mortality in Tsavo National Park, Kenya. *East African Wildlife Journal* 11, 339–368.

Cumming, D.H.M. (1981) The management of elephant and other large mammals in Zimbabwe. In: Jewel, P.J. & Holt, S. (eds.) *Problems in Management of Locally Abundant Wild Animals.* Academic Press, New York, pp. 91–118.

Davison, T. (1967) *Wankie. The Story of a Great Game Reserve.* Books of Africa, Cape Town, South Africa.

Dudley, J.P., Craig, G.C., Gibson, D.S.C., Haynes, G. & Klimowicz, J. (2001) Drought mortality of bush elephants in Hwange National Park, Zimbabwe. *African Journal of Ecology* 39, 187–194.

Dunham, K.M., Mackie, C.S., Musemburi, O.C., Zhuwau, C., Mtare, T.G., Taylor, R.D. & Chimuti, T. (2007) *Aerial Survey of Elephants and Other Large Herbivores in North-west Matabeleland, Zimbabwe.* Occasional Paper 19, WWF-SARPO, Harare, Zimbabwe.

Foley, C., Pettorelli, N. & Foley, L. (2008) Severe drought and calf survival in elephants. *Biology Letters* 4, 541–544.

Fritz, H., Duncan, P., Gordon, I.J. & Illius A.W. (2002) Megaherbivores influence trophic guilds structure in African ungulate communities. *Oecologia* 131, 620–625.

Guldemond R. & van Aarde R. (2008) A meta-analysis of the impact of African elephants on savanna vegetation. *Journal of Wildlife Management* 72, 892–899.

Illius, A.W. & O'Connor, T.G. (2000) Resource heterogeneity and ungulate population dynamics. *Oikos* 89, 283–294.

Junker, J., van Aarde, R.J. & Ferreira, S.M. (2008) Temporal trends in elephant *Loxodonta africana* numbers and densities in northern Botswana: is the population really increasing? *Oryx* 42, 58–65.

Kalwij, J.M., de Boer, W.F., Mucina, L., Prins, H.H.T., Skarpe, C. & Winterbach, C. (2010) Tree cover and biomass increase in a southern African savanna despite growing elephant population. *Ecological Applications* 20, 222–233.

Kohi, E.M., de Boer, W.F., Peel, M.J.S., Slotow, R., van der Waal, C., Heitkönig, I.M.A., Skidmore, A. & Prins, H.H.T. (2011) African elephants *Loxodonta africana* amplify browse heterogeneity in African savanna. *Biotropica* 43, 711–721.

Leggett, K. (2006) Effects of artificial waterpoints on the movement and behaviour of desert-dwelling elephants of north-western Namibia. *Pachyderm* 40, 24–34.

Loarie, S.R., van Aarde, R.J. & Pimm, S.L. (2009) Fences and artificial water affect African savannah elephant movement patterns. *Biological Conservation* 142, 3086–3098.

Lombard, A.T., Johnson, C.F., Cowling, R.M. & Pressey, R.L. (2001) Protecting plants from elephants: botanical reserve scenarios within the Addo Elephant National Park, South Africa. *Biological Conservation* 102, 191–203.

Loveridge, A.J., Hunt, J.E., Murindagomo, F. & Macdonald, D.W. (2006) Influence of drought on predation of elephant (*Loxodonta africana*) calves by lions (*Panthera leo*) in an African wooded savannah. *Journal of Zoology* 270, 523–530.

Makhabu, S.W., Skarpe, C. & Hytteborn, H. (2006) Elephant impact on shoot distribution on trees and on rebrowsing by smaller browsers. *Acta Oecologica* 30, 136–146.

Martin, J., Chamaillé-Jammes, S., Nichols, J.D., Fritz, H., Hines, J.E., Fonnesbeck, C.J., MacKenzie, D.I. & Bailey, L.L. (2010) Simultaneous modeling of habitat suitability, occupancy, and relative abundance: African elephants in Zimbabwe. *Ecological Applications* 20, 1173–1182.

O'Connor, T.G., Goodman, P.S. & Clegg, B. (2007) A functional hypothesis of the threat of local extirpation of woody plant species by elephant in Africa. *Biological Conservation* 136, 329–345.

Olson, D. (2004) *Elephant Husbandry Resource Guide*. Allen Press, Lawrence, USA.

Owen-Smith, N. (1996) Ecological guidelines for waterpoints in extensive protected areas. *South African Journal of Wildlife Research* 26, 107–112.

Rogers, C.M.L. (1993) *A Woody Vegetation Survey of Hwange National Park*. Department of National Parks and Wildlife Management, Harare, Zimbabwe.

Rutina, L.P., Moe, S.R. & Swenson, J.E. (2005) Elephant *Loxodonta africana* driven woodland conversion to shrubland improves dry-season browse availability for impalas *Aepyceros melampus*. *Wildlife Biology* 11, 207–213.

Smit, I.P.J. & Ferreira, S.M. (2010) Management intervention affects river-bound spatial dynamics of elephants. *Biological Conservation* 143, 2172–2181.

Smit, I.P.J., Grant, C.C. & Whyte, I.J. (2007a) Landscape-scale sexual segregation in the dry season distribution and resource utilization of elephants in Kruger National Park, South Africa. *Diversity and Distributions* 13, 225–236.

Smit, I.P.J., Grant, C.C. & Whyte, I.J. (2007b) Elephants and water provision: what are the management links? *Diversity and Distributions* 13, 666–669.

Spinage, C.A. (1973) A review of ivory exploitation and elephant population trends in Africa. *African Journal of Ecology* 11, 281–289.

Spinage, C.A. (1990) Botswana's problem elephants. *Pachyderm* 13, 15–19.

Stokke, S. & du Toit, J.T. (2002) Sexual segregation in habitat use by elephants in Chobe National Park, Botswana. *African Journal of Ecology* 40, 360–371.

Valeix, M., Chamaillé-Jammes, S. & Fritz, H. (2007a) Interference competition and temporal niche shifts: elephants and herbivore communities at waterholes. *Oecologia* 153, 739–748.

Valeix, M., Fritz, H., Dubois, S., Kanengoni, K., Alleaume, S. & Said, S. (2007b) Vegetation structure and ungulate abundance over a period of increasing elephant abundance in Hwange National Park, Zimbabwe. *Journal of Tropical Ecology* 23, 87–93.

Valeix, M., Fritz, H., Chamaillé-Jammes, S., Bourgarel, M. & Murindagomo, F. (2008) Fluctuations in abundance of large herbivore populations: insights into the influence of dry season rainfall and elephant numbers from long-term data. *Animal Conservation* 11, 391–400.

Valeix, M., Fritz, H., Sabatier, R., Murindagomo, F., Cumming, D. & Duncan, P. (2011) Elephant-induced structural changes in the vegetation and habitat selection by large herbivores in an African savanna. *Biological Conservation* 144, 902–912.

van Aarde, R.J. & Jackson, T.P. (2007) Megaparks for metapopulations: addressing the causes of locally high elephant numbers in southern Africa. *Biological Conservation* 134, 289–297.

van Aarde, R.J., Ferreira, S., Jackson, T., Page, B., de Beer, Y., Gough, K., Guldemond, R., Junker, J., Olivier, P., Ott, T. & Trimble, M. (2008) Elephant population biology and ecology. In: Scholes, R.J. & Mennell, K.G. (eds.) *Elephant Management: A Scientific Assessment for South Africa*. Witwatersrand University Press, Johannesburg, South Africa, pp. 84–145.

Verlinden, A. & Gavor, I.K.N. (1998) Satellite tracking of elephants in northern Botswana. *African Journal of Ecology* 36, 105–116.

Viljoen, P.J. (1989) Spatial distribution and movements of elephants (*Loxodonta africana*) in the northern Namib Desert region of the Kaokoveld, South West Africa/Namibia. *Journal of Zoology* 219, 1–19.

Williamson, B.R. (1975) Seasonal distribution of elephant in Wankie National Park. *Arnoldia* 7, 1–16.

Wittemyer, G. & Getz, W.M. (2007) Hierarchical dominance structure and social organization in African elephants, *Loxodonta africana*. *Animal Behaviour* 73, 671–681.

Wittemyer, G., Getz, W.M., Vollrath, F. & Douglas-Hamilton, I. (2007) Social dominance, seasonal movements, and spatial segregation in African elephants: a contribution to conservation behavior. *Behavioral Ecology and Sociobiology* 12, 1919–1931.

Young, E. (1970) Water as Faktor in die Ekologie van Wild in die Nasionale Krugerwildtuin. PhD Thesis, University of Pretoria, Pretoria, South Africa.

Young, K.D. & van Aarde, R.J. (2010) Density as an explanatory variable of movements and calf survival in savanna elephants across southern Africa. *Journal of Animal Ecology* 79, 662–673.

Young, K.D., Ferreira, S.M. & van Aarde, R.J. (2009) Elephant spatial use in wet and dry savannas of southern Africa. *Journal of Zoology* 278, 189–205.

Part IV
Controllers

In the heterogeneity framework we have adopted for this book, controllers are entities influencing the spatial distribution of the agent, the intensity of its action or the resultant effects on the substrate and organisms or processes responding to variation in that substrate (Pickett *et al.*, 2003). Controllers can be abiotic, such as soil water-logging controlling the distribution of termites, or biotic like grazing antelopes controlling the frequency and effect of fire in many savannas. In this section we describe soil type in the Chobe ecosystem as a controller both of the distribution of elephants (the agent), and of the effects by elephant impact on the vegetation (the substrate). In some areas, such as on alluvial soils along the Chobe River, controllers include populations of impala, *Aepyceros melampus*, which by browsing on seedlings control the regeneration of trees, and African buffalo, *Syncerus caffer*, which by grazing in large herds control the availability of feeding areas for elephants (Drawing by Marit Hjeljord).

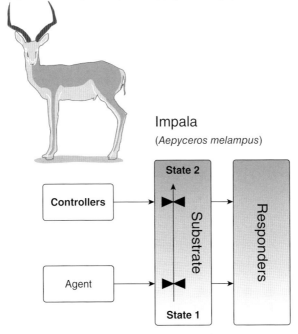

Impala
(*Aepyceros melampus*)

Elephants and Savanna Woodland Ecosystems: A Study from Chobe National Park, Botswana,
First Edition. Edited by Christina Skarpe, Johan T. du Toit and Stein R. Moe.
© 2014 John Wiley & Sons, Ltd. Published 2014 by John Wiley & Sons, Ltd.

References

Pickett, S.T.A., Cadenasso, M.L. & Benning, T.L. (2003) Biotic and abiotic variability as key determinants of savanna heterogeneity at multiple spatiotemporal scales. In: du Toit, J.T., Rogers, K.H. & Biggs, H.C. (eds.) *The Kruger Experience. Ecology and Management of Savanna Heterogeneity*. Island Press, Washington, Covelo, London, pp. 22–39.

9

Soil as Controller of and Responder to Elephant Activity

Christina Skarpe[1], Gaseitsiwe Masunga[2], Per Arild Aarrestad[3] and Peter G.H. Frost[4]

[1] Faculty of Applied Ecology and Agricultural Sciences, Hedmark University College, Norway
[2] Okavango Research Institute, University of Botswana, Botswana
[3] Norwegian Institute for Nature Research, Norway
[4] Science Support Service, New Zealand

Soil is both a biotic and an abiotic component of ecosystems and is the start and end of all trophic chains. Soil biota use the last remaining energy in the trophic chains and, once decomposed and mineralised, leave the chemical constituents of their organic tissues as ions, which can then be used again in primary production. The rate and pathways of decomposition and mineralisation, and the fate of the released ions, depend on interactions between organisms living on and in the soil and on abiotic factors such as soil moisture and temperature, linked to climate through soil properties such as texture, structure, porosity, mineralogy, bulk density and depth. These basic properties are determined by the parent material (bedrock and deposits), mineral weathering and organic matter decomposition (Ross, 1989). Secondary processes such as volcanism or redistribution of material by wind and water create heterogeneity and local, often distinct variation in soil productivity. At smaller temporal and spatial scales, the direct and indirect effects of large herbivores can also influence soil fertility and heterogeneity in savannas (Holdo *et al.*, 2007), as can individual trees (Campbell *et al.*, 1994) and soil organisms such as termites (Jouquet *et al.*, 2006; Sileshi *et al.*, 2010). There are no data on the influence of individual trees and soil organisms generally on soil properties and processes in the Chobe ecosystem, and only one brief and somewhat inconclusive study

Elephants and Savanna Woodland Ecosystems: A Study from Chobe National Park, Botswana,
First Edition. Edited by Christina Skarpe, Johan T. du Toit and Stein R. Moe.
© 2014 John Wiley & Sons, Ltd. Published 2014 by John Wiley & Sons, Ltd.

of the abundance and diversity of soil macrofauna (Dangerfield, 1997). For present purposes, we simply note the gap and encourage others to fill it.

Savannas have traditionally been classified according to their nutrient and water availability into either arid eutrophic (nutrient-rich) or moist dystrophic (nutrient-poor) systems. The coupling of rainfall and soil fertility is both functional, as high rainfall can cause soil nutrients to be leached from the biologically active surface soil layers, and coincidental because, in southern Africa at least, there tends to be a negative correlation at a regional scale between precipitation and the nutrient content of soil parent material, with a separation around an average annual rainfall of about 650 mm (Huntley, 1982). The correlation, however, is not absolute. Bell (1982) pointed out the sharp boundary of soil types with little difference in rainfall that goes through northern Tanzania, separating the nutrient-rich savannas of the Serengeti from the nutrient-poor miombo woodlands further west. Our study area within the nutrient-poor Chobe woodlands on Kalahari sand receive approximately the same annual rainfall as the nutrient-rich Serengeti plains. Moreover, the nutrient status of these woodlands contrasts strikingly with the small adjacent areas of alluvial soil that, by comparison, constitute nutrient-rich savanna. Small discrete patches of rich savanna within a matrix of nutrient-poor savanna woodland with the same climate and soil parent material have also been described by Blackmore et al. (1990), all of which raises the question: to what extent are the two types of savannas distinct classes or simply extremes on a continuum (Scholes, 1990)? We revisit this question later.

In our conception of the Pickett model (Pickett et al., 2003; Chapter 1) we see elephants as the agent of change in or maintenance of vegetation state. The vegetation in turn constitutes the substrate for elephant activity. Soil type is the most important controller in the Chobe ecosystem, influencing substrate (vegetation) quantity and quality, and through it, the activity of the agent (elephants), the agent's effect on the substrate and subsequently, the response of other organisms to the new state of the substrate. The Pickett model can be viewed at different scales, however, so that a controller at one scale can be an agent or a substrate at another. In this chapter we show that in Chobe the soil is both a controller of elephant activity, through its effects on the vegetation, and a responder to such activities.

The soils

The Soil Map of Botswana (FAO, 1990) shows three main soil units covering Chobe National Park, each corresponding broadly to one of the three main elephant habitats recognised in our study: ferralic arenosols (USDA Soil Taxonomy equivalent Ustic Quartzipsamments), underlying the *Baikiaea* and mixed woodlands; calcic luvisols (Typic Haplustalfs), forming the alluvial terraces that mostly support the *Capparis–Flueggea* shrublands; and calcic gleysols (Mollic Haplaquepts) on the grassy floodplains (Chapters 2 and 5). At a more detailed scale, as seen in the reconnaissance survey of the soils of Chobe District, including Chobe National Park (Remmelzwaal *et al.*, 1988), each of these regions comprises a complex of soil types that reflect the interplay of different factors associated with their origins and formation.

Yellowish brown to dark red arenosols cover most of the area south of the Chobe River beyond the alluvial terraces and form part of the extensive Kalahari sand sheet (Chapter 2). They are mostly deep to very deep, well to excessively drained, coarse to loamy fine sands. With less than 6% clay, few weatherable minerals, little soil organic matter (organic carbon content <0.5%), despite being covered by woodland, these arenosols are nutrient-poor (cation exchange capacity $<4\,cmol\,kg^{-1}$) and acid (pH 5.0–5.5) (Table 9.1). There is almost no surface runoff, a function of the relatively flat to gently undulating topography and deep, unimpeded drainage. Surface water for wildlife is generally lacking other than seasonally in small, shallow, largely treeless, grassy depressions ('pans') in which drainage is somewhat impeded and water accumulates for short periods during the wet season. The soils underlying these pans are mostly haplic luvisols (Typic Haplustalfs): deep to very deep but imperfectly drained, greyish brown to olive brown sands and loamy sands overlying sandy clay loams at depth (Remmelzwaal *et al.*, 1988). They are non-calcareous and marginally more fertile than the arenosols in which they are embedded (Table 9.1). The pans appear to be deflation hollows formed initially by wind and consolidated later by the gradually accumulation of finer soil particles carried downslope into the basin and fixed there during periods of wetness caused either by seasonal ponding of rainwater or groundwater seepage, or both (Thomas and Shaw, 1991). They are important ephemeral sources of water for wildlife, allowing some of the more water-dependent species to use the woodlands in the wet season.

The soils forming the alluvial terraces above the Chobe River are a complex of calcic and orthic luvisols (Typic and Udic Haplustalfs), calcic and haplic arenosols (Typic Ustochrepts and Ustic Quartzipsamments) and ferralic arenosols. The mosaic of soils on these terraces reflects their multiple origins, some originating as older alluvium deposited along the shoreline of the suspected former palaeolake Caprivi (Thomas and Shaw, 1991), others as reworked colluvium of different ages derived from the adjacent Kalahari sand sheet and eroded basal sediments. Calcic luvisols, the most widespread unit, are deep to very deep, dark greyish brown to yellow and reddish brown sandy clay loams and sandy clays. Clay contents increase with depth, producing soils that are imperfectly to moderately well drained. Soil pH (>7), cation exchange capacity ($>20\,cmol\,kg^{-1}$) and base saturation (100%) are all high, reflecting the nutrient-rich status of these soils (Table 9.1).

Table 9.1 Soil profile data for the north-eastern section of Chobe National Park and adjacent Kasane as recorded in the Africa Soil Profiles Database (Source: AfSPD: Leenaars, 2013). Profiles BW W3_0248 and BW W3_0316 are situated just outside Chobe National Park. The others come from within it.

	Ferralic arenosol Sandy plateau BW W3_0248			Haplic luvisol Sandy plateau BW W3_0640			Calcic luvisol Alluvial terrace BW W3_0316			Haplic arenosol Alluvial terrace BW W3_0566			Haplic arenosol Floodplain BW W3_0574			Eutric gleysol Floodplain BW W3_0602		
Soil unit / Landscape / AfSPD Reference No.																		
Location	−17.79861° S, 25.22389° E			−17.92222° S, 24.93194° E			−17.79222° S, 25.20222° E			−17.82556° S, 24.92778° E			−17.79583° S, 24.95833° E			−17.79722° S, 24.95972° E		
Soil horizon	A	B1	B2	A	Bt1	Bt2	A	Bt1	Bt2k	A	AC	C1	A	C1	C2	A	Bw	C1
Lower depth (cm)	40	80	200	20	53	90	10	50	85	25	57	73	20	60	125	20	65	110
Munsell colour (moist)	7.5YR 3/4	5YR 3/4	2.5YR 4/6	10YR 2/1	10YR 5/3	10YR 5/3	10YR 2/1	10YR 2/1	10YR 3/2	10YR 3/2	10YR 3/2	10YR 3/3	10YR 3/2	10YR 6/3	10YR 7/3	10YR 2/1	10YR 3/1	10YR 8/1
Sand (%)	91	94	91	–	91	87	68	65	51	93	94	94	93	96	–	16	47	97
Silt (%)	4	1	3	–	2	1	15	11	13	3	3	2	4	2	–	24	15	2
Clay (%)	5	5	6	–	7	12	17	25	36	4	3	4	3	2	–	60	38	1
Texture	Sa	Sa	Sa	Sa	LSa	LSa	SaL	SaCL	SaC	Sa	Sa	Sa	Sa	Sa	–	C	C	Sa
pH H$_2$O	5.6	5.3	5.4	7.0	6.4	6.9	6.8	7.3	8.0	5.7	7.9	7.5	4.7	4.8	4.9	5.4	6.6	6.9
EC (dS m^{-1})	0.1	0.1	0.1	–	0.1	0.1	0.2	0.2	0.2	0.1	0.1	0.1	0.1	0.1	0.1	0.1	0.3	–
Ca^{++} (cmol kg^{-1})	–	–	–	2.8	1.4	–	–	–	–	2.4	4.6	2.8	0.6	0.5	–	10.2	12.7	0.5

Table 9.1 (*Continued*)

Mg^{++} (cmol kg^{-1})	–	–	–	1.3	0.9	2.9	–	–	–	0.4	0.8	0.7	0.2	0.2	0.1	2.7	3.5
Na^{+} (cmol kg^{-1})	–	–	–	–	–	–	–	–	–	–	0.2	–	–	–	0.1	1.9	2.4
K^{+} (cmol kg^{-1})	–	–	–	0.3	1.6	1.8	–	–	–	0.2	0.1	0.1	0.1	0.1	–	0.2	0.1
Sum exchangeable bases (cmol kg^{-1})	2.1	0.7	0.4	4.4	3.9	–	22.4	24.4	56.7	3.0	5.7	3.6	0.9	0.8	–	15	18.7
Soil CEC (cmol kg^{-1})	2.7	1.8	1.6	4.8	4.6	8.5	21.3	22.5	22.0	4.5	4.0	4.3	1.9	–	1.3	16.2	17.5
Base saturation (%)	78	39	25	92	85	–	100	100	100	67	100	84	47	–	–	93	100
Organic carbon (%)	0.2	0.1	–	–	0.1	0.1	1.5	0.7	0.5	0.4	0.2	0.2	–	–	–	1.2	0.4

EC, exchange capacity; CEC, cation exchange capacity.

The soils forming the floodplain are similarly diverse, a result of the complex interplay of river flow, sediment carrying capacity, sediment loads and alternating periods of deposition and entrainment. The soils of the lower Chobe floodplain are unique in being the product of the transport, distribution and sorting of material from two entirely opposite directions, the normal but variable downstream flow of the Chobe River and backflow from the Zambezi River during its flood. Over years, as the main river channel has meandered its way across the floodplain, it has deposited sediments of different sizes according to the position and speed of flow of the water across it, producing both vertically and horizontally stratified sediments. The principal soil types are calcic and eutric gleysols (Mollic Haplaquepts) and haplic and eutric arenosols (Ustic Quartzipsamments). The gleysols are predominantly black to very dark grey, sandy clay to clays, some of which show a substantial increase in clay with depth, whereas others overlie lenses of almost pure sand (see profile BW W3_0602 in Table 9.1). The soils are generally deep to very deep but poorly to imperfectly drained, a consequence of both their generally high clay content and low-lying position in the landscape. The variation in soil structure is matched by diversity in soil chemistry, the soils being slightly acidic to substantially alkaline (pH 5.4–9.0), occasionally strongly sodic (exchangeable sodium percentage >40%) and moderately to extremely fertile (cation exchange capacity 16–33 $cmol\,kg^{-1}$, sum of exchangeable bases 15.0 to >60.0 $cmol\,kg^{-1}$, both increasing with depth: Table 9.1; Remmelzwaal et al., 1988). The physical and chemical attributes of the arenosols appear broadly similar to those of arenosols elsewhere in the region, but must function differently because of their lower lying, wetter positions in the landscape. In this chapter we refer to both the terrace and floodplain soils as alluvial soils.

To supplement the extremely limited soil profile data available for our study area we collected surface soil samples in the 70 sample plots distributed across the floodplain, in the shrublands on alluvial soil, and in the woodlands on Kalahari sand (Masunga, 2008; Chapter 5). Soil concentrations of calcium, magnesium, sodium, potassium, phosphorus and carbon, as well as cation exchange capacity, increased from the Kalahari sand to the shrublands on the raised alluvial terrace to the floodplain, whereas pH varied less but was lowest on the floodplain and highest in the shrublands (Table 9.2). Vegetation composition and physiognomy are strongly related to soil factors (Masunga, 2008; Aarrestad et al., 2011; Chapter 5).

Soil as a controller of elephant activities and impact

The most striking feature in our study area is the contrast between the ecosystem on relatively nutrient-rich, fine textured alluvial soil and that on nutrient-poor Kalahari sand. This difference in soil accounts for the fundamentally different ecosystem dynamics in the two areas during the last 150 or so years, concurrent with the fall and rise of the

Table 9.2   Cation exchange capacity (CEC), exchangeable calcium (Ca), magnesium (Mg), sodium (Na), potassium (K), extractable phosphorus (P) and organic carbon (C) levels in the A horizons of the dominant soils underlying the woodlands, shrublands and floodplain of Chobe National Park, along with the number of samples analysed (*n*). Standard deviations are shown in parentheses.

	Woodland: mainly ferralic arenosol	Shrubland: mainly calcic luvisol	Floodplain: mainly calcic gleysol
n	141	38	35
pH	5.2 (0.4)	6.2 (0.7)	4.8 (0.3)
CEC (cmol kg^{-1})	2.60 (1.10)	5.80 (2.40)	32.60 (15.80)
Ca (cmol kg^{-1})	1.52 (1.02)	6.59 (4.61)	21.01 (11.54)
Mg (cmol kg^{-1})	0.32 (0.13)	0.63 (0.23)	3.33 (2.92)
Na (cmol kg^{-1})	0.03 (0.02)	0.13 (0.10)	3.04 (5.23)
K (cmol kg^{-1})	0.20 (0.24)	0.59 (0.63)	0.84 (1.15)
P (mg 100 g^{-1})	16.2 (13.6)	65.5 (93.1)	43.3 (23.4)
C (%)	0.5 (0.2)	0.7 (0.4)	4.6 (3.0)

Chobe elephant population. It also underlies the past and present contrasts in structure, function and composition of the plant and animal communities (Chapters 4, 5, 12 and 13).

Before elephants were nearly eliminated by ivory hunters at the end of the 19th century the raised alluvial flats close to the Chobe River had open vegetation, in contrast to the woodlands on Kalahari sand (Selous, 1881). With the sharp decline in herbivory by elephants and ungulates, caused by ivory hunting and rinderpest, respectively, the raised alluvial flats became covered by *Acacia* woodland in the early 20th century (Simpson, 1974), so that both the nutrient-poor Kalahari sand and the alluvium had woodlands, but of quite different types (Chapter 4). When the elephants increased again from about the mid-1900s onwards, they fed preferentially in the nutrient-rich woodlands on alluvial soil, opening the dense woody vegetation by trampling and browsing and successively killing the large trees (Chapter 4). Because the nutritive quality of much of the plant material on alluvial soil was high enough to support medium-sized ruminants, foraging ungulates could further reduce plant biomass. Moe *et al.* (2009) have shown how seedling-eating antelopes, primarily impala, *Aepyceros melampus*, act as controllers of the elephant-induced transition in state from tall woodland to fairly open shrubland (about 15% woody cover) on alluvial soil (Chapter 10).

In contrast, there is little evidence of elephants inducing a shift from woodland to shrubland or grassland on the nutrient-poor Kalahari sands, where the nutritive quality of much of the plant biomass is too low to be eaten extensively by elephants or to support medium-sized ruminants (Chapter 4). This suggests that herbivory might be an important factor explaining the distinct difference in plant biomass density between nutrient-rich and nutrient-poor savanna, depending on whether a considerable proportion of the plant biomass is of high enough quality to support mesoherbivores (Bell, 1982; Wallgren, 2008; O'Kane *et al.*, 2011).

Soil controls the vegetation

Nutrient-poor savannas are typically dominated by tufted nutrient-poor grasses and large trees with simple or pinnate mesophyllous leaves, often in the subfamily Caesalpinioideae, whereas nutrient-rich savannas generally are dominated by nutrient-rich grazing-tolerant grasses and relatively small trees with bipinnate nanophyllous leaves, often in the sub-family Mimosoideae (Huntley, 1982; Bell, 1982; Scholes, 1990). We sampled woody and herbaceous vegetation in 70 plots of 400 m², distributed in the four elephant-habitat types identified earlier (Chapter 5). The vegetation in our study area and its relation to soil properties is described in Chapter 5. The woody vegetation on Kalahari sand is dominated by *Baikiaea plurijuga* (Caesalpinioideae) with an undergrowth of tufted grasses and tall forbs (Aarrestad *et al.*, 2011; Chapter 5). The vegetation on the alluvium in the early 1900s was described as *Acacia* woodland or *Acacia* tree savanna (Simpson, 1974; Chapter 4), and was thus dominated by species in the subfamily Mimosoideae. The present woody vegetation on the alluvial terraces, however, is dominated by shrubs and scramblers of other families (Aarrestad *et al.*, 2011; Chapter 5), a community shift likely to have been caused by intense elephant browsing and controlled by soil properties. Soil nutrient availability and browsing regime strongly influence which tolerance and avoidance traits maximise competitive ability and subsequently lead to dominance by a species in the community (Stamp, 2003; Chapter 12). Many herbivory-related plant traits differ between the woody species on the Kalahari sand and those on the alluvium, principally that the leaves of many species on Kalahari sand have high concentrations of quantitative, carbon-based, defence compounds, in contrast to species on the alluvial terraces, many of which are spinescent (Rooke, 2003; Chapter 12). These features correspond with those characterising nutrient-poor and nutrient-rich savannas, respectively (Scholes, 1990).

Mammal communities and soil

We found differences between the mammal community confined largely to alluvial soils and that using both the alluvium and the areas on Kalahari sand, in spite of the proximity of the two areas and the small extent of the patch with nutrient-rich savanna. Using the Distance survey method (Buckland *et al.*, 2001), we counted mammals along road transects on alluvial soil and Kalahari sand within about 10 km of the Chobe River (Chapter 13), and on the lacustrine flats and adjacent areas with Kalahari sand at the then-dry Savuti Marsh about 150 km southwest of the river. Total mammal biomass density was about 10 times higher on the alluvial and lacustrine soils compared with the Kalahari sand area (calculated from data in Wallgren, 2008). As expected from the Bell hypothesis (Bell, 1982) the herbivore biomass on Kalahari sand consisted primarily of elephants, which because of their large body size and hind-gut fermentation, can tolerate low quality forage. Elephants also made up a large proportion of the mammal biomass on the richer soils but did little foraging there. The concentration of elephants in nutrient rich alluvial areas is understandable given that in both Chobe and Savuti the permanent water resources (the river and drilled boreholes, respectively) were situated in these areas. Elephants spend much time at these water sources to drink and socialise. An additional reason for the concentration of elephants on the Chobe floodplain and adjacent alluvial terraces can be their frequent use of soil licks containing clay, hypothesised to assist in detoxifying secondary plant metabolites (Johns and Duquette, 1991) and providing essential minerals, particularly sodium which occurs abundantly in patches of sodium-enriched soils (Remmelzwaal et al., 1988). Moreover, in nearby Hwange National Park, Zimbabwe, elephants are known to use mineral licks, apparently to meet their sodium requirements, something that has been noted for other herbivores as well (Holdo et al., 2002).

Determination of habitat preferences based on the transect counts in the Botswana-Norway Institutional Co-operation and Capacity Building (BONIC) study area (Chapter 13) showed that most ungulate species, independent of size, preferred habitats on alluvial soil and avoided the Baikiaea woodland on Kalahari sand. Small-bodied species, weighing less than about 150 kg, such as impala, Aepyceros melampus, red lechwe, Kobus leche, puku, Kobus vardonii, bushbuck, Tragelaphus scriptus, and common warthog, Phacochoerus africanus, were virtually confined to the habitats on alluvium where the nutritional quality of the plants is generally high, no doubt reflecting the relatively high nutrient status of the soils (Ben-Shahar and MacDonald, 2002). As with elephant, these herbivores might also be benefitting directly from the presence of soil licks on alluvium, allowing them to supplement any mineral deficiencies in their diet. A few large grazers weighing more than about 200 kg, such as the plains zebra, Equus quagga, a hindgut fermenter, and roan antelope, Hippotragus equinus, a tall-grass specialist, avoided the alluvium, preferring habitats on sand. Surprisingly, this was also the case with the smallest browsers, weighing less than about 25 kg: steenbok, Raphicerus campestris, and common duiker, Sylvicapra grimmia (Chapter 13).

Soil as a responder to elephant activities

At a local scale, interactions among vegetation, herbivores, fires and flooding regimes are important modifiers of soil properties and nutrient dynamics. In African savannas, spatial and temporal heterogeneity in soil nutrients is associated with the activities of herbivores, from termites to elephants (McNaughton, 1988; Pickett et al., 2003; Jouquet et al., 2006; Sileshi et al., 2010). Herbivores graze, browse, excrete and die, and thereby directly influence nutrient cycling and availability. In open ecosystems, migratory or highly mobile herbivores with differential habitat use can contribute to the flux and redistribution of nutrients among habitats (Senft et al., 1987; Hobbs, 1996; Holdo et al., 2007). Large herbivores can also have many indirect effects on soil properties, for example, by causing changes in plant community composition, soil water regime or fire frequency and intensity (McNaughton, 1992; Hobbs, 1996).

Large herbivores and nutrient cycling

Large herbivores can enhance or retard the rate of soil nutrient cycling and influence the cycling of particular elements. Acceleration of nutrient cycling and availability is mainly found in ecosystems with relatively high initial soil fertility. They are often grass-grazer systems, where grazing animals remove a large proportion of the available biomass, and the grasses respond by fast regrowth of nutrient-rich tissue. In such systems much of the nutrient flux from plants to soil goes via animal dung and urine, and the small amount of plant litter produced is nutrient-rich and decomposes and mineralises rapidly (Ruess and Seagle, 1994; McNaughton et al., 1997). In nutrient-poor ecosystems, particularly those dominated by browsers, herbivory can retard nutrient cycling. By selectively removing nutrient-rich plants and plant parts, herbivores can promote an increase in the production of carbon-based secondary metabolites, either by triggering production of induced defences in browsed plants or by causing a shift in species composition towards those with more chemical defences. In both cases the result will be a reduction in litter quality leading to slower decomposition and nutrient cycling (Pastor et al., 1993; Ritchie et al., 1998; Persson et al., 2009). In browser-dominated systems typically only a small proportion of the most nutritious plant biomass is eaten, so that the positive effect on nutrient cycling arising from the transformation of plant biomass to easily biodegradable urine and faeces does not offset the negative effect of the decrease in litter quality. Elephants often remove a comparatively large proportion of biomass when browsing on shrubs and small trees, either by breaking branches or stems, or by biting off large twigs. Removing a lot of biomass from a tree can trigger substantial regrowth, potentially resulting in carbon-limitation in the plant and allocation of available carbon to growth at the expense of production of carbon-based secondary metabolites, resulting in a lower $C:N$ ratio in the regrowing plant tissues (Rooke and Bergström, 2007; Scogings et al., 2011, 2013). This response would moderate or reverse the reduction in litter quality. Thus, browsing by elephants might be less likely to retard nutrient cycling than browsing by a highly selective small herbivore.

In the BONIC area, elephants forage mostly in the woodlands on nutrient-poor sand, although they do not browse the dominant, slow-growing, chemically defended, tall canopy trees there (Owen-Smith and Chafota, 2012; Chapter 12). Instead, they feed on grasses in the wet season and on small trees and shrubs in the dry season. Many of these species respond to severe browsing by regrowing rapidly. If there are two over-lapping nutrient cycles operating in nutrient-poor woodland savanna, as suggested by Frost (1985), elephants could enhance the relatively fast, shallow cycle by consuming a large proportion of the grasses, forbs and the fast-growing shrubs and small trees in the woodlands. Because the elephants do not browse canopy trees such as *Baikiaea plurijuga*, they obviously do not interfere with the slow, deep nutrient cycling of this species. We do not know the biomass of the different vegetation compartments or the nutrient flux rates in the Chobe woodlands but it is almost certain that large trees contain most of the phytomass. Because the turnover of tree biomass is slower than that of herbaceous vegetation and relatively fast-growing shrubs, much of the annual nutrient flux in this ecosystem would go via the faster pathway (Frost, 1985), something which browsing by elephants would enhance by inducing fast regrowth of relatively nutrient-rich biomass and converting large quantities of plant matter to dung and urine. In such a system, the often deep-rooted, seldom-browsed trees are the conservative component, gathering nutrients both radially near the surface and from deep down in these well-drained soils, and sequestering them in long-lived biomass. Such nutrients are mobilised relatively slowly through annual production and leaf fall, occasional herbivory and sometimes fire. In contrast, the shallower-rooted, faster growing and more palatable shrubs and grasses form the more productive component, one linked more broadly with herbivores and fire, and responding to their pressures through rapid regrowth and nutrient mobilisation. Nutrients circulating through both the 'slow' and the 'fast' components converge during decomposition and mineralisation, at which point movement can occur between the two cycles, depending on the timing and site of ion release, the nutrient requirements of the plants and their uptake capacity at that time. These two cycles – one slow and conservative, the other fast and reactive – could well be a key feature enhancing in the resilience of this system to disturbance.

Large herbivores and nutrient translocation

Elephants used the habitats in our study area differently and disproportionately to their areal extent. Älvgren (2009), using day and night-time transect counts and Distance statistics (Buckland *et al.*, 2001), calculated elephant densities in different habitats. For the BONIC study area within about 10 km of the river, he found an overall dry season (July–September) elephant density of about 5.6 animals km^{-2}. The woodlands on Kalahari sand, covering about 90% of the area, had a density of about 4.8 elephants km^{-2}, whereas the floodplain, covering 2% of the area on average had 28 elephants km^{-2}.

Elephant density in the shrublands, which covered the remaining 8% of the area, averaged 8.4 animals km^{-2}. Elephants, depending on the Chobe River for drinking during the dry season, use the area within about 10 km of the river (Stokke and du Toit, 2002; Älvgren, 2009), commuting between foraging habitats and the river. Thus, densities observed in the different habitats are a measure of habitat use and not of densities in a strict sense. The proportion of time elephants spent on average in the different habitats was estimated by weighting the average (day and night) density in the habitat by the proportion of that habitat in the area investigated (Figure 9.1; Älvgren, 2009). On average, elephants spent about 78% of their time in the woodlands on Kalahari sand, 10% on the floodplain and about 12% in the shrublands. The proportion of time spent in a habitat by elephants or other large herbivores does not necessarily reflect forage intake but, as defecation is fairly evenly distributed over the day (Coe, 1972), it broadly reflects the amount of dung and urine deposited. With a conservative estimate of 50 kg dry dung and 50 L urine produced per day per elephant, across all ages (Coe, 1972), this suggests that the woodlands in the dry season receive about 240 kg dung km^{-2} day^{-1} or about 43 tons km^{-2} during the 6-month dry season. For the floodplains, with their higher density of elephants, and using the same assumptions as above, we calculate that they could receive about 1400 kg dry $dung^{-1}$ km^{-2} day^{-1}. Combining the habitats on alluvium, shrublands and floodplain, gives more than 600 kg km^{-2} day^{-1} or more than 100 tons during the 6-month dry season. About the same amount of urine (litres) is deposited as the dry mass (kg) of dung. Elephant dung has comparatively low

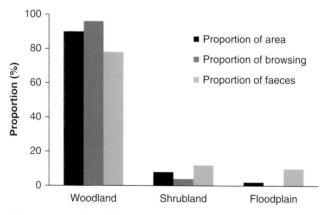

Figure 9.1 The proportion (%) of the study area covered by, respectively, woodland on Kalahari sand, shrubland on alluvium, and floodplain; the proportion of browsing by elephants in the respective habitats (% of total bites and branches broken by elephants) and the proportion of faeces deposited in each habitat. (Source: Data are from Älvgren, 2009 and unpublished data from the browsing study in the BONIC project (Chapter 12).)

Table 9.3 Mean concentrations of major and trace elements in elephant dung in late wet, early dry and late dry season. Samples are separate dung heaps (*n*). Standard errors are shown in parentheses. (Source: Adapted from Masunga *et al.*, 2006 and A. Brodén, unpublished.) Different superscripts within each of the first two rows indicate significant differences between late wet and early dry seasons ($p < 0.05$).

	Late wet season ($n = 4$)	Early dry season ($n = 2$)	Late dry season ($n = 5$)
Major elements ($g\,kg^{-1}$)			
N	1.62 (0.09)[a]	1.24 (0.21)[a]	1.04 (0.19)
P	0.28 (0.02)[a]	0.11 (0.03)[b]	0.08 (0.01)
K	1.66 (0.18)[a]	0.82 (0.03)[a]	–
Ca	1.02 (0.11)[a]	1.04 (0.20)[a]	–
Mg	0.23 (0.00)[a]	0.15 (0.00)[b]	–
Na	0.03 (0.01)[a]	0.03 (0.02)[a]	–
S	0.11 (0.00)[a]	0.09 (0.00)[b]	–
C	39.0 (2.0)[a]	36.5 (0.4)[a]	45.0 (6.2)
C/N	24.0 (1.0)[a]	30.0 (5.0)[a]	44.5 (10.6)
Trace elements ($mg\,kg^{-1}$)			
Fe	670.0 (78.4)[a]	864.0 (0.0)[a]	–
Cu	10.9 (0.6)[a]	7.8 (0.8)[b]	–
Mn	367.0 (61.3)[a]	372.5 (100.0)[a]	–
Zn	38.3 (4.7)[a]	19.2 (2.2)[a]	–
Mo	1.5 (0.0)[a]	1.5 (0.0)[a]	–
Bo	10.5 (1.2)[a]	14.0 (1.9)[a]	–
Al	473.7 (66.0)[a]	960.0 (0.0)[a]	–

nutrient concentration in the dry season, with slightly above 1% nitrogen and about 0.1% phosphorus (Masunga *et al.*, 2006; Table 9.3). Nevertheless, the amount of nitrogen deposited in elephant dung in the habitats on alluvial soil during the dry season is about 20% of that recommended annually for agricultural crops. To this should be added the input from urine, which we surmise has a nitrogen concentration of about 0.5% on a mass basis, mainly in the form of urea (assuming that it has about the same chemical composition as that of other herbivores).

Thus the distribution of elephant dung and urine was unequal between habitats and did not correspond with each habitat's contribution to elephant foraging. The treeless floodplains, comprising only 2% of the study area, received about 10% of the dung while contributed nothing to browsing by elephants and little to their grazing either, especially in the dry season when the grasses are too short to provide an acceptable intake rate. The alluvial soils overall constituted about 10% of the study area and made up some 4% of elephant browse (based on the number of twig bites and breaks recorded in the plots; Chapter 12). In turn, they received about 22% of the dung and urine. Even if the data in these back-of-an-envelope calculations are rough, there is clearly considerable net nutrient transfer by the elephants from the woodlands on Kalahari sand to the shrublands and, in particular, to the floodplain on alluvium, sharpening the contrast in resource availability between the two areas.

Large herbivores and fire

Besides affecting ecosystem productivity directly through influencing plant litter quality and quantity, and by redistributing nutrients via dung and urine, large herbivores can also affect soil nutrients and plant productivity indirectly (Hobbs, 1996; Augustine *et al.*, 2003). Grazers reduce the amount of dry grass available as fuel, thereby efficiently reducing savanna fire frequencies and intensities (McNaughton, 1992). Holdo *et al.* (2007) suggest that in a grazer-dominated ecosystem such as the Serengeti the direct effect of grazer-enhanced nitrogen turnover is small compared with the indirect effect that grazers can have on nitrogen losses through fire. Fire and herbivory are hypothesised to have opposite effects on nitrogen availability and cycling in savanna (Holdo *et al.*, 2007). Fire often, but not always, leads to depletion of nitrogen through volatilisation (Hobbs *et al.*, 1991; Ojima *et al.*, 1994; Dell *et al.*, 2005), whereas herbivory, particularly grazing, at least under some circumstances, enhances nitrogen cycling and availability (McNaughton *et al.*, 1997; de Mazancourt *et al.*, 1998, 1999). As discussed above, the 'grazing optimisation' hypothesis (McNaughton, 1979) is not directly transferable to browser-dominated systems, more so perhaps to those with a predominance of megabrowsers such as elephants, than to those dominated by highly selective small browsers.

Fire control, particularly in the northern areas that are most visited by tourists, is part of the management of Chobe National Park. However, the strong gradient in fire frequency with distance from the river suggests that fire is also controlled by natural causes. We recorded the presence or absence of fire scars on *Baikiaea plurijuga* trees with different diameters at breast height (DBH: 1.37 m above ground), along transects running perpendicular to the river at 1 km intervals. Ten trees of different sizes were sampled at each site, the distribution of size classes being similar across sites. A total of 270 trees were sampled. The incidence of fire scars increased both with tree size ($p < 0.001$) and with distance from the river ($p < 0.001$). Among the largest trees, those with a DBH greater than 50 cm, 72% had fire scars, uncorrelated with distance from the river ($p = 0.250$). Among trees with DBH smaller than 50 cm, 60% had fire scars, the frequency of which was positively correlated with distance from the river ($p < 0.001$) with little variation across these smaller size classes (Figure 9.2).

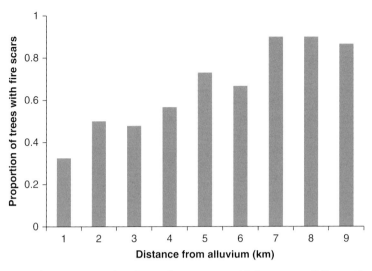

Figure 9.2 **The proportion of *Baikiaea plurijuga* trees with fire scars at different distances (km) along a gradient from the boundary between alluvial soil close to the Chobe River and the Kalahari sand extending into the woodlands on Kalahari sand.**

The strong correlation between the incidence of fire scars and distance from the river suggests that fire frequency increases along a gradient of declining herbivory from permanent water and nutrient-rich soils on alluvium to the sandy interior where limited surface water and nutrient-poor soils and vegetation limit herbivory. The lack of correlation between fire scars and distance from the river in the largest trees might depend on the gradient being of fairly recent origin, or simply on these trees being old enough to have encountered fire, even if infrequent. Grazing pressure is most intense on the alluvium (Chapters 12 and 13), and the reduction in fire frequency probably started here, through grazing reducing fuel loads and inhibiting fires. Further south on the Kalahari sand, grazing is primarily by elephants and buffalo. Because of their large size and high density, elephants are likely to be a main agent reducing grass fuel loads and hence the frequency of fires in this ecosystem. In other situations, however, this can be offset if elephants initially, and intermittent fires subsequently, open up the woodland canopy, allowing more grass to grow increasing fire frequency (Holdo, 2007). As *Baikiaea plurijuga* only grows on sand, we can say little on the fire history on the alluvium, except that fires were recorded to be common there in the mid-20th century (Child, 1968; Simpson, 1974). Presumably they ceased when densities of, for example, buffalo and elephant, increased shortly thereafter (Chapter 13). If a reduction in fire frequency limits the losses of nitrogen from the ecosystem, as suggested by Holdo *et al.* (2007), the likely herbivore-induced gradient in fire frequency can enhance the difference in nutrient status between the alluvial areas close to the river and the expanses of nutrient-poor sand further from the river.

As shown in this chapter the contrast between the alluvium and the Kalahari sand is the most important controller of ecosystem processes in our area, distinctly separating the patches of rich savanna from the expanses of poor savanna on Kalahari sand. In the former, plant material is comparatively nutrient-rich and herbivores have strong impacts on vegetation structure and biomass and have induced a transition in state from woodland to open shrubland during the last 150 or so years. On the nutrient-poor Kalahari sand, plant biomass tends to be too nutrient-poor for ungulate herbivores, and no transition in state of the vegetation has taken place. We have seen that the difference in nutrient availability is reflected also in the density and composition of herbivore communities utilising the vegetation on the two soil types. Further, we have shown evidence that herbivores, particularly elephants, can enhance the difference in nutrient status between the alluvium and the sand by redistributing nutrients and by impacting fire regime.

References

Aarrestad, P.A., Masunga, G.S., Hytteborn, H., Pitlagano, M.L., Marokane, W. & Skarpe, C. (2011) Influence of soil, tree cover and large herbivores on field layer vegetation along a savanna landscape gradient in northern Botswana. *Journal of Arid Environment* 75, 290–297.

Älvgren, J. (2009) Space-time pattern of habitat utilization for the Chobe riverfront elephant population. MSc Thesis, Uppsala University, Uppsala, Sweden.

Augustine, D.J., McNaughton, S.J. & Frank, D.A. (2003) Feedbacks between soil nutrients and large herbivores in a managed savanna ecosystem. *Ecological Applications* 13, 1325–1337.

Bell, R.H.V. (1982) The effect of soil nutrient availability on community structure in African ecosystems. In: Huntley, B.J. & Walker, B.H. (eds.) *Ecology of Tropical Savannas*. Springer-Verlag, Berlin, pp. 193–216.

Ben-Shahar, R. & MacDonald, D.W. (2002) The role of soil factors and leaf protein in the utilization of mopane plants by elephants in northern Botswana. *BMC Ecology* 2, article 3. [online] http//www.biomedcentral.com/1472-6785/2/3/.

Blackmore, A.C., Mentis, M.T. & Scholes, R.J. (1990) The origin and extent of nutrient-enriched patches within a nutrient-poor savanna in South Africa. *Journal of Biogeography* 17, 463–470.

Buckland, S.T., Anderson, D.R., Burnham, K.P., Laake, J.L., Borchers, D.L. & Thomas, L. (2001) *Introduction to Distance Sampling – Estimating Abundance of Biological Populations.* Oxford University Press, Oxford.

Campbell, B.M., Frost, P., King, J.A., Mawanza, M. & Mhlanga, L. (1994) The influence of trees on soil fertility on two contrasting semi-arid soil types at Matopos, Zimbabwe. *Agroforestry Systems* 27, 1–14.

Child, G. (1968) *An Ecological Survey of North-eastern Botswana.* Food and Agriculture Organization of the United Nations, Rome.

Coe, M. 1972. Defaecation by African elephants (*Loxodonta africana* (Blumenbach)). *East African Wildlife Journal* 10, 165–174.

Dangerfield, J.M. (1997) Abundance and diversity of soil macrofauna in northern Botswana. *Journal of Tropical Ecology* 13, 527–538.

de Mazancourt, C., Loreau, M. & Abbadie, L. (1998) Grazing optimization and nutrient cycling: when do herbivores enhance plant production? *Ecology* 79, 2242–2252.

de Mazancourt, C., Loreau, M. & Abbadie, L. (1999) Grazing optimization and nutrient cycling: potential impact of large herbivores in savanna system. *Ecological Applications* 9, 784–797.

Dell, C.J., Williams, M.A. & Rice, C.W. (2005) Partitioning of nitrogen over five growing seasons in tallgrass prairie. *Ecology* 86, 1280–1287.

FAO (1990) *Soil Map of the Republic of Botswana*. Soil Mapping and Advisory Services Project FAO/BOT/85/011. FAO and Government of Botswana, Gaborone, Botswana.

Frost, P.G.H. (1985) Organic matter and nutrient dynamics in a broadleafed African savanna. In: Tothill, J.C. & Mott, J.J. (eds.) *Ecology and Management of the World's Savannas*. Australian Academy of Sciences, Canberra, pp. 200–206.

Hobbs, N.T. (1996) Modification of ecosystems by ungulates. *Journal of Wildlife Management* 60, 695–713.

Hobbs, N.T., Schimel, D.S., Owensby, C.E. & Ojima, D.S. (1991) Fire and grazing in the tallgrass prairie: contingent effects on nitrogen budgets. *Ecology* 72, 1374–1382.

Holdo, R.M. (2007) Elephants, fire, and frost can determine community structure and composition in Kalahari woodlands. *Ecological Applications* 17, 558–568.

Holdo, R.M., Dudley, J.P. & McDowell, L.R. (2002) Geophagy in the African elephant in relation to availability of dietary sodium. *Journal of Mammalogy* 83, 652–664.

Holdo, R.M., Holt, R.D., Coughenour, M.B. & Richie, M.E. (2007) Plant productivity and soil nitrogen as a function of grazing, migration and fire in an African savanna. *Journal of Ecology* 95, 115–128.

Huntley, B.J. (1982) Southern African savannas. In: Huntley, B.J. & Walker, B.H. (eds.) *Ecology of Tropical Savannas*. Springer-Verlag, Berlin, pp. 101–119.

Johns, T. & Duquette, M. (1991) Detoxification and mineral supplementation as functions of geophagy. *The American Journal of Clinical Nutrition* 53, 448–456.

Jouquet, P., Dauber, J., Lagerlo, J., Lavelle, P. & Lepage, M. (2006) Soil invertebrates as ecosystem engineers: intended and accidental effects on soil and feedback loops. *Applied Soil Ecology* 32, 153–164.

Leenaars, J.G.B. (2013) *Africa Soil Profiles Database, Version 1.1. A Compilation of Georeferenced and Standardized Legacy Soil Profile Data for Sub-Saharan Africa (with dataset)*. ISRIC Report 2013/03, Africa Soil Information Service (AfSIS) Project. ISRIC – World Soil Information, Wageningen, The Netherlands.

Masunga, G.S. (2008) Ecosystem processes, nutrients, plant and fungal species diversity in Chobe National Park, Botswana. PhD Thesis, Norwegian University of Life Sciences, Ås, Norway.

Masunga, G.S., Andresen, Ø., Taylor, J.E. & Dhillion, S.S. (2006) Elephant dung decomposition and coprophilous fungi in two habitats of semi-arid Botswana. *Mycological Research* 110, 1214–1226.

McNaughton, S.J. (1979) Grazing as an optimization process: grass-ungulate relationships in the Serengeti. *American Naturalist* 113, 691–703.

McNaughton, S.J. (1988) Mineral nutrition and spatial concentrations of African ungulates. *Nature* 334, 343–345.

McNaughton, S.J. (1992) The propagation of disturbance in savannas through food webs. *Journal of Vegetation Science* 3, 303–314.

McNaughton, S.J., Banyikwa, F.F. & McNaughton, M.M. (1997) Promotion of the cycling of diet-enhancing nutrients by African grazers. *Science* 178, 1798–1800.

Moe, S.R., Rutina, L.P., Hytteborn, H. & du Toit, J.T. (2009) What controls woodland regeneration after elephants have killed the big trees? *Journal of Applied Ecology* 46, 223–230.

O'Kane, C.A., Duffy, K.J., Page, B.R. & Macdonald, D.W. (2011) Are the long-term effects of mesobrowsers on woodland dynamics substitutive or additive to those of elephants? *Acta Oecologica* 37, 393–398.

Ojima, D.S., Schimel, D.S., Parton, W.J. & Owensby, C.E. (1994) Long-term and short-term effects of fire on nitrogen cycling in tallgrass prairie. *Biogeochemistry* 24, 67–84.

Owen-Smith, N. & Chafota, J. (2012) Selective feeding by a megaherbivore, the African elephant (*Loxodonta africana*). *Journal of Mammalogy* 93, 698–705.

Pastor, J., Dewey, N., Naiman, R.J., McInnes, P.F. & Cohen, Y. (1993) Moose browsing and soil fertility in the boreal forests of Isle Royale National Park. *Ecology* 74, 467–480.

Persson, I., Nilsson, M.B., Pastor, J., Eriksson, T., Bergström, R. & Danell, K. (2009) Depression of below ground respiration rates at simulated high moose population densities in boreal forest. *Ecology* 90, 2724–2733.

Pickett, S.T.A., Cadenasso, M.L. & Benning, T.L. (2003) Biotic and abiotic variability as key determinants of savanna heterogeneity at multiple spatiotemporal scales. In: du Toit, J.T., Rogers, K.H. & Biggs, H.C. (eds.) *The Kruger Experience. Ecology and Management of Savanna Heterogeneity*. Island Press, Washington, DC, pp. 22–39.

Remmelzwaal, A., Van Waveren, E. & Baert, G. (1988) *The Soils of Chobe District*. Soil Mapping and Advisory Services (Botswana) Field Document 8. FAO/UNDP/Government of Botswana, Gaborone.

Ritchie, M.E., Tilman, D. & Knops, J.M.H. (1998) Herbivore effects on plant and nitrogen dynamics in oak savanna. *Ecology* 79, 165–177.

Rooke, T. (2003) Defences and responses: woody species and large herbivores in African savannas. PhD Thesis, Swedish University of Agricultural Sciences, Umeå, Sweden.

Rooke, T. & Bergström R. (2007) Chemical responses and herbivory after simulated leaf browsing in *Combretum apiculatum*. *Plant Ecology* 189, 201–212.

Ross, S. (1989). *Soil Processes: A Systematic Approach*. Routledge, London, UK.

Ruess, R.W. & Seagle, S.W. (1994) Landscape patterns in soil microbial processes in the Serengeti National Park, Tanzania. *Ecology* 75, 892–904.

Scholes, R.J. (1990) The influence of soil fertility on the ecology of southern African savannas. *Journal of Biogeography* 17, 415–419.

Scogings, P.F., Hjältén, J. & Skarpe, C. (2011) Secondary metabolites and nutrients of woody plants in relation to browsing intensity in African savannas. *Oecologia* 167, 1063–1073.

Scogings, P.F., Hjältén, J. & Skarpe, C. (2013) Does large herbivore removal affect secondary metabolites, nutrients and shoot length in woody species in semi-arid savannas? *Journal of Arid Environments* 88, 4–8.

Selous, F.C. (1881) *A Hunter's Wanderings in Africa*. Richard Bentley & Son, London, UK.

Senft, R.L., Coughenour, M.B., Bailey, D.W., Rittenhouse, L.R., Sala, O.E. & Swift, D.M. (1987) Large herbivore foraging and ecological hierarchies. *Bioscience* 37, 789–799.

Sileshi, G.W., Arshad, M.A., Konaté, S. & Nkunika, P.O.Y. (2010) Termite-induced heterogeneity in African savanna vegetation: mechanisms and patterns. *Journal of Vegetation Science* 21, 923–937.

Simpson, C.D. (1974) Ecology of the Zambezi Valley Bushbuck *Tragelaphus scriptus ornatus* Pocock. PhD Thesis, Texas A&M University, College Station, Texas.

Stamp, N. (2003) Out of the quagmire of plant defense hypotheses. *The Quarterly Review of Biology* 78, 23–55.

Stokke, S. & du Toit, J.T. (2002) Sexual segregation in habitat use by elephants in Chobe National Park, Botswana. *African Journal of Ecology* 40, 360–371.

Thomas, D.S.G. & Shaw, P.A. (1991) *The Kalahari Environment*. Cambridge University Press, Cambridge, UK.

Wallgren, M. (2008) Mammal community structure in a world of gradients: effects of resource availability and disturbance across scales and biomes. PhD Thesis, Swedish University of Agricultural Sciences, Umeå, Sweden.

10

Impala as Controllers of Elephant-Driven Change within a Savanna Ecosystem

Stein R. Moe[1], Lucas Rutina[2], Håkan Hytteborn[3,4]
and Johan T. du Toit[5]

[1]Department of Ecology and Natural Resource Management, Norwegian
University of Life Sciences, Norway
[2]Okavango Research Institute, University of Botswana, Botswana
[3]Department of Plant Ecology and Evolution, Evolutionary Biology
Centre, Uppsala University, Sweden
[4]Department of Biology, Norwegian University of Science and
Technology, Norway
[5]Department of Wildland Resources, Utah State University, USA

Introduction

The impala, *Aepyceros melampus*, is the most numerous and widespread ungulate across eastern and southern African savannas. With a global population estimated at about two million animals it can be matched, among African ungulates, only by springbok, *Antidorcas marsupialis*, which has a much more limited range (IUCN, 2011). Although some duiker (Tribe Cephalophini) and dik-dik (Tribe Neotragini) have wider ranging populations, they are fragmented, with individuals occurring at comparatively low local densities. Despite the substantial densities and wide distribution of impala, it is only relatively recently that their functional significance has been recognised in savanna ecosystems (Prins and van der Jeugd, 1993; Moe *et al.*, 2009; O'Kane *et al.*, 2012). This is surprising because they are conspicuous by their numbers to any visitor to a wildlife area in eastern and southern Africa.

It is uncommon to be a common species. Universally, across all taxa, the distributions of species' relative abundances show that there are only a few common species; most are rare (McGill *et al.*, 2007). Despite their evident functional importance, common

Elephants and Savanna Woodland Ecosystems: A Study from Chobe National Park, Botswana,
First Edition. Edited by Christina Skarpe, Johan T. du Toit and Stein R. Moe.

species have received less attention than they deserve (Gaston, 2010). To understand ecosystem structure and dynamics we need more knowledge of how common species affect ecosystem dynamics and, ultimately, ecosystem resilience (Gaston and Fuller, 2007; Gaston, 2010). In this context, impala are especially interesting in the Chobe ecosystem, where they are now common but were much less so just a few decades ago (Sheppe and Haas, 1976), when the elephant *Loxodonta africana* population was also low. Here we use the Chobe ecosystem as a case-study to explore the possibility that the recovery of the elephant population (Chapter 6) has created habitat that favours impala. We hypothesise that, following changes in plant structure and species composition along the Chobe river caused by increasing numbers of elephant, the impala population grew substantially and is currently preventing the shrubland from reverting to its previous woodland state (i.e. the impala population is acting as a controller, *sensu* Pickett *et al.*, 2003; Chapter 1), and will continue to do so, even if elephant numbers were now to decrease.

For the past 50 years, during the post-descriptive phase of ecological research, ecological experiments and theory focused primarily on predation, competition and external stress as principal determinants of the dynamics, composition and diversity of natural communities. Recent work has challenged this emphasis, arguing that positive interactions or facilitation have been underestimated (Bruno *et al.*, 2003; Ellison *et al.*, 2005; Altieri *et al.*, 2007). Incorporating facilitation in ecological theory may well change traditional views of important concepts such as the relationship between fundamental and realised niche (Bruno *et al.*, 2003). Whereas the realised niche has previously been regarded as being encompassed by the fundamental niche, it could subsume the fundamental niche if facilitation more than offsets the effects of those factors tending to shrink a species' niche, even to the point of expanding beyond the current fundamental niche (Bruno *et al.*, 2003).

Recent research shows that interactions between foundation species (*sensu* Ellison *et al.*, 2005) and other species are often strong and facilitative (Altieri and Witman, 2006; Altieri *et al.*, 2007; Ellison *et al.*, 2005), and that these interactions are far more important for ecosystem structure and function than has been previously recognised (Bruno *et al.*, 2003; Altieri *et al.*, 2007). Following this avenue of research opportunities, researchers are increasingly focusing on facilitation in their studies of herbivores in African savannas (e.g. Fritz *et al.*, 2002; Rutina *et al.*, 2005; Fornara and du Toit, 2007; Valeix *et al.*, 2011). In this chapter, we consider how the increasing numbers of impala are influencing woody vegetation dynamics along the Chobe riverfront, primarily through their control of seedling establishment (Moe *et al.*, 2009). This in turn could be changing food availability for elephant along the Chobe riverfront. The conceptual framework we use is that of Pickett *et al.* (2003), in which elephants function as agents responsible for the change in state from woodland to shrubland, with impala serving as controllers of the return from shrubland to woodland.

Impala and seedlings

On a continental scale the balance between herbaceous and woody vegetation in savanna ecosystems is regulated by rainfall and edaphic factors (Scholes and Archer, 1997). African savannas with less than 650 mm mean annual precipitation (MAP) have been termed 'stable', implying that grass cover will be maintained, regardless of disturbance regimes, because soil water availability is too low to allow canopy closure (Sankaran et al., 2005). At higher rainfall, however, there is potential for a closed canopy but the relative cover of woody and herbaceous plants is determined by local disturbances such as fire (Bond, 2008) and herbivory (Asner et al., 2009). Fire, which was frequent right up to the Chobe riverfront in the past, is now effectively absent there (Mosugelo et al., 2002) because increased herbivory by impala has removed almost all fine fuels.

Elephants are able to kill mature trees either by pushing them over or by debarking them (Ben-Shahar, 1993). Elephants also kill seedlings and saplings. On the Savuti floodplain in Chobe they killed 36% of *Acacia erioloba* seedlings less than 1.5 m high (Barnes, 2001). Conversely, substantial regeneration of *Acacia sieberiana* in Uganda has been attributed to the long-term removal of elephants (Hatton and Smart, 1984). Although *Acacia* seedlings are probably nutritious for elephants, they are dispersed, inconspicuous and difficult to harvest in bulk. With elephants in Chobe browsing predominantly in the 1–3 m zone above ground (Stokke and du Toit, 2000) we hypothesise that they cannot account for the almost complete current absence of regenerating *Acacia* and other formerly abundant trees along the Chobe riverfront. Profound changes in relative plant dominance during the past decades have been documented along the riverfront (Simpson, 1975, Chapter 4). These differences are apparent between heavily and lightly disturbed sites, particularly on alluvium covered with Kalahari sand (Figure 10.1). There are almost no regenerating or mature *Acacia* trees any longer (Chapter 5, Figure 10.1). *Garcinia livingstonei*, the dominant woody species where elephants have limited access to the riverside on Kalahari sand overlying alluvial soil, is virtually absent from heavily impacted areas of the riparian fringe, where densities of up to 20 elephants km^{-2} have been estimated (Teren and Owen-Smith, 2010). The three woody species now dominating these heavily impacted areas are *Croton megalobotrys*, *Combretum mossambicense* and *Capparis tomentosa*. *Croton megalobotrys* is avoided by impala whereas the other two species are preferred (Makhabu, 2005; Moe et al., 2009). Impala typically browse these species during the peak dry season from June to September (Moe et al., 2009), a period where seedlings are vulnerable because of drought stress (Hawkes and Sullivan, 2001). We found the density of tree seedlings less than 1.0 m high in heavily impacted sites along the riverfront to be 162.5 ($\pm$45.4 s.e.) seedlings ha^{-1} in sites with alluvial soils covered by

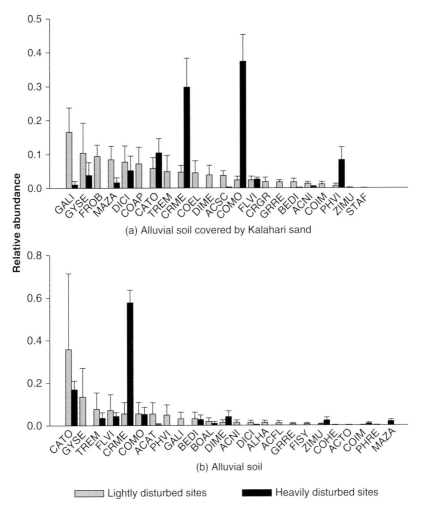

Figure 10.1 **Relative abundance (number of stems/km² ± s.e.) of woody plant species on lightly disturbed (i.e. in close proximity to human settlements) and heavily disturbed sites on two different soil types along the Chobe riparian woodland, northern Botswana.** *Acacia ataxacantha* (ACAT), *Acacia schweinfurthii* (ACSC), *Acacia fleckii* (ACFL), *Acacia nigrescens* (ACNI), *Acacia tortilis* (ACTO), *Albizia harveyi* (ALHA), *Berchemia discolor* (BEDI), *Boscia albitrunca* (BOAL), *Capparis tomentosa* (CATO), *Combretum apiculatum* (COAP), *Combretum elaeagnoides* (COEL), *Combretum hereroense* (COHE), *Combretum imberbe* (COIM), *Combretum mossambicense* (COMO), *Croton gratissimus* (CRGR), *Croton megalobotrys* (CRME), *Dichrostachys cinerea* (DICI), *Diospyros mespiliformis* (DIME), *Ficus sycomorus* (FISY), *Flueggea virosa* (FLVI), *Friesodielsia obovata* (FROB), *Garcinia livingstonei* (GALI), *Grewia retinervis* (GRRE), *Gymnosporia senegalensis* (GYSE), *Markhamia zanzibarica* (MAZA), *Philenoptera violacea* (PHVI), *Phyllanthus reticulatus* (PHRE), *Sterculia africana* (STAF), *Trichilia emetica* (TREM), *Ziziphus mucronata* (ZIMU).

Kalahari sand, which are less susceptible to evaporative water loss, and 18.8 (±6.6 s.e.) seedlings ha^{-1} on pure alluvium, where the clay-rich soils are comparatively more arid.

To understand how impala affect seedling survival, we conducted an experiment with four species of woody plants (Moe *et al.*, 2009). Two of the once-common species, *Garcinia livingstonei* and *Faidherbia albida*, now are decreasing, with only a few individuals of *Faidherbia albida* presently scattered along the riverfront. The two other species, *Combretum mossambicense* and *Croton megalobotrys*, are currently among the most common woody species along the Chobe riverfront, especially on heavily disturbed sites (Figure 10.1). We randomly selected 12 sites, six in areas heavily impacted by elephants and six in less disturbed areas (see Moe *et al.*, 2009 for details). For each of the two disturbance regimes, three sites were on Kalahari sand overlying alluvial soil and three were on exposed alluvium. At each site we planted five seedlings (grown in a nursery for 2 weeks) of each of the four species in three treatments: (i) complete exclosure (1.2 m × 2.0 m, fenced with diamond mesh on all four sides to a height of 1.5 m and on top), allowing access by insects and small mammals only; (ii) semi-permeable exclosure (same size as the complete exclosure, but sides open to 0.5 m above ground), allowing access by gallinaceous birds, primates and lagomorphs; (iii) open plots, where all browsers had access. After fencing, the seedlings were monitored for 9 months (Moe *et al.*, 2009).

Seedling mortality was relatively low except for the open plots where seedlings of all species were extensively consumed (Figure 10.2), showing that large browsers exert a particularly strong influence on seedling mortality in the area. In addition, *Faidherbia albida* experienced high predation in the semi-permeable exclosures, indicating that foraging by smaller mammals like primates and rodents, and possibly also gallinaceous birds, can affect survival. Because the seedlings in this experiment had been germinated in a nursery and transplanted into prepared holes, the soil around their roots was loose and they could be uprooted easily by elephants. To determine the cause of seedling mortality more precisely, we documented whether the stem was cut, leaving a sharp stump (ascribed to rodents); torn, leaving a rough stump (ungulate browsing); or uprooted (elephant damage). During the 9 months of the trial, no uprooting was recorded. There was also no significant variation across both heavily and lightly disturbed sites in seedling mortality ascribed to ungulates, which had accounted for 81% of the seedlings consumed in the open plots in the first 4 months after planting (Rutina, 2004).

After monitoring the seedlings for 9 months the survivors were uprooted, all vegetation within each plot was manually removed and the plots were monitored for another year to study natural regeneration (Moe *et al.*, 2009). The results showed the same general trend as in the predation experiment. We observed substantial seedling establishment in the complete exclosures (58 individuals from 8 species) but little in the semi-permeable and the open plots (12 individuals each, from 5 and 4 species, respectively). *Diospyros mespiliformis*, with a total of 30 new individuals, regenerated particularly vigorously in the complete exclosures. We also found seven

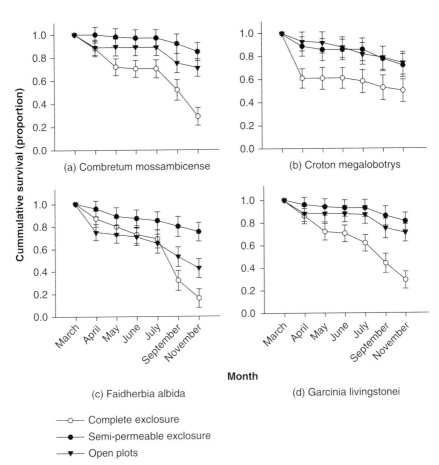

Figure 10.2   **Cumulative survival of seedlings of each tree species in each of three different treatments through the year. Note how survival of *C. mossambicense, F. albida* and *G. livingstonei* declined sharply in the open plots (open circles) after June/July, which is when impala typically begin increasing their browse intake to offset the reduced availability of green grass in the dry season. (Source: Moe *et al.*, 2009. Reproduced with permission of John Wiley & Sons.)**

new individuals of *Combretum mossambicense* in the complete exclosure, but only one and none in the semi-permeable and open treatments respectively. Seedlings of *Capparis tomentosa* established in all treatments with four, one and four seedlings in the complete, semi-permeable and open treatments, respectively, as did *Croton megalobotrys* with eight, six and six sprouting seedlings in the closed, semi-permeable

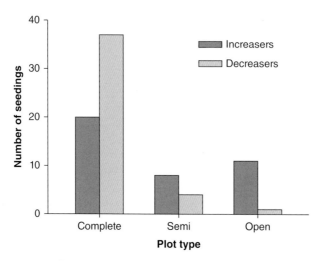

Figure 10.3 **Total numbers of woody seedlings naturally establishing after 15 months (December 2001–March 2003) in each of three different experimental plot types across 12 sites along the Chobe riverfront, categorised by plant functional type as 'decreasers' (*Garcinia livingstonei, Diospyros mespiliformis, Friesodielsia obovata*) or 'increasers' (*Combretum mossambicense, Croton megalobotrys, Capparis tomentosa, Dichrostachys cinerea, Flueggea virosa*). The distribution of seedlings across plot types differed significantly between plant functional types ($\chi^2 = 14.6$, d.f. = 2, $p < 0.001$).**

and open treatments, respectively. By classifying seedlings into those species that are either increasing or decreasing in dominance along the Chobe riverfront, we found that excluding large herbivores differentially affected the natural regeneration of 'increasers' versus 'decreasers'. As expected, the 'decreasers' dominated the naturally establishing seedlings in the complete exclosures and the 'increasers' dominated in the open plots (Figure 10.3).

The relative abundances of increaser and decreaser woody species appear to be related to the food preferences of impala. *Croton megalobotrys*, for example, is an increaser that is avoided by impala (Makhabu, 2005). *Croton megalobotrys, Combretum mossambicense* and *Capparis tomentosa* dominate the disturbed sites along the riverfront (Figure 10.1). The fact that the last two of these increaser species are favoured by impala (Makhabu, 2005) shows that intense browsing does not reduce habitat quality for impala. On the contrary, because both these species are shrubs, they are readily accessible to impala. In contrast, many of the woody species that have been reduced or virtually eliminated are not available to impala once they grow above about 1.5 m. For example, along the riverfront, *Acacia schweinfurthii, Acacia fleckii, Acacia nigrescens* and *Acacia tortilis* have disappeared or been substantially reduced in abundance (Figure 10.1). By consuming seedlings, impala can either remove these species altogether, or arrest them in a shrub-like state.

Because of elephants' high energy demands, one would expect them to regulate woodland dynamics more than do impala. When evaluating browsers' impacts on woody species assemblages, however, it is important to consider the complete developmental pathway. Impala feed on trees at the seedling stage when plants are small and vulnerable to mortality from browsing. Elephants, in contrast, feed predominantly on adult trees better able to withstanding browsers. Selective feeding by impala can therefore prevent certain species reaching heights preferred by feeding elephants.

We have estimated population densities of about 71 impala km^{-2} along the Chobe riverfront. In this area, mean seedling dry mass after 15 months of natural growth from seeds in exclosure plots was 0.636 g m^{-2} (s.e. $= 0.244$; $n = 12$ sites) whereas densities of seedlings less than 1 m high (maximum feeding height for impala) in the area were 183,021 seedlings km^{-2}. This gives a seedling biomass density of 636 kg km^{-2} or a seedling biomass of approximately 9 kg per impala. Obviously impala do not browse only on seedlings below 1 m, but also on larger trees with canopies available below that height. Nevertheless, our data show that impala have the potential to retain the present vegetation along the riverfront by limiting seedling recruitment and pruning saplings.

When surveying animal densities along the riverfront and relating ungulate densities to seedling predation on each plot, we found a strong positive relationship between local impala density and seedling predation, but nothing similar with other herbivores (Moe et al., 2009). Thus, although the Chobe riverfront sustains a wide variety of herbivores, including elephants, impala seem to be the main consumers of tree seedlings in this ecosystem.

Seedling predation across eastern and southern African savannas

Many studies have shown how seedling establishment is regulated across African savannas. Results are conflicting and context-dependent. Important factors for seedling recruitment are bottom-up limitations of nutrients and water (Scheiter and Higgins, 2007), competition with herbaceous vegetation for soil moisture, nutrients and light (Kambatuku et al., 2011), top-down factors like fire and herbivory (Goheen et al., 2010; Scogings et al., 2012) and interactions among these factors (e.g. van Langevelde et al., 2003; Midgley et al., 2010).

Fire can have both positive and negative effects on woody seedling establishment. Occasional fires temporarily reduce herbaceous vegetation, thereby opening windows of opportunity for woody seedling establishment if browsing is low. Frequent fires alone can reduce woody regeneration (Roques et al., 2001) as well as increase it (Aleper et al., 2008), depending on the number of consecutive years of burning and the susceptibility of seedlings to fire.

Herbivory, particularly by impala, is known to affect seedling recruitment in eastern and southern Africa. Impala are clearly major seedling predators along the Chobe riverfront. Results from studies elsewhere in sub-Saharan Africa have yielded conflicting results on whether browsers control seedling recruitment. Experimental studies from central Kenya found invertebrates (Shaw et al., 2002) and rodents (Goheen et al., 2004) to be important predators of Acacia seedlings. Net survival of seedlings (size 3–7 cm when planted) was approximately twice as great in areas where large mammals were present compared to where they were excluded, apparently because of an increase of rodents and invertebrates in plots excluding large mammals (Goheen et al., 2004). Similarly, in a 10-year exclosure study, Riginos and Young (2007) found that wild herbivores had a net positive effect on the growth of Acacia drepanolobium saplings, and they attributed this to the impact of elephants on the densities of other, competing, woody species.

Although these studies from Kenya (Shaw et al., 2002; Goheen et al., 2004; Riginos and Young, 2007; Goheen et al., 2010) have reached different conclusions from those in Chobe (Barnes, 2001; Moe et al., 2009), several other studies also show the importance of browsing for seedling mortality. Using long-term data from Lake Manyara, Tanzania, Prins and van der Jeugd (1993) observed that the impala population occasionally collapsed as a result of anthrax epidemics, and that such population crashes were associated with subsequent pulses of Acacia regeneration. Sharam et al. (2006), studying the recruitment of Euclea divinorum in Serengeti National Park, concluded that the low recruitment rate resulted from a combination of fire, browsing (mainly impala) and competition from grass.

When large herbivores were excluded from large vegetated termite (Macrotermes sp.) mounds and adjacent areas in Lake Mburo National Park, Uganda, impala exerted a strong limiting effect on recruitment of the most common woody species in the area (Støen et al., 2013). Similarly, impala and dik-dik (Madoqua kirkii) limited recruitment of a number of woody species in central Kenya (Augustine and McNaughton, 2004). Both species richness and density of woody plants increased 5 years after excluding impala and other large herbivores in Kruger National Park in South Africa (Scogings et al., 2012). Van der Waal et al. (2011) showed that tree seedlings compete poorly with grass under fertile conditions in semi-arid areas, and that impala indirectly affect seedling recruitment by concentrating dung and urine, thereby creating localised nutrient-rich patches that promote grass growth and aggregation of herbivores, leading to lower tree seedling survival. Lagendijk et al. (2011) found that the mixed feeder, nyala Tragelaphus angasii limited seedling establishment in sand forests in South Africa, both where nyala were present alone and together with elephants. This shows that species other than impala can play a role in seedling predation in African savannas, but over a much smaller range.

How does Chobe compare with other African savanna ecosystems? First, fire, which used to be frequent right up to the Chobe riverfront, is now effectively absent from impala-dominated areas near the riverfront (Mosugelo *et al.*, 2002; Skarpe *et al.*, 2004), because with the increase in herbivore density almost all fine fuels have been removed. This in turn implies competitive release for woody seedlings (Riginos, 2009). Low herbaceous cover also provides sub-optimal conditions for potential rodent predators (Goheen *et al.*, 2004). Because seedling survival was high in exclosures in Chobe (Moe *et al.*, 2009) small consumers such as insects and rodents seem to be relatively unimportant. Nevertheless, both insect and rodent populations fluctuate markedly, and these groups might therefore have more effect in years when their populations peak. Seedlings growing on nutrient-rich soils might be more able to compensate for herbivory (Hawkes and Sullivan, 2001), so that survival could vary with resource availability. Our experiments in Chobe, however, were done both in relatively nutrient poor areas, where the alluvial soil is covered by a thin layer of Kalahari sand, and on the more nutrient-rich alluvial floodplain soil, and there was no corresponding interaction between soil type and survival (Moe *et al.*, 2009).

Ungulate control of tree-seedling survival clearly differs across savanna ecosystems, something that can be attributed largely to different densities of browsers or to the ratio between browsers and grazers in the various areas. Most importantly, where impala are present, they invariably limit or reduce growth of tree seedlings (Prins and van der Jeugd, 1993; Barnes, 2001; Augustine and McNaughton, 2004; Sharam *et al.*, 2006; Moe *et al.*, 2009; Scogings *et al.*, 2012; Støen *et al.*, 2013). Conversely, where impala are absent, as in all the studies from Kenya (Shaw *et al.*, 2002; Goheen *et al.*, 2004; Goheen *et al.*, 2010; Riginos and Young, 2007), seedling recruitment increases partly as an effect of grazing, even if other browsers are present at low densities. In grazer-dominated systems, competition from grass (Riginos, 2009) is reduced and consequently seedling survival increases (Kambatuku *et al.*, 2011). Reduced grass cover also leads to reduced densities of rodents (Saetnan and Skarpe, 2006; Vial *et al.*, 2011), thereby decreasing seedling predation (Shaw *et al.*, 2002; Goheen *et al.*, 2004).

Impala prevent woodland regeneration

Throughout this book we argue that the increasing elephant population in the Chobe ecosystem is driving substantial shifts in vegetation structure and composition. Within the local ungulate assemblage, the impala population is by far the main responder to these vegetation changes, increasing from low levels prior to the 1960s (Sheppe and Haas, 1976) to population densities over 50 animals km^{-2} along the riverfront currently. Intense and selective consumption of seedlings and pruning of saplings by impala is now preventing the Chobe riverfront vegetation from returning to its original woodland state. By virtue of their high densities, therefore, impala have now become controllers of the transition of the riverfront vegetation from shrubland back to woodland.

Following the rinderpest and the ivory hunt during the late 19th century, elephant and impala populations were both mere fractions of their former and present numbers. We know impala were less affected by the rinderpest virus than most other artiodactyls (Chapter 6) but we also know the impala population density in Chobe was substantially lower in the first half of the 20th century than it is at present (Sheppe and Haas, 1976). We can speculate that the rapid collapse of other prey populations during the rinderpest panzootic caused hyperpredation on impala, as has been found in predator-prey systems elsewhere (Moleón et al., 2008). Added to this would have been habitat changes caused by the virtual absence of elephants and other browsers, resulting in negative feedback on the impala population. All evidence points to the densities of elephants and all ungulate species, including impala, being at unprecedentedly low levels in the first few decades of the 20th century. This would have permitted a pulse of seedling establishment, leading to the development of a more-or-less continuous woodland cover all the way down to the riverfront along with a small strip of riparian forest along the riverfront itself (Mosugelo et al., 2002). Subsequently, the increasing elephant population changed this vegetation ('substrate' sensu Pickett et al., 2003), by converting woodland (State 1) into a shrubland (State 2). The woodland has not recovered, and shows no signs of doing so, because of a virtual absence of recruits to woodland tree populations. Our studies have confirmed that where impala occur in relatively high densities they are important consumers of tree seedlings (Moe et al., 2009), thereby strongly influencing long-term vegetation dynamics. The same interactions have recently been reported for the Hluhluwe-iMfolozi Park in South Africa (O'Kane et al., 2012). Other studies from East Africa have also shown that impala retard growth rates of trees. Sharam et al. (2006) showed experimentally how impala not only decreased the survival of Euclea divinorum seedlings by 70%, but also reversed seedling growth to negative. Similarly, in the Serengeti, a range of small browsers, including impala, reduced Acacia seedlings to less than 0.3 m in height in the Serengeti (Belsky, 1984).

The importance of impala as seedling consumers is also reflected in the size-class distributions of woody species along the Chobe riverfront, particularly where Kalahari sand overlies alluvial soil. A typical plant size-class distribution with normal recruitment contains many small individuals and fewer large trees. At riverfront sites with overlying Kalahari sand and with high densities of impala, 25% of woody plants were less than 1 m tall, the maximum feeding height of impala (Rutina et al., 2005). Impala can kill seedlings within this height range, thereby depleting this size class. In contrast, at sites on similar soils, but close to human settlement, where impala numbers were lower, 80% of woody plants were below 1 m in height. On pure alluvium with high impala densities, just 5% of woody plants were less than 1 m in height, compared with 28% in the same height class in areas close human settlements with lower impala densities. In all cases, trees and shrubs taller than 1 m had either managed to escape being eaten during their transition through this zone, or had established earlier, when impala were much less abundant. Overall, our evidence from the Chobe riverfront indicates that impala can arrest the regeneration of elephant-impacted woodland and that, functionally, they control vegetation dynamics by preventing shrubland (State 2) from returning to woodland (State 1).

A guild-based approach to predicting the effects of ungulates on tree establishment

As outlined above, different ungulate assemblages affect tree establishment differently across African savannas. In an attempt to synthesise contemporary studies across sub-Saharan Africa we propose an approach based on ungulate guilds for predicting the effect of herbivores on seedling establishment (Figure 10.4). The effect of grazers is always indirect and their effect on seedling establishment is positive. Grass can compete strongly with trees (Riginos, 2009). When grazers significantly reduce herbaceous cover, rodent densities are generally reduced (Goheen *et al.*, 2004; Saetnan and Skarpe, 2006). Rodents can either directly and negatively affect seedlings through consumption or indirect affect the seedling population by consuming seeds (Goheen *et al.*, 2010). By reducing herbaceous biomass, grazers also reduce the importance of fire in an ecosystem. Fire generally has a negative impact on seedling survival (Roques *et al.*, 2001), although Aleper *et al.* (2008) showed that two consecutive years of burning had a positive effect on *Acacia sieberiana* regeneration, but that 3 years of burning reduced regeneration.

When feeding on grass, mixed feeders will affect seedling establishment positively in the same way as grazers. But mixed feeders also consume and potentially kill seedlings, especially in the dry season when the plants are water stressed. When the biomass of mixed feeders (e.g. impala in Chobe) is high relative to that of seedlings and browse, their potential impact can be considerable (Moe *et al.*, 2009). Mixed feeders can also affect seedlings indirectly, either negatively through seed predation (Goheen *et al.*, 2010) or positively through seed dispersal (Milton and Dean, 2001; Miller, 1996). Compared with grazers and mixed feeders, browsers affect tree regeneration through fewer pathways. Their effect is either positive through seed dispersal or negative through predation of seeds and seedlings. Overall, given the spatio-temporal heterogeneity of African savannas, windows of opportunity for seedling establishment, open and close across the landscape, depending on the relative strengths of each interaction within this complex system, thereby creating considerable potential for path-dependency in vegetation dynamics.

Following our conceptual model (Figure 10.4) we propose that the composition of ungulate guilds and the relative population biomass densities of guild members is important in determining the balance between trees and grass. In most savannas and

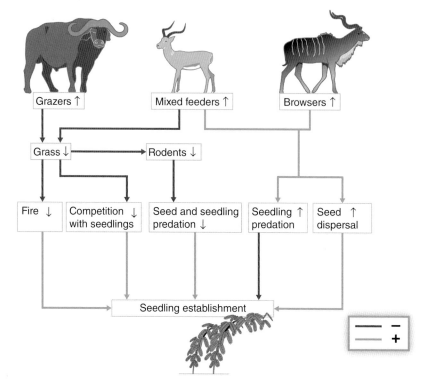

Figure 10.4   Conceptual model of the potential effects of grazers, mixed feeders and browsers on seedling establishment, showing the complexity of multiple interactions. Arrows shows the cascading effects caused by increased population densities of grazers, mixed feeders and browsers, respectively. Minus signs indicate negative effects on the subsequent box whereas plus signs indicate positive effects. For example, an increase in grazers reduces grass abundance. Reduced grass cover in turn decreases competition with seedlings for soil moisture while also reducing rodent populations, thereby lowering seed predation. Although the negative effect of rodents on seedling numbers is overshadowed by the stronger negative effects of mixed feeders (impala) and browsers, ungulates in general also have a positive effect on the dispersal of seeds through their dung. Drawing by Marit Hjeljord.

grasslands around the world, species-rich assemblies of wild grazing and browsing ungulates have been replaced by livestock, thus moving these systems towards strong dominance by a few species, typically grazers. Such changes have resulted in shrub encroachment (e.g. Coetzee *et al.*, 2007; McGranahan, 2008) with substantial negative implications for animal production.

Cascading effects of an ecosystem controller

The Kalahari sand ecosystem close to the Chobe riverfront was once woodland, supporting a high richness, evenness and diversity of woody species (Figure 10.1). Control of this ecosystem was essentially bottom-up, with the *Baikiaea* woodland maintaining a deep-slow nutrient cycle (Chapter 3), but with the recovery of the elephant population this is reverting to a faster-shallower cycle. In particular, elephants have facilitated the impala population, a mixed-feeder, whose density and ecological influence have increased substantially, with cascading effects through the ecosystem, both facilitating (Chapter 17) and competing (Dipotso *et al.*, 2007) with other species.

Several direct and indirect effects of high impala population densities can be predicted. We know from other studies that impala, because of their large numbers, are among the most commonly consumed prey for many medium-sized predators such as leopard *Panthera pardus* (Hayward *et al.*, 2006) and wild dog *Lycaon pictus* (Woodroffe *et al.*, 2007). Both occur in Chobe and, as such, impala can sustain populations of rare predators. Impala are grazers for most of the year. High densities of grazers reduce grass cover, which in turn reduces the abundance of small mammals (Saetnan and Skarpe, 2006) and thereby affects the abundance of raptors and small mammalian predators. Furthermore, impala pellets are consumed by coprophagous insects such as dung beetles (Edwards, 1991), which facilitate nutrient cycling. Impala are also common hosts for endoparasites, such as nematodes (Anderson, 1985), as well as for ectoparasitic ticks, which are consumed by oxpeckers (Robertson and Jarvis, 2000). Thus, although elephants function as the main biotic agent of change in the Chobe system, our research points to impala being the major controllers along the riverfront, directly governing certain ecosystem processes such as woodland regeneration, and indirectly controlling the abundance of a wide range of organisms, from parasites to predators.

References

Aleper, D., Lye, K.A. & Moe, S.R. (2008) Response of *Acacia sieberiana* to repeated experimental burning. *Rangeland Ecology and Management* 61, 182–187.

Altieri, A.H. & Witman, J.D. (2006) Local extinction of a foundation species in a hypoxic estuary: integrating individuals to ecosystem. *Ecology* 87, 717–730.

Altieri, A.H., Silliman, B.R. & Bertness, M.D. (2007) Hierarchical organization via a facilitation cascade in intertidal cordgrass bed communities. *American Naturalist* 169, 195–206.

Anderson, I.G. (1985) Nematode–impala interactions and their relationship to the environment. *South African Journal of Science* 81, 697–697.

Asner, G.P., Levick, S.R., Kennedy-Bowdoin, T., Knapp, D.E., Emerson, R., Jacobson, J., Colgan, M.S. & Martin, R.E. (2009) Large-scale impacts of herbivores on the structural diversity of African savannas. *Proceedings of the National Academy of Sciences* 106, 4947–4952.

Augustine, D.J. & McNaughton S.J. (2004) Regulation of shrub dynamics by native browsing ungulates on East African rangeland. *Journal of Applied Ecology* 41, 45–58.

Barnes, M.E. (2001) Effect of large herbivores and fire on the regeneration of *Acacia erioloba* woodlands in Chobe National Park, Botswana. *African Journal of Ecology* 39, 340–350.

Belsky, A.J. (1984) Role of small browsing mammals in preventing woodland regeneration in Serengeti National Park, Tanzania. *African Journal of Ecology* 22, 271–279.

Ben-Shahar, R. (1993) Patterns of elephant damage to vegetation in northern Botswana. *Biological Conservation* 65, 249–256.

Bond, W.J. (2008) What limits trees in C4 grassy biomes? *Annual Review of Ecology and Systematics* 39, 641–659.

Bruno, J.F., Stachowicz, J.J. & Berness, M.D. (2003) Inclusion of facilitation into ecological theory. *Trends in Ecology and Evolution* 18, 119–125.

Coetzee, B.W.T., Tincani, O.L., Wodu, Z. & Mwasi, S.M. (2007) Overgrazing and bush encroachment by *Tarchonantus camphoratus* in a semi-arid savanna. *African Journal of Ecology* 46, 449–451.

Dipotso, F.M., Skarpe, C., Kelaeditse, L. & Ramotadima, M. (2007) Chobe bushbuck in an elephant-impacted habitat along the Chobe River. *African Zoology* 42, 261–267.

Edwards, P.B. (1991) Seasonal variation of the dung of African grazing mammals, and its consequences for coprophagous insects. *Functional Ecology* 5, 617–628.

Ellison, A.M., Bank, M.S., Clinton, B.D., Colburn, E.A., Elliott, K., Ford, C.R., Foster, D.R., Kloeppel, B.D., Knoepp, J.D., Lovett, G.M., Mohan, J., Orwig, D.A., Rodenhouse, N.L., Sobczak, W.V., Stinson, K.A., Stone, J.K., Swan, C.M., Thompson, J., Von Holle, B. & Webster, J.R. (2005) Loss of foundation species: consequences for the structure and dynamics of forested ecosystems. *Frontiers in Ecology and the Environment* 3, 479–486.

Fornara, D.A. & du Toit, J.T. (2007) Browsing lawns? Responses of *Acacia nigrescens* to ungulate browsing in an African savanna. *Ecology* 88, 200–209.

Fritz, H., Duncan, P., Gordon, I.J. & Illius A.W. (2002) Megaherbivores influence trophic guilds structure in African ungulate communities. *Oecologia* 131, 620–625.

Gaston, K.J. (2010) Valuing common species. *Science* 327, 154–155.

Gaston, K.J. & Fuller, R.A. (2007) Commonness, population depletion and conservation biology. *Trends in Ecology and Evolution* 23, 14–19.

Goheen, J.R., Keesing, F., Allan, B.F., Ogada, D. & Ostfeld, R.S. (2004) Net effects of large mammals on *Acacia* seedling survival in an African savanna. *Ecology* 85, 1555–1561.

Goheen, J.R., Palmer, T.M., Keesing, F., Riginos, C. & Young, T.P. (2010) Large herbivores facilitate savanna tree establishment via diverse and indirect pathways. *Journal of Animal Ecology* 79, 372–382.

Hatton, J.C. & Smart, N.O.E. (1984) The effect of long-term exclusion of large herbivores on soil nutrient status in Murchison Falls National Park, Uganda. *African Journal of Ecology* 22, 23–30.

Hawkes, C.V. & Sullivan, J.J. (2001) The impact of herbivory on plants in different resource conditions: a meta-analysis. *Ecology* 82, 2045–2058.

Hayward, M.W., Henschel, P., O'Brian, J., Hofmeyr, M., Balme, G. & Kerley, G.I.H. (2006) Prey preferences of the leopard (*Panthera pardus*). *Journal of Zoology* 270, 298–313.

IUCN (2011) *IUCN Red List of Threatened Species.* [online] http://www.iucnredlist.org.

Kambatuku, J.R., Cramer, M.D. & Ward D. (2011) Savanna tree-grass competition is modified by substrate type and herbivory. *Journal of Vegetation Science* 22, 225–237.

Lagendijk, D.D., Mackey, R.L., Page, B.R. & Slotow, R. (2011) The effects of herbivory by a mega- and mesoherbivore on tree recruitment in sand forest, South Africa. *PLoS ONE* 6(3): e17983. doi:10.1371/journal.pone.0017983.

Makhabu, S.W. (2005) Resource partitioning within a browsing guild in a key habitat, the Chobe Riverfront, Botswana. *Journal of Tropical Ecology* 21, 641–649.

McGill, B.J., Etienne, R.S., Gray, J.S., Alonso, D., Anderson, M.J., Benecha, H.K., Dornelas, M., Enquist, B.J., Green, J.L., He, F., Hurlbert, A.H., Magurran, A.E., Marquet, P.A., Maurer, B.A., Ostling, A., Soykan, C.U., Ugland, K.I. & White, E.P. (2007) Species abundance distributions: moving beyond single prediction theories to integration within an ecological framework. *Ecology Letters* 10, 995–1015.

McGranahan, D.A. (2008) Managing private, commercial rangelands for agricultural production and wildlife diversity in Namibia and Zambia. *Biodiversity and Conservation* 17, 1965–1977.

Midgley, J.J., Lawes, M.J. & Chamaillé-Jammes, S. (2010) Savanna woody plant dynamics: the role of fire and herbivory, separately and synergistically. *Australian Journal of Botany* 58, 1–11.

Miller, M.F. (1996) Dispersal of *Acacia* seeds by ungulates and ostriches in an African savanna. *Journal of Tropical Ecology* 12, 345–346.

Milton, S.J. & Dean, W.R.J. (2001) Seed dispersed in dung of insectivores and herbivores in semi-arid southern Africa. *Journal of Arid Environments* 47, 465–483.

Moe, S.R., Rutina, L.P., Hytteborn, H. & du Toit, J.T. (2009) What controls woodland regeneration after elephants have killed the big trees? *Journal of Applied Ecology* 46, 223–230.

Moleón, M., Almaraz, P. & Sánchez-Zapata, J.A. (2008) An emerging infectious disease triggering large-scale hyperpredation. *PLoS ONE* 3(6), e2307. doi:10.1371/journal.pone.0002307.

Mosugelo, D.K., Moe, S.R., Ringrose, S. & Nellemann, C. (2002) Vegetation changes during a 36-year period in northern Chobe National Park, Botswana. *African Journal of Ecology* 40, 232–240.

O'Kane, C.A.J., Duffy, K.J., Page, B.R. & Macdonald, D.W. (2012) Heavy impact on seedlings by the impala suggests a central role in woodland dynamics. *Journal of Tropical Ecology* 28, 291–297.

Pickett, S.T.A., Cadenasso, M.L. & Benning, T.L. (2003) Biotic and abiotic variability as key determinants of savanna heterogeneity at multiple spatiotemporal scales. In: du Toit, J.T., Rogers, K.H. & Biggs, H.C. (eds.) *The Kruger Experience. Ecology and Management of Savanna Heterogeneity.* Island Press, Washington, DC, pp. 22–40.

Prins, H.H.T. & van der Jeugd, H.P. (1993) Herbivore population crashes and woodland structure in East Africa. *Journal of Ecology* 81, 305–314.

Riginos, S. (2009) Grass competition suppresses savanna tree growth across multiple demographic stages. *Ecology* 90, 335–340.

Riginos, S. & Young, T.P. (2007) Positive and negative effects of grass, cattle, and wild herbivores on Acacia sapling in East African savanna. *Oecologia* 153, 985–995.

Robertson, A. & Jarvis, A.M. (2000) Oxpeckers in North-eastern Namibia: recent population trends and the possible negative impacts of drought and fire. *Biological Conservation* 92, 241–247.

Roques, K.G., O'Connor, T.G. & Watkinson, A.R. (2001) Dynamics of shrub encroachment in an African savanna: relative influences of fire, herbivory, rainfall and density dependence. *Journal of Applied Ecology* 38, 268–280.

Rutina, L.P. (2004) Impalas in an elephant-impacted woodland: browser-driven dynamics of the Chobe riparian zone, Northern Botswana. PhD Thesis, Agricultural University of Norway, Ås, Norway.

Rutina, L.P., Moe, S.R. & Swenson, J.E. (2005) Elephant *Loxodonta africana* driven woodland conversion to shrubland improves dry-season browse availability for impalas *Aepyceros melampus*. *Wildlife Biology* 11, 207–213.

Saetnan, E.R. & Skarpe, C. (2006) The effect of ungulate grazing on small mammal community in Southeastern Botswana. *African Zoology* 41, 9–16.

Sankaran, M., Hanan, N.P., Scholes, R.J., Ratnam, J., Augustine, D.J., Cade, B.S., Gignoux, J., Higgins, S.I., Le Roux, X., Ludwig, F., Ardo, J., Banyikwa, F., Bronn, A., Bucini, G., Caylor, K.K., Coughenour, M.B., Diouf, A., Ekaya, W., Feral, C.J., February, E.C., Frost, P.G.H., Hiernaux, P., Hrabar, H., Metzger, K.L., Prins, H.H.T., Ringrose, S., Sea, W., Tews, J., Worden, J. & Zambatis, N. (2005) Determinants of woody cover in African savannas. *Nature* 438, 846–849.

Scheiter, S. & Higgins, S.I. (2007) Partitioning of root and shoot competition and the stability of savannas. *American Naturalist* 170, 587–601.

Scholes, R.J. & Archer, S.R. (1997) Tree–grass interactions in savannas. *Annual Review of Ecology and Systematics* 28, 517–544.

Scogings, P.F., Johansson, T., Hjalten, J. & Kruger, J. (2012) Response of woody vegetation to exclusion of large herbivores in semi-arid savannas. *Austral Ecology* 37, 56–66.

Sharam, G., Sinclair, A.R.E. & Turkington, R. (2006) Establishment of broad-leaved thickets in Serengeti, Tanzania: the influence of fire, browsers, grass competition, and elephants. *Biotropica* 38, 599–605.

Shaw, M.T., Keesing, F. & Ostfeld, R.S. (2002) Herbivory on Acacia seedlings in an East African savanna. *Oikos* 98, 385–392.

Sheppe, W. & Haas, P. (1976) Large mammal populations of the lower Chobe River, Botswana. *Mammalia* 40, 223–243.

Simpson, C.D. (1975) A detailed vegetation study on the Chobe river in north-east Botswana. *Kirkia* 10, 185–227.

Skarpe, C., Aarrestad, P.A., Andreassen, H.P., Dhillion, S., Dimakatso, T., du Toit, J.T., Halley, D.J., Hytteborn, H., Makhabu, S., Mari, M., Marokane, W., Masunga, G., Modise, D., Moe, S.R., Mojaphoko, R., Mosugelo, D., Motsumi, S., Neo-Mahupeleng, G., Ramotadima, M., Rutina, L., Sechele, L., Sejoe, T.B., Stokke, S., Swenson, J.E., Taolo, C., Vandewalle, M. & Wegge, P. (2004) The return of the giants: ecological effects of an increasing elephant population. *Ambio* 33, 276–282.

Støen, O.-G., Okullo, P., Eid, T. & Moe, S.R. (2013) Termites facilitate and ungulates limit savanna tree regeneration. *Oecologia* 172, 1085–1093.

Stokke, S. & du Toit, J.T. (2000) Sex and size related differences in the dry season feeding patterns of elephants in Chobe National Park, Botswana. *Ecography* 23, 70–80.

Teren, G. & Owen-Smith, N. (2010) Elephants and riparian woodland changes in the Linyanti region, northern Botswana. *Pachyderm* 47, 18–25.

Valeix, M., Fritz, H., Sabatier, R., Murindagomo, F., Cumming, D. & Duncan, P. (2011) Elephant-induced structural changes in the vegetation and habitat selection by large herbivores in an African savanna. *Biological Conservation* 144, 902–912.

van der Waal, C., Kool, A., Meijer, S.S., Kohi, E., Heitkönig, I.M.A., de Boer, W.F., van Langevelde, F., Grant, R.C., Peel, M.J.S., Slotow, R., de Knegt, H.J., Prins, H.H.T. & de Kroon, H. (2011) Large herbivores may alter vegetation structure of semi-arid savanna through soil nutrient mediation. *Oecologia* 165, 1095–1107.

van Langevelde, F., Van de Vijver, C.A.D.M., Kumar, L., van de Koppel, J., de Ridder, N., van Andel, J., Skidmore, A.K., Hearne, J.W., Stroosnijder, L., Bond, W.J., Prins, H.H.T. & Rietkerk, M. (2003) Effects of fire and herbivory on the stability of savanna ecosystems. *Ecology* 84, 337–350.

Vial, F., Macdonald, D.W. & Haydon, D.T. (2011) Limits to exploitation: dynamic food web models predict the impact of livestock grazing on Ethiopian wolves *Canis simensis* and their prey. *Journal of Applied Ecology* 28, 340–347.

Woodroffe, R., Lindsey, P.A., Romanach, S.S. & Ole Ranah, M.K. (2007) African wild dogs (*Lycaon pictus*) can subsist on small prey: implication for conservation. *Journal of Mammalogy* 88, 181–193.

11

Buffalo and Elephants: Competition and Facilitation in the Dry Season on the Chobe Floodplain

Duncan J. Halley[1], Cyril Taolo[2] and Stein R. Moe[3]

[1]Norwegian Institute for Nature Research, Norway
[2]Department of Wildlife and National Parks, Botswana
[3]Department of Ecology and Natural Resource Management, Norwegian University of Life Sciences, Norway

Introduction

Large herbivore communities in semi-arid African ecosystems are often resource limited (Cumming *et al.*, 1997; Coe *et al.*, 1976; Botkin *et al.*, 1981; East, 1984; Owen-Smith, 1990), although the degree to which predation can modify this conclusion is still much discussed (Fritz, 1997; Fritz and Duncan, 1994; Grange and Duncan, 2006; Owen-Smith and Mills, 2006). Interspecific competition is therefore likely to play an important role in shaping these communities (Farnsworth *et al.*, 2002; Kleynhans *et al.*, 2011). In addition, the dependence of herbivore dietary tolerance on body size translates into important size-related differences between savanna ungulates in terms of habitat specificity and the share of community resources exploited, and therefore the degree of dietary overlap within which competition can occur (du Toit and Cumming, 1999).

It has been hypothesised that interactions within a guild of size-differentiated herbivores will change from being driven by competition at low vegetation density to facilitation at high vegetation density (van de Koppel and Prins, 1998). A small-bodied grazing animal selectively removes young leaves or flower buds, reducing the quality of the remaining grass for larger bulk-grazing species; both they and mesoherbivore bulk grazers can crop a sward so short that the intake rate of larger species is

Elephants and Savanna Woodland Ecosystems: A Study from Chobe National Park, Botswana,
First Edition. Edited by Christina Skarpe, Johan T. du Toit and Stein R. Moe.

constrained. This suggests that on nutrient-rich soils, where intense herbivory keeps vegetation density low, smaller grazers can competitively exclude larger ones. Interaction between species with similar food requirements has been studied extensively (e.g. McNaughton, 1976; Sinclair and Norton-Griffiths, 1982; Bell, 1986; Owen-Smith, 1988; du Toit and Owen-Smith, 1989; Gordon and Illius, 1989; du Toit, 1990; Cromsigt *et al.*, 2009; Macandza *et al.*, 2012a,b). The results of most of these studies indicate that species with similar food requirements, although competing for forage, can coexist despite overlaps in fundamental niches provided the overlap in potential resource use is incomplete (Putnam, 1996). Fritz *et al.* (2002) analysed wildlife censuses from 31 conserved African ecosystems and concluded that elephants *Loxodonta africana* negatively affected populations of browsers and mixed feeders, but had no influence on grazers. This suggests a difference in the dynamics of competition between elephants and other grazers, on the one hand; and browsers on the other.

Facilitation can also occur where one species improves resource availability for another by improving accessibility of food (van de Koppel and Prins, 1998), stimulating regrowth (McNaughton, 1976), and/or improving food quality (McNaughton *et al.*, 1988). Competition and facilitation are not mutually exclusive, and can vary in time and space and in the species involved (van de Koppel and Prins, 1998; Taolo, 2003).

African elephant and buffalo, *Syncerus caffer*, comprise a large proportion of the ungulate biomass in dystrophic ecosystems like Chobe (du Toit and Owen-Smith, 1989). The large-bodied, relatively unselective African elephant removes large quantities of grass and browse with significant impact on plant dynamics and hence the biodiversity of savanna ecosystems (Guldemond and van Aarde, 2008). Although much of the impact of elephants on their environment is associated with their activity as browsers, they can consume large quantities of good quality grasses even in the dry season. There is increasing evidence, however, that smaller herbivores, or the interaction between smaller herbivores and elephants, also strongly influence shifts in savanna vegetation (Belsky, 1984; Prins and van der Jeugd, 1993; Rutina, 2004; Moe *et al.*, 2009).

At Chobe, the African buffalo is the largest grazing mammal apart from elephants and hippopotamus, *Hippopotamus amphibius*, in terms of individual size, and second only to elephants in overall biomass and forage consumption. Its body size allows it to digest low quality forage prevalent in dystrophic savannas (Prins, 1996) and therefore to overlap significantly in its foraging niche with elephants.

Niche separation among large herbivores results from differences in digestive systems (Hofmann and Stewart, 1972; Hofmann, 1989), in body size and associated metabolic requirements (Bell, 1971; Jarman, 1974), and in the morphology of the foraging apparatus affecting the use of different plant parts (Gordon *et al.*, 1996). All these features differ markedly between buffalo and elephants. Despite their differing feeding strategies, buffalo and elephants compete for the same resource where their dietary and spatial ranges overlap sufficiently (de Boer and Prins, 1990). Although elephant mainly browse during the dry season (Estes, 1991), they are known to prefer green graminoids associated with water (Kalamera, 1989; Prins and Douglas-Hamilton, 1990). Buffalo are ruminant bulk grazers, but they also prefer green graminoids found in riverine habitats during the dry season (Sinclair, 1977). At most, browsing appears to be only a very small part of the diet of buffalo (Leuthold, 1972; Beekman and Prins, 1989), even when they live in or near scrub and savanna woodland, as in Chobe (Halley and Minagawa, 2005). Conditions in Chobe therefore entail considerable potential for competition between buffalo and elephant for grazing resources, with the floodplain maintaining growing swards of grass throughout the dry season while there is little or no primary production elsewhere (Mosugelo et al., 2002).

Although elephants and buffalo can compete with one another by removing large quantities of grazing (de Boer and Prins, 1990), elephants can also facilitate buffalo by removing substantial amounts of browse, thereby increasing grass production (Dublin et al., 1990). In addition, feeding and trampling by elephants can induce regrowth of new, higher quality, herbaceous shoots (Vesey-Fitzgerald, 1969). Fritz et al. (2002) showed that the biomass of mesograzers (4–450 kg body mass) such as buffalo responds positively to rainfall and soil nutrient content, reflecting a positive correlation between grass production and rainfall, and a higher nitrogen concentration in grasses growing on nutrient-rich than on nutrient-poor soils. They concluded that the metabolic biomass of mesograzers was unaffected by megaherbivores, an observation that is consistent with the argument that smaller grazing herbivores are competitively dominant over large herbivores on shorter and scarcer grass resources (Illius and Gordon, 1987). Thus the impact of the two species on each other is complex.

In this chapter we investigate elements of the dry-season foraging interactions between elephants and buffalo in Chobe. Competition theory predicts that (i) elephant will avoid buffalo grazed patches; (ii) buffalo will avoid elephant grazed patches. Facilitation theory predicts (iii) elephants will be attracted to buffalo grazed patches; (iv) buffalo will be attracted to elephant grazed patches. If there is neither competition nor facilitation, then (v) neither species will respond to grazing by the other.

Spatial and temporal overlap between elephant and buffalo in Chobe

In the Chobe ecosystem, dystrophic Kalahari sands juxtapose with the eutrophic alluvial soils along the Chobe River (Chapter 9). The ecosystem is dominated by a gradient of elephant impact towards the river whereas the influence of fire is more pronounced in the opposite direction (Ben-Shahar, 1993, 1998; Mosugelo et al., 2002; Chapter 9). The Chobe River is the only source of water in the dry season (apart from a few recently established pumped pans), resulting in an annual aggregation of ungulates and elephants within the extreme northern tip of the protected area that produces one of the highest concentrations of wildlife in Africa (Chapter 3). The floodplain of the river is the only area in which new forage growing throughout the dry season (Mosugelo *et al.*, 2002).

The elephant population increased markedly in Chobe over the 20 years prior to this study, resulting in local dry season densities of elephants of 20 elephants km^{-2} along the Chobe riverfront (Melton, 1985; Gibson *et al.*, 1998; DWNP – Department of Wildlife and National Parks, Botswana, 2000; Teren and Owen-Smith, 2010). Buffalo have also increased substantially in numbers in recent decades (Taolo, 2003). After elephants, buffalo are the most numerous large herbivore species (2747 $\pm$ 96 animals in the river-front area in the dry season during the study period: Taolo, 2003), and the heaviest in total biomass. The loss of large tracts of floodplain outside the Chobe National Park and across the Chobe River in Namibia to other land uses, however, has prevented buffalo along the Chobe River from reaching the numbers apparently present when the first written accounts were made. Selous (1881: 156) described the population size then as 'astonishing'.

The population is divided in the dry season into two large herds and several smaller herds, which show a fission-fusion pattern of local foraging movements. Although populations at Chobe have increased, the maximum size of individual herds on the floodplain in the dry season has remained fairly stable over the last two decades. Sheppe and Haas (1976) estimated that the largest herd to be around 1000 animals. Current figures indicate that this number remains substantially unchanged (Halley *et al.*, 2003). This upper size is most likely the result of a trade-off between predation risk and intraspecific competition for food (Sinclair, 1977; Prins, 1996).

In the wet season there is a short-distance migration to savanna woodlands south of the river (Halley *et al.*, 2003). Significant interchange of both bulls and cows occurs between herds, including long-distance movements (e.g. to the Okavango region), between dry seasons (Halley *et al.*, 2002).

The Chobe River provides not only water during the dry season. High quality grass on the floodplain is intensively utilised by local herbivores. Sedudu Island (2.5 km^2) lies on the floodplain of the Chobe River, with the northern channel of the river forming the international boundary with Namibia and the boundary of the National Park. The flat, low-lying island falls within the floodplain vegetation type as defined by Mosugelo *et al.* (2002). The level of the Chobe River varies during the course of the year and from year to year. Most of the flood waters come from the upper Zambezi basin in Angola, with the flood usually peaking between March and May, when most of the floodplain, including Sedudu Island, is under water. This makes it impossible for most herbivores, except hippopotamus, to use the vegetation for grazing. During the dry season, the floodwaters recede, allowing a variety of animals to graze and drink on the island.

Sedudu Island is open grassland dominated by *Cynodon dactylon*, frequently interrupted by isolated patches of *Diplachne fusca* and species of *Panicum*, *Digitaria*, *Eragrostis*, *Echinochloa* and *Sporobolus* (Chapter 5). Patches of *Cyperus papyrus*, *Phragmites australis* and *Typha capensis* fringe the water channels. Woody vegetation does not occur on the island. The soils on the floodplain are characterised by calcic gleysols and eutric arenosols, deep to very deep, poorly to imperfectly drained, with very dark sand clay to clay interspersed on dead river courses. Apart from buffalo and elephant, other herbivores that occur commonly on the island include hippopotamus and lechwe, *Kobus lechwe*.

The simultaneous occurrence of herds of buffalo and elephant on Sedudu Island provided an opportunity to evaluate interactions between the two species. Although buffalo and elephant overlap in terms of macrohabitat use during the dry season, the overlap in diet shrinks considerably at this time of the year (Omphile, 1997). Browse constitutes the bulk of elephant diet, although they also consume similar floodplain grasses to those used by buffalo, including *Cynodon dactylon* (Omphile, 1997). Buffalo are predominantly grazers with only a small element of browse in their diet (Leuthold, 1972; Mloszewski, 1983; Beekman and Prins, 1989; Halley and Minagawa, 2005). The biomass of other large herbivores on the island was insignificant compared to the study species. Buffalo herds were present on the island throughout the day and night for most of the dry season although the actual numbers fluctuated from about 50 to 1400 animals. Elephant herds swam across the river to the island each day but left the floodplain for the woodlands in the early evening.

We concentrated our study on the feeding interactions between elephants and buffalo in the resource-limited dry season. During this period grass food resources are confined to the floodplain and high densities of both buffalo and elephants occur adjacent to the Chobe River. Thus, any competitive or facilitative interactions would be expected to be high during the dry season.

Sedudu Island was delineated into six grazing patches based on the pattern of use by buffalo. The patches were *Cynodon dactylon*-dominated swards visible to the eye and were characterised by shorter grass than in adjacent unused areas. These patches ranged in size from 0.27 to 0.63 km². Buffalo and elephant on the island were observed from the mainland by telescope once a day during the 2000 and 2001 dry seasons (July to November). The date, time, patch, number of animals and activity (grazing, resting and drinking) of the animals was recorded. Each visit to the patches was recorded as a discrete event. Contingency tables were used to determine if two events were randomly distributed and mutually independent.

The relative grazing pressure exerted on the patches by each species during bouts of grazing was estimated using methods adapted from de Boer and Prins (1990). Because the reported effects in the current study were relative, that is, *buffalo* activity was assessed only in relation to how much *elephant* grazing pressure was more, or less, than average for each patch and vice versa, we disregarded several of the assumptions made by de Boer and Prins (1990). We did not try to assess either the absolute amount of forage consumed by each species, or how much more or less buffalo consumed relative to elephants. Average estimates of residence were obtained separately for each of the six patches because their sizes differed (Taolo, unpublished data). Relative grazing pressure (RGP, animal-hour/patch) was estimated as

$$RGP = N_i G_i A_i \qquad (11.1)$$

where N_i, number of animals present in observation period; i; G_i, proportion of animals grazing during period i; and A_i, duration of observation period i (h).

The median grazing pressure of buffalo was calculated by this method to be 600; of elephants, 16. A day on which a patch was grazed by buffalo or elephant was defined as day 0. The relative grazing pressures by the other species in the patch on the 10 days preceding day 0 were then tabulated. The average relative grazing pressure was calculated for each day (i.e. day −10, day −9, etc.) by fixing the data series to day 0. The cumulative differences between values calculated for days when relative grazing pressures were higher or lower than the median grazing pressure on day 0 for each species were tested using two-tailed paired sample t-tests for means.

Interference competition

There was no association between the visits by the two species to a patch at the scale of a single day. The herd size of one species on a particular patch was not influenced by the absence or presence of the other species (Table 11.1).

Thus there appears to be no evidence of one species displacing the other, or one or both species being attracted to the other on the same day. Although interference competition between several ungulate species has been shown (e.g. Faas and Weckerly, 2010; Colman *et al.*, 2012), exploitation competition between large herbivores is

Table 11.1 The relationship of presence or absence of buffalo and elephant on average group sizes of the other species on Sedudu Island (significance tested using two-tailed paired sample t-test; N.S., not significant).

	Buffalo	Elephant
Buffalo		
Present		12
Absent		15
		$t = 1.274, n = 121$ N.S.
Elephant		
Present	370	
Absent	317	
	$t = 0.921, n = 144$ N.S.	

generally more common (Illius and Gordon, 1987). When resources are limited, such as during our study, food resources are typically more concentrated, and the level of aggression between conspecifics and heterospecifics could be expected to increase. We found no such evidence of displacement between elephant and buffalo, however.

Exploitation competition

To determine if grazing pressure of one species, rather than its mere presence, affected grazing by the other within a single day, visits of buffalo were classified as being to sites of less than or greater than the median grazing pressure of elephant and vice versa. A contingency table analysis showed that these associations were effectively random (Table 11.2).

Although we found no evidence of direct interference competition or exploitation competition within the same day, the two species responded to the grazing pressure exerted by the other over a timescale of days, and in different ways. Elephant responded negatively to high recent buffalo grazing pressure whereas buffalo responded positively to high recent elephant grazing. Relatively low grazing pressure by buffalo up to 2 days prior to a bout of grazing by elephant (day 0) was associated with a greater grazing pressure (i.e. higher than the median of all visits) by elephant on day 0 (Table 11.3). Conversely, high previous grazing pressure by elephant up to 10 days before, but not on the day before, was associated with higher than median grazing pressure by buffalo on day 0 (Table 11.4).

Elephant grazing pressure on a particular patch was thus lower if the buffalo grazing pressure on the same patch was previously relatively high (hypothesis 1, competition). In contrast, buffalo grazing pressure was higher on patches that had experienced high grazing pressure by elephant up to 10 days beforehand (hypothesis 4, facilitation).

Table 11.2 **Comparison of buffalo and elephant grazing pressure in the same patch on the same day on Sedudu Island using Cochran's corrected chi square (expected values given in parentheses): $\chi^2 = 1.622$, d.f. $= 2$, NS.**

		Elephant grazing pressure		
		0 – median	>median	Total
Buffalo grazing pressure	0 – median	9 (7.8)	6 (7.2)	15
	>median	14 (15.2)	15 (13.8)	29
	Total	23	21	44

Table 11.3   The effect of prior grazing pressure of buffalo on later elephant grazing pressure of the same patch. The day elephant grazed on a patch is defined as day 0. Lower-than and higher-than median refer to the median grazing pressure (animal-hour/patch) for elephant. The values are the number of grazing buffalo-hour/patch on days before the elephant grazed at higher or lower than median intensity. Significance of the cumulative difference from day 0 was determined by paired sample t-tests for means. Higher buffalo grazing pressures the day before, and (cumulatively) 2 days before, were significantly associated with below median elephant grazing pressures on day 0.

Elephant grazing on day 0.	Buffalo grazing pressure, days before elephant grazed patch (day 0)									
	−1	−2	−3	−4	−5	−6	−7	−8	−9	−10
Higher than median	124	187	164	243	241	169	103	99	53	124
Lower than median	226	234	157	94	153	189	145	54	77	152
Significance of cumulative difference from day 0	<0.01	<0.05	NS	NS	NS	NS	NS	NS	NS	NS

Table 11.4   The effect of prior grazing pressure of elephant on later buffalo grazing pressure of the same patch. The day buffalo grazed on a patch is defined as day 0. Lower-than and higher-than median refer to the median grazing pressure (animal-hour/patch) for buffalo. The values are the number of grazing elephants-hour/patch on days before the buffalo grazed at higher or lower than median intensity. Significance of the cumulative difference from day 0 was determined by paired sample t-tests for means. Higher elephant grazing pressures from 2 days before to 10 days before (cumulatively) day 0 were significantly associated with above median buffalo grazing pressures on day 0.

Buffalo grazing on day 0	Elephant grazing pressure, days before buffalo grazed patch (day 0)									
	−1	−2	−3	−4	−5	−6	−7	−8	−9	−10
Higher than median	4.5	2.6	2.8	2.8	3.9	3.2	5.6	4.0	3.8	4.9
Lower than median	3.0	1.9	1.8	2.6	1.9	2.1	1.8	0.8	2.7	1.6
Significance of cumulative difference from day 0	NS	<0.1	<0.05	<0.05	<0.01	<0.01	<0.01	<0.01	<0.01	<0.01

Vesey-Fitzgerald (1969) suggests two mechanisms by which elephant grazing might facilitate grazing by buffalo. First, by pulling up the prostrate stolons of *Cynodon dactylon* while feeding, elephants increase the amount of grass available to the buffalo. Second, trampling by elephant breaks down the grass sward, contributing to production of nutritious green tillers. Regrowth of tillers takes time, which might be why there was no immediate effect of elephant grazing on buffalo numbers, and why the facilitation effect lasted some time. Facilitation has been reported for wildebeest by zebra (Bell, 1970; McNaughton, 1976), although the conclusion that wildebeest benefits from the zebra in the grazing succession was not upheld when the population dynamics of these herbivores was examined (Sinclair and Norton-Griffiths, 1982).

De Boer and Prins (1990) found competition between buffalo and elephant in East Africa in some ways similar to that in Chobe, with elephant reacting negatively to grazing by buffalo, but with no positive or negative effect of elephant on buffalo. Larger animals have a disadvantage when the quantity of food items is limited, because smaller animals, requiring less, can more easily satisfy their quantity requirements (Illius and Gordon, 1987). Elephant are much larger than buffalo and can only compete successfully with buffalo when food is sufficiently abundant for elephant to maintain their higher intake rate, something that they cannot do on the shorter grasses that dominate Chobe River floodplain. To graze these swards, where the grasses are shortened and flattened primarily by grazing buffalo, an elephant must switch its usual method of grazing to one of pulling up grass with its trunk and then kicking it loose with its foreleg (Prins, 1996; pers. obs.). This substantially reduces the rate of intake. Time-limited elephants must therefore include a large proportion of relatively lower quality bulk in their diet in the dry season to escape competition with specialised grazers such as buffalo for higher-value but quantitatively and spatially limited graminoid food.

Browse is not normally so limited, except where the bushes are short. This is because, although grass is essentially a two-dimensional resource at meso- and megaherbivore scales, browse is three-dimensional – browse above 2 m is not accessible to most meso-herbivores. Only elephants and giraffe, *Giraffa camelopardalis*, both megaherbivores, can do this. The quality of coarser browse (e.g. twigs and branches) is also so low that only mammals able to process large amounts can get any benefit from it, which effectively restricts this resource to elephants.

Nevertheless, buffalo, like elephants, are susceptible to the influence of grass height on the rate of intake (Smallegange and Brunsting, 2002). To minimise the effects of competition and yet meet their dietary requirements, Chobe riverfront buffalo regularly move throughout the floodplain within their range, crossing river channels and lagoons on the Botswana side of the border and even, to a limited extent, venturing onto the domestic cattle-dominated pastures on the Namibian side (Halley *et al.*, 2003). These floodplain herds fragment and fuse throughout the dry season (Halley *et al.*, 2003), a pattern noted elsewhere. Sinclair (1977) and Prins (1996) explained this fission and fusion behaviour as a trade-off between intraspecific competition, which is inherently more severe with increasing herd size, and predation risk, which is lower in larger herds.

With the onset of the wet season, both buffalo and elephant move away from the Chobe River and deeper into the interior of the protected area where they encounter abundant long grasses. At this time, elephants switch to feeding much more on grass, the increased overlap in diet with buffalo indicating greater resource availability.

Intensive burning and cattle grazing in the early to mid-20th century converted the Chobe floodplain, over a period of almost 40 years, from mostly tall reeds and sedges to a vegetation type dominated by *Cynodon dactylon* and other grass species (Child, 1968). Following the exclusion of human settlement and livestock from the national park, this shift in the composition of the floodplain was temporarily arrested (Simpson, 1975). But with increasing numbers of elephant, buffalo and other herbivores, such as impala, and their resultant impact on riverine habitats in recent decades, the trend towards dominance by seral grass species has continued. Grazing and trampling now maintain the dominance of shorter grasses on the floodplain, which favours buffalo over elephant.

Buffalo as a controller of elephant grazing

In summary, we found no interference competition between buffalo and elephant on the same patch and day. Over a period of a week or more, however, elephant reacted negatively to prior high buffalo grazing pressure. Buffalo, in contrast, reacted positively to elephant grazing pressure. Although elephants did graze commonly on the floodplain, especially when they came to drink at the river, they appeared to be unable to compete with buffalo. As such, buffalo are controlling elephant foraging choices in the dry season, whereas elephant grazing facilitates later grazing by buffalo.

The increasing dominance of short, stoloniferous seral grass species such as *Cynodon dactylon* on the Chobe floodplain further favours buffalo grazing at the expense of elephant (Child, 1968, Simpson, 1975, 1978). In the dry season, the only growing grass is restricted to the floodplain and so is limited in both areal extent and quantity. Elephants require more food per individual than ungulates due to their faster passage rates and much larger body size (Owen-Smith, 1988); in order to fulfil dietary requirements, time limited elephants therefore must switch to a greater proportion of browse during the dry season. The dietary segregation between elephant increases to its maximum at this time of the year. In this way, the buffalos are controlling the elephant foraging choices in the dry season, while elephant grazing facilitates later buffalo grazing.

This system also appears to be an example of a persistent change in state caused by human disturbances to the ecosystem: long grass, with elephants up to the early 20th century, shifted to short grass as a result of a marked reduction in elephant densities caused by hunting, and subsequently maintained by buffalo despite the recovery in elephant numbers (A.R.E. Sinclair, pers. comm.). The longer grass itself might have developed as a result of the 19th century rinderpest epidemic, which reduced ungulates to extremely low levels while not affecting elephants.

There is one caveat to our conclusions. Elephant numbers on Sedudu Island during the course of this study were not high relative to the overall elephant population. In years with exceptionally severe or prolonged dry seasons elephants could possibly deplete their browse resources, forcing them to forage more on the floodplain at densities which could place greater competitive pressures on buffalo.

References

Beekman, J.H. & Prins, H.H.T. (1989) Feeding strategies of sedentary large herbivores in East Africa with emphasis on the African buffalo, *Syncerus caffer*. *African Journal of Ecology* 27, 129–147.

Bell, R.H.V. (1970) The use of the herb layer by grazing ungulates in the Serengeti. In: Watson, A. (ed.) *Animal Populations in Relation to their Food Source*. Blackwell Scientific Publications, Oxford, pp. 111–124.

Bell, R.H.V. (1971) A grazing ecosystem in the Serengeti. *Scientific American* 225, 86–93.

Bell, R.H.V. (1986) Carrying capacity and off-take quotas. In: Bell, R.H.V. & McShane-Caluzi, E. (eds.) *Conservation and Wildlife Management in Africa: The Proceedings of a Workshop Organized by the U.S. Peace Corps at Kasungu National Park, Malawi*. U.S. Peace Corps, Washington, DC, pp. 145–181.

Belsky, A.J. (1984) Role of small browsing mammals in preventing woodland regeneration in Serengeti National Park, Tanzania. *African Journal of Ecology* 22, 271–279.

Ben-Shahar, R. (1993) Patterns of elephant damage to vegetation in northern Botswana. *Biological Conservation* 65, 249–256.

Ben-Shahar, R. (1998) Elephant density and impact on Kalahari woodland habitats. *Transactions of the Royal Society of South Africa* 53, 149–155.

Botkin, D.B., Mellilo, J.M. & Wu, L.S.-Y. (1981) How ecosystem processes are linked to large mammal population dynamics. In: Fowler, C.W. & Smith, T.D. (eds.) *Dynamics of Large Mammal Populations*. John Wiley & Sons, New York, pp. 373–387.

Child, G. (1968) *An Ecological Survey of North-eastern Botswana*. Food and Agriculture Organization of the United Nations, Rome.

Coe, M.J., Cumming, D.H.M. & Phillipson, J. (1976) Biomass and production of large African herbivores in relation to rainfall and primary production. *Oecologia* 22, 341–354.

Colman, J.E., Tsegaye, D., Pedersen, C., Eidesen, R., Arntsen, H., Holand, O., Mann, A., Reimers, E. & Moe, S.R. (2012) Behavioral interference between sympatric reindeer and domesticated sheep in Norway. *Rangeland Ecology and Management* 65, 299–308.

Cromsigt, J.P.G.M., Prins, H.H.T. & Olff, H. (2009) Habitat heterogeneity as a driver of ungulate diversity and distribution patterns: interaction of body mass and digestive strategy. *Diversity and Distributions* 15, 513–522.

Cumming, D.H.M., Fenton, M.B., Rautenbach, I.L., Taylor, R.D., Cumming, G.S., Cumming, M.S., Dunlop, J.M., Ford, G.S., Hovorka, M.D., Johnston, D.S., Kalcounis, M.C., Mahlanga,

Z. & Portfors, C.V. (1997) Elephants, woodlands and biodiversity in southern Africa. *South African Journal of Science* 93, 231–236.

de Boer, W.F. & Prins, H.H.T. (1990) Large herbivores that strive mightily but eat and drink as friends. *Oecologia* 82, 264–274.

du Toit, J.T. (1990) Feeding height stratification among African browsing ruminants. *African Journal of Ecology* 28, 55–61.

du Toit, J.T. & Cumming, D.H.M. (1999) Functional significance of ungulate diversity in African savannas and the ecological implications of the spread of pastoralism. *Biodiversity and Conservation* 8, 1643–1661.

du Toit, J.T. & Owen-Smith, N. (1989) Body size, population metabolism, and habitat specialisation among large African herbivores. *The American Naturalist* 133, 736–740.

Dublin, H.T., Sinclair, A.R.E. & McGlade, J. (1990) Elephants and fire as causes of multiple stable states in the Serengeti-Mara woodlands. *Journal of Animal Ecology* 59, 1147–1164.

DWNP – Department of Wildlife and National Parks, Botswana (2000) *Dry Season Aerial Survey.* Division of Wildlife Research, Gaborone, Botswana.

East, R. (1984) Rainfall, soil nutrient status and biomass of large African savanna mammals. *African Journal of Ecology* 22, 245–270.

Estes, R.D. (1991) *The Behavior Guide to African Mammals: Including Hoofed Mammals, Carnivores, Primates.* University of California Press, Berkeley, USA.

Faas, C.J. & Weckerly, F.W. (2010) Habitat interference by axis deer on white-tailed deer. *Journal of Wildlife Management* 74, 698–706.

Farnsworth, K.D., Focardi, S. & Beecham, J.A. (2002) Grassland–herbivore interactions: how do grazers coexist? *The American Naturalist* 159, 24–39.

Fritz, H. & Duncan, P. (1994) On the carrying capacity for large ungulates of African savanna ecosystems. *Proceedings of the Royal Society of London B* 256, 77–82.

Fritz, H. (1997) Low ungulate biomass in west African savannas: primary production or missing megaherbivores or large predator species. *Ecography* 20, 417–421.

Fritz, H., Duncan, P., Gordon, I.J. & Illius A.W. (2002) Megaherbivores influence trophic guilds structure in African ungulate communities. *Oecologia* 131, 620–625.

Gibson, D.C., Craig, C.G. & Masogo, R.M. (1998) Trends of the elephant population in northern Botswana from aerial survey data. *Pachyderm* 25, 14–27.

Gordon, I.J. & Illius, A.W. (1989) Resource partitioning by ungulates on the Isle of Rhum. *Oecologia* 79, 383–389.

Gordon, I.J., Illius, A.W. & Milne, J.D. (1996) Sources of variation in the foraging efficiency of grazing ruminants. *Functional Ecology* 10, 219–226.

Grange, S. & Duncan, P. (2006) Bottom-up and top-down processes in African ungulate communities: resources and predation acting on the relative abundance of zebra and grazing bovids. *Ecography* 29, 899–907.

Guldemond R. & van Aarde R. (2008) A meta-analysis of the impact of African elephants on savanna vegetation. *Journal of Wildlife Management* 72, 892–899.

Halley, D.J. & Minagawa, M. (2005) African buffalo diet in a woodland and bush dominated biome, as determined by stable isotope analysis. *African Zoology* 40, 160–163.

Halley, D.J., Vandewalle, M., Mari, M. & Taolo, C. (2002) Herd-switching and long-distance dispersal in female African buffalo *Syncerus caffer. African Journal of Ecology* 40, 97–99.

Halley, D.J., Taolo, C. & Mari, M. (2003) Herd dynamics, movements, and habitat use of African buffalo *Syncerus caffer* at the Chobe riverfront, Botswana. Paper I in: Taolo, C. L. (2003). Population ecology, seasonal movement and habitat use of African Buffalo (*Syncerus caffer*) in Chobe National Park, Botswana. PhD Thesis, Norwegian University of Science and Technology, Trondheim, Norway.

Hofmann, R.R. (1989) Evolutionary steps of ecophysiological adaptation and diversification of ruminants: a comparative review of their digestive system. *Oecologia* 78, 443–457.

Hofmann, R.R. & Stewart, D.R.M. (1972) Grazer or browser: a classification based on the structure and feeding habits of East African ruminants. *Mammalia* 36, 226–240.

Illius, A.W. & Gordon, I.J. (1987) The allometry of food intake in grazing ruminants. *Journal of Animal Ecology* 56, 989–999.

Jarman, P.J. (1974) The social organisation of antelope in relation to their ecology. *Behaviour* 48, 215–267.

Kalamera, M.C. (1989) Observations on feeding preference of elephants in the Acacia tortilis woodland of Lake Manyara National Park, Tanzania. *African Journal of Ecology* 27, 325–334.

Kleynhans, E.J., Jolles, A.E., Bos, M.R.E. & Olff, H. (2011) Resource partitioning along multiple niche dimensions in differently sized African savanna grazers. *Oikos* 120, 591–600.

Leuthold, W. (1972) Home range, movements, and food of a buffalo herd in Tsavo National Park. *East African Wildlife Journal* 10, 237–243.

Macandza, V.A., Owen-Smith, N. & Cain, J.W. (2012a) Dynamic spatial partitioning and coexistence among tall grass grazers in an African savannah. *Oikos* 121, 891–898.

Macandza, V.A., Owen-Smith, N. & Cain, J.W. (2012b) Habitat and resource partitioning between abundant and relatively rare grazing ungulates. *Journal of Zoology* 287, 175–185.

McNaughton, S.J. (1976) Serengeti migratory wildebeest: facilitation of energy flow by grazing. *Science* 191, 92–94.

McNaughton, S.J., Ruess, W. & Seagle, S.W. (1988) Large mammals and process dynamics in African ecosystems. *Bioscience* 38, 794–800.

Melton, D.A. (1985) The status of elephants in northern Botswana. *Biological Conservation* 31, 317–334.

Mloszewski, J. (1983) *The Behaviour and Ecology of the African Buffalo.* Cambridge University Press, Cambridge, UK.

Moe, S.R., Rutina, L.P., Hytteborn, H. & du Toit, J.T. (2009) What controls woodland regeneration after elephants have killed the big trees? *Journal of Applied Ecology* 46, 223–230.

Mosugelo, D.K., Moe, S.R., Ringrose, S. & Nellemann, C. (2002) Vegetation changes during a 36 year period in northern Chobe National Park, Botswana. *African Journal of Ecology* 40, 232–240.

Omphile, U.J. (1997) Seasonal diet selection and quality of large savanna ungulates in Chobe National Park, Botswana. PhD Thesis, University of Wyoming, Laramie, Wyoming, USA.

Owen-Smith, N. (1990) Demography of a large herbivore, the greater kudu *Tragelaphus strepsiceros*, in relation to rainfall. *Journal of Animal Ecology* 59, 893–913.

Owen-Smith, R.N. (1988) *Megaherbivores: The Influence of Very Large Body Size on Ecology.* Cambridge University Press, Cambridge, UK.

Owen-Smith, R.N. & Mills, M.G.L. (2006) Manifold interactive influences on the population dynamics of a multispecies ungulate assemblage. *Ecological Monographs* 76, 73–92.

Prins, H.H.T. (1996) *Ecology and Behaviour of the African Buffalo: Social Inequality and Decision Making*. Chapman and Hall, London, UK.

Prins, H.H.T. & Douglas-Hamilton, I. (1990) Stability in a multi-species assemblage of large herbivores in East Africa. *Oecologia* 83, 392–400.

Prins, H.H.T. & van der Jeugd, H.P. (1993) Herbivore population crash and woodland structure in East Africa. *Journal of Ecology* 81, 305–314.

Putnam, R.J. (1996) Ungulates in temperate forest ecosystems: perspectives and recommendations for future research. *Forest Ecology and Management* 88, 205–214.

Rutina, L.P. (2004) Impalas in an elephant-impacted woodland: browser-driven dynamics of the Chobe riparian zone, Northern Botswana. PhD Thesis, Agricultural University of Norway, Ås, Norway.

Selous, F.C. (1881) *A Hunter's Wanderings in Africa*. Richard Bentley & Son, London, UK.

Sheppe, W. & Haas, P. (1976) Large mammal populations of the lower Chobe River, Botswana. *Mammalia* 40, 223–243.

Simpson, C.D. (1975) A detailed vegetation study on the Chobe river in north-east Botswana. *Kirkia* 10, 185–227.

Simpson, C.D. (1978) Effects of elephant and other wildlife on vegetation along the Chobe River, Botswana. *Occasional Papers, The Museum Texas Technical University* 48, 1–15.

Sinclair, A.R.E. (1977) *The African Buffalo: A Study of Resource Limitation of Populations*. University of Chicago Press, Chicago.

Sinclair, A.R.E. & Norton-Griffiths, M. (1982) Does competition or facilitation regulate migrant ungulate populations in the Serengeti? A test of hypotheses. *Oecologia* 53, 364–369.

Smallegange, I.M. & Brunsting, A.M.H. (2002) Food supply and demand, a simulation model of the functional response of grazing ruminants. *Ecological Modelling* 149, 179–192.

Taolo, C.L. (2003) Population ecology, seasonal movement and habitat use of African buffalo (*Syncerus caffer*) in Chobe National Park, Botswana. PhD Thesis, Norwegian University of Science and Technology, Trondheim, Norway.

Teren, G. & Owen-Smith, N. (2010) Elephants and riparian woodland changes in the Linyanti region, northern Botswana. *Pachyderm* 47, 18–25.

van de Koppel, J. & Prins, H.H.T. (1998) The importance of herbivore interactions for the dynamics of African savanna woodlands: an hypothesis. *Journal of Tropical Ecology* 14, 565–576.

Vesey-Fitzgerald, D.F. (1969) Utilisation of the habitat by buffalo in Lake Manyara National Park. *East African Wildlife Journal* 7, 131–145.

Part V
Responders

In the heterogeneity framework provided by the model by Pickett *et al.* (2003) and described in Chapter 1, responders are organisms or processes that respond to heterogeneity in the substrate as imposed by the agent. There may be direct and indirect responders, as when certain ungulate populations respond directly to elephant-driven changes in vegetation conditions, and when predator populations respond indirectly to such changes that influence the availability of ungulate prey types. For the heterogeneity framework to be meaningful, responders should be specified in relation to variation in a specified substrate at a specified spatial scale, as a responder in one context can be an agent or a controller in another. Soil, for example, is a controller of variation in vegetation at the landscape scale, but is also a responder to changes in the state of the vegetation at any one site. In this section, we discuss some responders in the context of vegetation heterogeneity as influenced by the elephant population along the Chobe River. Responders include populations of ungulates such as bushbuck, *Tragelaphus scriptus*, gallinaceous birds and lion, *Panthera leo* (Drawing by Marit Hjeljord).

Elephants and Savanna Woodland Ecosystems: A Study from Chobe National Park, Botswana,
First Edition. Edited by Christina Skarpe, Johan T. du Toit and Stein R. Moe.
© 2014 John Wiley & Sons, Ltd. Published 2014 by John Wiley & Sons, Ltd.

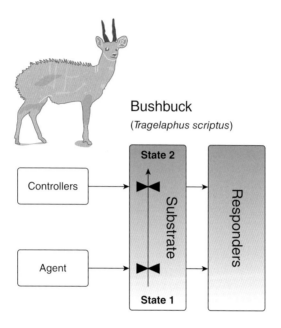

Bushbuck
(*Tragelaphus scriptus*)

References

Pickett, S.T.A., Cadenasso, M.L. & Benning, T.L. (2003) Biotic and abiotic variability as key determinants of savanna heterogeneity at multiple spatiotemporal scales. In: du Toit, J.T., Rogers, K.H. & Biggs, H.C. (eds.) *The Kruger Experience. Ecology and Management of Savanna Heterogeneity*. Island Press, Washington, Covelo, London, pp. 22–39.

12

Plant – Herbivore Interactions

Christina Skarpe[1], Roger Bergström[2], Shimane Makhabu[3],
Tuulikki Rooke[4], Håkan Hytteborn[5,6] and Kjell Danell[7]

[1] Faculty of Applied Ecology and Agricultural Sciences, Hedmark
University College, Norway
[2] Forestry Research Institute of Sweden, Uppsala Science Park, Sweden
[3] Department of Basic Sciences, Botswana College of Agriculture,
Botswana
[4] Research and Assessment Department, Swedish Environmental
Protection Agency, Sweden
[5] Department of Plant Ecology and Evolution, Evolutionary Biology
Centre, Uppsala University, Sweden
[6] Department of Biology, Norwegian University of Science and
Technology, Norway
[7] Department of Wildlife, Fish, and Environmental Studies, Swedish
University of Agricultural Sciences, Sweden

To a casual observer, the importance of large herbivores for ecosystem structure and dynamics can seem more obvious in African savannas than in many other ecosystems because of their high abundance, diversity and species richness of ungulates. African savannas have also had a long uninterrupted history of mammalian herbivory, leading to the evolution of plant traits adapted to herbivory and to reciprocal traits in herbivores. Mammalian herbivory is, together with fire, a key process in structuring savanna ecosystems, influencing primary production, morphology and chemistry of plants, vegetation structure and composition, nutrient cycling and distribution and fire regime.

Elephants and Savanna Woodland Ecosystems: A Study from Chobe National Park, Botswana,
First Edition. Edited by Christina Skarpe, Johan T. du Toit and Stein R. Moe.
© 2014 John Wiley & Sons, Ltd. Published 2014 by John Wiley & Sons, Ltd.

In nutrient-poor savannas such as those on Kalahari sand in the Chobe National Park, Botswana, elephants, *Loxodonta africana*, are a main agent creating spatial and temporal variation in the vegetation and ecosystems (Chapters 1 and 3–5). Within this framework, elephants and smaller herbivores interact with individual plants and plant populations, exploiting and modifying heterogeneity at many scales (Hobbs, 1996; Augustine and McNaughton, 1998; Danell *et al.*, 2003). Such interactions can lead to the development of feeding loops, which have been recorded from many savanna ecosystems (du Toit *et al.*, 1990, Makhabu and Skarpe, 2006; Fornara and du Toit, 2007). In such loops, the herbivores first respond to the characteristics of available plants in the habitat and feed selectively on plant species, plant individuals (genets) and plant parts (ramets). Second, the plants can respond to herbivore attack both quantitatively and qualitatively by adjusting features such as growth rates and resource allocation, patterns of tillering or branching or by changing nutritive chemistry and chemical or mechanical deterrents. Third, the herbivores respond to these changes by selecting for or against the affected plants, by increasing or decreasing food intake from the plant or by selecting or avoiding specific plant parts (Danell *et al.*, 2003; Figure 12.1).

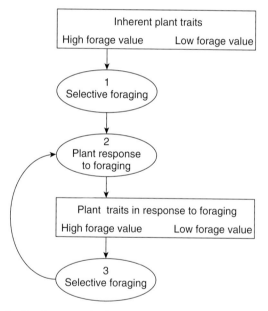

Figure 12.1 **The feeding loop. In the first step in the loop herbivores feed selectively in response to plant inherent traits. In the second step browsed plants can respond to being eaten, and, subsequently, in the third step selectively foraging herbivores respond to the traits induced by plant responses.**

A partially migratory system

In the Chobe ecosystem most water-dependent mammals select dry season home ranges within reach of permanent water in the Chobe River (Ben-Shahar, 1993; Omphile and Powell, 2002), a distance of 1 km or less in small species with limited mobility and high water requirements, and more than 10 km in highly mobile elephant bull groups (Stokke and du Toit, 2002; Chapter 7) and in species with low water requirements (Cain *et al.*, 2012; Chapter 13). As a result, areas adjacent to the Chobe River are heavily used so that by the end of the dry season, in September–October, grazing and browsing resources on the floodplains and in the shrublands close to the river are more or less depleted. Once the rains start and surface water becomes widely available in seasonal pans and vleis many large herbivores, including many elephants and African buffalo, *Syncerus caffer*, move away from the vicinity of the river into the woodlands on Kalahari sand where forage resources are less heavily exploited. Some woody species will have sprouted new leaves ahead of the rains. Other plants, especially grasses, produce a flush of young, relatively nutrient-rich forage in response to a pulse of nutrient mineralisation following the first rains (Scholes and Walker, 1993). This increase in nutrients and plant-available moisture is faster in the coarse sands than in the compact alluvial soils close to the river (Walker and Noy-Meir, 1982).

The dispersal of herbivores away from the river in the wet season is probably a major reason for the survival of heavily browsed or grazed plants and hence for the sustainability of the system. Browsing and grazing of vegetation on the alluvium close to the river is most intense when the plants are dormant, and then becomes relaxed during the growing period. Intermittent grazing in systems of migratory or highly mobile herbivores provides food plants with a recovery period, and could be one reason for the 'success' and abundance of many migratory herbivore species (Fryxell, 1995; Augustine and McNaughton, 1998). Another factor contributing to the sustainability of the floodplain and shrubland ecosystems on the Chobe River, with their seasonally high animal densities, might be the net import of nutrients via dung and urine from elephants foraging mainly in the woodland on sand but spending much time near the river drinking and socialising (Chapter 9).

Plant characteristics vary with resource availability and herbivory

Plants have evolved various morphological, chemical and life history traits in response to environmental conditions. Some of these traits reduce the loss of fitness in the presence of large herbivores either by reducing the loss of biomass to herbivores or by enhancing the rate of regrowth following its loss. The two strategies are often referred to as avoidance and tolerance, respectively (Briske, 1996; Strauss and Agrawal, 1999; Skarpe and Hester, 2008). Most plants employ both strategies and, although each involve traits that are believed to incur costs to a plant, the evidence of trade-off between the strategies is weak (Stamp, 2003; Agrawal, 2011). Nonetheless, species that have evolved in resource-poor environments tend to have high concentrations of carbon-based quantitative defence compounds – for example, various phenolic compounds – because growth is often limited by a shortage of nutrients, whereas photosynthesis provides carbon, which can be invested in defences with little opportunity cost (Herms and Mattson, 1992; Stamp, 2003). In contrast, fast-growing species, evolved in resource-rich environments, require considerable carbon for growth and therefore tend to have traits that either confer tolerance to herbivory or provide qualitative non-carbon-based defences, for example, alkaloids (Coley et al., 1985). Consequently, we would expect that, in Chobe, plant strategies related to herbivory will differ between species growing on nutrient-poor Kalahari sand, which are likely to have carbon based quantitative defences, and those living on richer alluvial soils, which we expect will rely on tolerance traits or invest in qualitative defences.

Although resource availability places different constraints on the evolution of plants' strategies in relation to mammalian herbivory, the strongest determinant is the history of herbivory, mammalian as well as by insects. Tolerance or avoidance traits only evolve in plant populations and become or remain common in vegetation in situations where the fitness cost of the trait is less than the cost of herbivory that it prevents (Vourc'h et al., 2001). Megaherbivores such as elephants seem fairly insensitive to morphological defences, for example, spines of different types. It has been suggested (Shrader et al., 2012) that elephants, because of their low mass-specific metabolic requirements and low digestive efficiency, can tolerate digestibility-reducing compounds such as tannins and phenolics better than smaller ruminating species can, whereas, as hind-gut fermenters, elephants might be less capable than ruminants in handling potentially toxic compounds such as alkaloids (Alexander, 1993).

We recorded morphological and chemical herbivory-related traits of 51 tree species from Chobe including chemistry of mature leaves: that is, the concentration of acid detergent fibre, carbon and nitrogen, all in percentages of dry mass; and total phenolics and tannin activity in milligram tannic acid (TA) per gram dry mass of leaf (Rooke, 2003). Sampling was done separately in the four wooded elephant habitat types – Baikiaea woodland, mixed woodland, Combretum shrubland and Capparis shrubland – recognised in the approximately 400 km^2 core study area (Chapter 5). In addition, we recorded available and utilised browse on all trees more than 0.5 m high in the 400 m^2 plots distributed in these elephant habitat types (Chapter 5), including the 58 plots in the woodlands and shrublands but excluding the 12 plots on the treeless floodplains. Available browse for elephants was quantified as the number of twigs with 8 mm diameter occurring within 2.5 m of the ground, meaning that they offered the opportunity for a herbivore to crop a bite with 8 mm diameter, the average bite diameter for elephants (Stokke and du Toit, 2000). In turn, available browse for ruminants was measured as the number of last-generation annual shoots, also up to 2.5 m above ground. Utilised browse was recorded as the number of bites. Bites by elephant and ungulates could generally be clearly separated.

As hypothesised above, we found tannin activity, concentration of total phenolics, and acid detergent fibre to be lower in shrubland plants on relatively nutrient-rich alluvial soil than in plants in the woodlands on nutrient-deficient sand. This pattern was consistent both when calculated as the average for all species in a habitat and when taking their relative frequencies into account (Figure 12.2). Nitrogen concentration in the tree leaves did not vary between species on nutrient-poor sand and the richer alluvium. We did not analyse for qualitative defences such as alkaloids or glycosides, but a couple of species on the sand were totally (*Erythrophleum africanum*) or strongly (*Baikiaea plurijuga*) avoided by mammalian browsers, including elephants. Both these species are reported to contain powerful qualitative defences (Jansson and Cronlund, 1976; Wink, 2000; Felpin and Lebreton, 2003). Such defences would not normally be expected in slow-growing trees on the nutrient-poor sand, but might have been evolutionarily and ecologically selected by a long history of elephant browsing. Qualitative defences are expected to play a more prominent role in the vegetation on the nutrient-rich alluvium, where, for example, the currently expanding shrub or scrambler *Capparis tomentosa* is rich in alkaloids (Ahmed and Adam, 1980). In addition, some forbs in the heavily grazed and browsed vegetation on alluvial soil, for example, *Tribulus terrestris*, *Eclipta alba* and *Heliotropium ovalifolium* and the floodplain grass *Vetiveria nigritana*, contain alkaloids that act as deterrents to large herbivores (Creeper *et al.*, 1999; Bremner *et al.*, 2003).

The frequency of spinescence, a morphological defence against browsing by mammalian herbivores, increased in the woody vegetation from almost 0% in Kalahari sand at the crest of the sand ridge to about 40% of the species and 60% of the trees in the shrublands on the alluvium (Rooke, 2003; Figure 12.2). In the heavily grazed and browsed vegetation on alluvium some forbs, for example, *Tribulus terrestris* and *Acanthospermum hispidum*, were also found to be spiny or have spiny fruits, as deterrents to herbivores or as means of dispersal or both.

In contrast to the high frequency and diversity of defensive traits in dicotyledonous plants grasses are typically tolerant of grazing, showing adaptive traits such as clonal growth, which allows vegetative reproduction, redistribution of resources between ramets and rapid replacement of biomass lost to herbivory. Along the vegetation gradient from the crest of the sand ridge to the floodplain, tall, tufted, obligatory seed-reproducing grasses, which are poorly adapted to heavy grazing (Díaz *et al.*, 2007), dominated the relatively lightly grazed upper, sandy slopes, whereas most grasses in the heavily grazed vegetation on alluvium were clonal and had prostrate growth forms (Chapters 4 and 5).

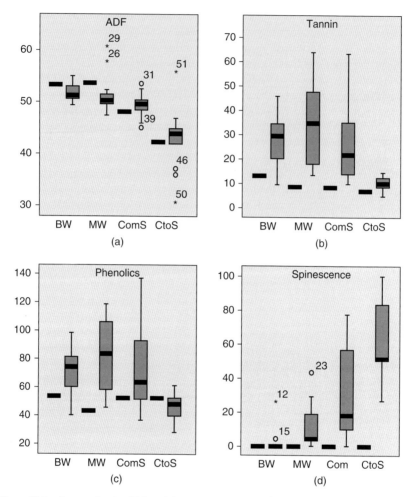

Figure 12.2   **Average levels of (a) acid detergent fibre (mg g⁻¹ dry matter); (b) tannin activity (mg TA g⁻¹ dry matter) (c) total phenolics (mg TA g⁻¹ dry matter) and (d) relative frequency of spinescence (%) in leaves of woody plants in *Baikiaea* woodlands (BW), mixed woodlands (MW), *Combretum* shrublands (ComS) and *Capparis tomentosa* shrublands (CtoS). Values are means of all woody plants in 15 (13 in CtoS) plots of 400 m² in each vegetation type. The graph shows median, quartiles and extreme values. Medians for the average of all woody species in the plots are included to the left for comparison. (Source: Rooke, 2003.)**

Large herbivores feed selectively in response to plant traits

In the first link in the feeding loop described above, selectively foraging herbivores respond to differences in nutritive value among plant species and between individual plants and plant parts (Figure 12.1). These choices are influenced in turn by the season, age and position of the shoot or leaf within a plant, all of which affect resource quality and availability. Selectivity also varies widely with animal species depending on size, digestive system and feeding guild (Chapter 13), with age, sex and reproductive status of the animal, all adding complexity to the system.

This complexity might be one reason for the weak relationships we found between specific traits of the woody species and the recorded intensity of foraging on twigs and shoots by elephants and ruminant browsers. Further, we only recorded feeding on twigs and shoots, whereas both ruminants (Cooper *et al.*, 1988; Makhabu, 2005) and elephants (Makhabu, 2005; Owen-Smith and Chafota, 2012) often browse on other plant parts, particularly leaves. Nevertheless, selectivity indices (proportion of a species in the total number of bites divided by proportion of the species in the total number of available shoots or twigs in the habitat) were calculated for 23 tree species contributing more than 1% of the total number of trees in the plots in the four habitats and for which data on herbivory-related traits had been recorded. In the *Baikiaea* woodland neither elephant nor ungulate selectivity indices showed any significant correlation with the concentrations of acid detergent fibre, nitrogen and total phenolics; tannin activity or with the nitrogen : total phenolics and nitrogen : tannin activity ratios. The only significant correlation was a strong negative relationship between ungulate selectivity indices and carbon concentration ($p = 0.003$). The equivalent correlation for elephant selectivity was not significant ($p = 0.70$). In the *Capparis* shrubland on alluvium ungulate selectivity indices were positively related to the concentration of acid detergent fibre ($p = 0.02$) and showed a weak negative regression on total phenolics ($p = 0.09$), and equally weak positive regressions on nitrogen : total phenolics and nitrogen : tannin activity ratios (p values about 0.06). In contrast, elephant selectivity indices in the vegetation in the *Capparis* shrubland showed no significant regression on any of the recorded plant traits, other than weak tendencies to positive relationships with total phenolics and tannin activity (p values about 0.1).

Herbivores select food based on the total food quality of plant material, whereas the correlations discussed above are with individual chemical compounds. The low significance of the relationships between animal selectivity and the recorded plant traits strongly suggests that some important determinants of selectivity had been omitted in the analyses. It is unlikely that elephants select for high concentration of phenolics or tannins. The regressions, if valid, are more likely a result of the concentration of total phenolics and tannin activity being negatively correlated with one or more powerful deterrents that we did not investigate. *Combretum elaeagnoides* and *Combretum apiculatum*, the species contributing most to elephant browse in our area, as measured by the total number of bites or breaks in the plots across all habitats, had high tannin activity (63 and 120 mg TA g^{-1} respectively) and high concentrations of phenolics (145 and 172 mg TA g^{-1} respectively). Species in *Baikiaea* woodland that were highly selected by elephants, including *Bauhinia petersiana* and *Baphia massaiensis*, and species that were greatly avoided, such as *Erythrophleum africanum*, *Baikiaea plurijuga* and *Combretum mossambicense* had low tannin activity (1–14 mg TA g^{-1}) and low concentration of total phenolics (25–58 mg TA g^{-1}). Some of the avoided species, including *Erythrophleum africanum* and *Baikiaea plurijuga*, are known to contain powerful alkaloids (Jansson and Cronlund, 1976; Ahmed and Adam, 1980; Wink, 2000; Felpin and Lebreton, 2003). Our results, like those of Owen-Smith and Chafota (2012), therefore suggest that elephants do not prefer browse species with high concentrations of phenolics, although we found that such species constituted a large proportion of their diet, seemingly because the preferred species provided insufficient dietary intake. Our results are also consistent with the notion put forward by Shrader *et al.* (2012) that elephants tolerate high concentrations of digestibility-reducing compounds and might be more sensitive to qualitative, toxic defences. Ruminant browsers to some extent have evolved the means to handle such compounds (Alexander, 1993). This is obvious in the shrublands along the Chobe River where *Capparis tomentosa* is the species most browsed by giraffe, *Giraffa camelopardalis*, and is also much eaten by greater kudu, *Tragelaphus strepsiceros* and impala, *Aepyceros melampus* (Makhabu, 2005), despite reportedly containing a toxin that is potent enough to kill domestic goats and calves that feed on the leaves (Ahmed and Adam, 1980). The species is avoided by elephants, however.

Browsing elephants and ruminants largely select different tree species in the shrublands on the Chobe alluvium (Makhabu, 2005), but often browse the same tree species in the woodlands on the sand. Our records of available and utilised browse for both elephants and the combined ruminant browsing guild show that two of the tree species that have increased most with the elephant-induced vegetation change on the alluvium (Chapter 4), *Capparis tomentosa* and *Combretum mossambicense*, together constituted 75–80% of available browse for both elephants and ruminants and contributed 86% of ruminant bites but only 10% of elephant bites and breaks. Of the common increasing species, only *Flueggea virosa* was much browsed by both elephants and ruminants (Makhabu, 2005), whereas *Combretum elaeagnoides* was the species most eaten by elephants on alluvial- and mixed soil. On the sand, the difference in species selection between ruminants and elephants was less pronounced than on the alluvium. The dominant woodland species, *Baikiaea plurijuga*, constituted about 50% of available browse in these woodlands, but was avoided by both elephant and ruminants. Browsers instead targeted smaller species, some of which might have increased in abundance through the impact of elephants, fire or logging, individually or in some combination (Chapter 4). The woodland species that contributed most to the browse for both elephants and ruminants included *Markhamia zanzibarica* and *Combretum apiculatum*. The shrubs *Baphia massaiensis* and *Bauhinia petersiana* were browsed by both ruminants and elephants, with elephants, in particular, evidently preferring these species, which ranked third and first, respectively, of the 11 species included in the analysis for the *Baikiaea* woodland.

The discrepancy in elephant selectivity indices between Table 7.1 and our results can have a number of reasons. Stokke and du Toit in Chapter 7 and Stokke (1999) calculated selection indices with trees as units, whereas our unit was the available twigs. Thus, small shrubs such as *Baphia massaiensis* and *Bauhinia petersiana* contribute few available twigs compared with, say, a *Capparis tomentosa* shrub often with hundreds of twigs, but which still counts as one tree. Furthermore, many of the highly preferred species in Stokke and du Toit's study are rare in the vegetation and were probably sought out by the elephants. They were either not encountered in our plots or constituted less than 1% of the trees there, and so were not included in our analysis. An inverse relationship between the relative frequency of a species and its utilisation by browsers has often been recorded (Owen-Smith and Novellie, 1982; Skarpe *et al.*, 2000).

Spinescence does not deter browsers in the same way as many secondary compounds do, but it often decreases intake rates through a combination of increasing handling time, reducing bite size and slowing the biting rate (Cooper and Owen-Smith, 1986; Gowda, 1996; Skarpe *et al.*, 2012). Both long straight spines and short hooked spines reduce intake rate in medium-sized browsers, but the functional mechanism is different (Gowda, 1996; Skarpe *et al.*, 2012). The smallest browsers, picking leaves between the spines, and the largest, such as elephant, are largely unresponsive to spinescence. In the shrublands on the Chobe alluvium, 4 of the 10 woody species that were most commonly browsed by ruminants were spinescent. Moreover, neither browsing pressure (% utilisation) nor preference was related to spinescence, contrary to findings elsewhere in Botswana (Skarpe *et al.*, 2000). The spinescent and evergreen *Capparis tomentosa* was one of the most intensively browsed species. By the end of the dry season, 80–90% of available leaves had been browsed on some *Capparis* shrubs in the most heavily used section of the shrubland, primarily by impala picking leaves from between the sharp, hooked spines.

Grazers select for swards of optimal height and with high bulk density and leaf : stem ratio (Wilmshurst *et al.*, 1999; Fryxell *et al.*, 2005). In the dry season in Chobe, low swards, cropped to less than 20 cm and with high leaf : stem ratios, were found primarily on the floodplains, but also in a few other alluvial or lacustrine flats south of our main study area. These floodplains were grazed intensively by small and large grazers including puku, *Kobus vardonii*, red lechwe, *Kobus leche*, buffalo and hippopotamus, *Hippopotamus amphibius*, as well as impala. Some of the large grazing species, elephants and buffalo, also grazed the tall, tufted grasses in the woodlands on sand, alongside specialist tall-grass grazers such as sable antelope, *Hippotragus niger*, roan antelope, *Hippotragus equinus* and tsessebe, *Damaliscus lunatus* (Chapter 13).

Plant responses to large herbivores

Plant responses to herbivore foraging form the second link in the feeding loop (Figure 12.1). In savannas, such as Chobe National Park, overall plant responses are complex, not least because of the many species of plants and large herbivores with different feeding modes. Furthermore, the plants are subject to seasonal differences in the kinds and intensity of herbivory, for instance during dry and wet seasons, and vary in their capacity to respond. For example, compare an impala nibbling a grass shoot or a tree leaf, a greater kudu biting a top or lateral shoot, and an elephant breaking or felling a tree. Clearly, both the impact and the plant's response will be different.

Plant responses to large herbivore feeding occur at both the population level and the level of the individual plant and plant parts (Danell *et al.*, 2003). At the level of an individual plant many morphological characteristics that influence subsequent herbivore selectivity and feeding, such as plant architecture, allocation of plant biomass above and below ground, shoot and leaf sizes and types, as well as spinescence, can be altered after a plant has responded to herbivory (Rooke *et al.*, 2004; Makhabu *et al.*, 2006; Hrabar *et al.*, 2009). Makhabu *et al.* (2006) found tree height to be reduced and shoot distribution in the canopies changed following elephant browsing of 6 tree species in the shrubland. An experiment with simulated elephant browsing of 5 tree species in the study area, involving clipping shoots at 8 mm diameter and cutting saplings at 0.5 m above ground, resulted in fewer but larger canopy shoots and an increase in both the number and size of basal shoots on the 'browsed' trees compared to unbrowsed controls (Skarpe and Makhabu, unpublished). Similar results have been shown by Bergström *et al.* (2000) and Bond and Midgley (2001).

Plant responses usually include changes in the chemistry of the tissues of the plants, although this was not studied in the Botswana-Norway Institutional Co-operation and Capacity Building (BONIC) project. Nutrients, fibre and secondary compounds have all been shown to change in plants responding to herbivory, although the direction and extent of change varies among species, plant size, age and growth rate, site productivity, seasons and the intensity and mode of herbivory (Rooke and Bergström, 2007; Wessels *et al.*, 2007; Scogings *et al.*, 2011). It has been hypothesised that severe pruning by large herbivores, for example, by elephants, can led to carbon shortage in the plant and, hence, to reduced allocation of carbon to defences, whereas defoliation by a small herbivore may instead trigger induced defences (Scogings *et al.*, 2011). Some plant responses to herbivory are rapid, appearing in hours or days; others take longer to emerge and need an intervening growth period to develop. Simulated browsing of five deciduous species in Chobe showed that plant-level responses were still detectable by elephants 3 years after treatment (Makhabu and Skarpe, 2006). Similarly long-lasting individual plant responses have been noted in boreal ecosystems (Danell *et al.*, 1994, 2003).

At the plant population level, large herbivores can affect dispersal, plant recruitment and mortality (Hester *et al.*, 2006; Midgley *et al.*, 2010). Herbivory can also affect reproductive tissues, with changes in seed production or seed size (Bergström and Danell, 1987; Midgley *et al.*, 2010) or to the allocation of resources to growth rather than to reproduction (Skarpe and Hester, 2008). The importance of all these impacts for plant population dynamics is still poorly known, however. Herbivory on seedlings, elephant pushing over trees or digging for roots and exceptionally high grazing or browsing pressures can all contribute to plant mortality (Midgley *et al.*, 2010). There is often an interplay between many different factors; for example, herbivores, fire, frost and landscape characteristics variously affecting patterns of tree mortality (Vanak *et al.*, 2012). Often herbivory does not kill the plant, but change the competitive hierarchy between species, potentially leading to changes in vegetation composition. For example, because elephants are relatively insensitive to quantitative defences, heavy browsing by elephants would be expected to lead to a competitive advantage to plant species with tolerance rather than defensive traits, and with qualitative rather than quantitative defences.

Herbivore responses to plant responses

The third link in the feeding loop is the response of herbivores to previously attacked plants (Figure 12.1). Foraging herbivores exploit the increased heterogeneity within and between plants caused by herbivory (Danell *et al.*, 2003), with previously grazed or browsed plants or particular plant parts sometimes experiencing either increased or reduced probability of being eaten compared with previously unaffected plants. In widely different ecosystems, previously browsed trees experience higher probability of being browsed compared with previously unbrowsed conspecifics (Löyttyniemi, 1985; Danell *et al.*, 1985; du Toit *et al.*, 1990; Skarpe *et al.*, 2000; Makhabu *et al.*, 2006). Whereas a browser's initial selection of a particular plant might reflect pre-existing differences between plants within a species, subsequent browsing depends more on herbivory-induced changes in plant palatability. This has been shown experimentally by the response of browsers to plants subjected to different randomly allocated levels of simulated browsing (Bergström *et al.*, 2000; Makhabu and Skarpe, 2006). Among woody plants in the Chobe ecosystem, the long-term effects of accumulated browsing, particularly alteration of tree morphology by elephants, were strong predictors of browsing by elephants and ruminants. In five out of six studied tree species, both impala and kudu selectively browsed trees with moderate or high accumulated elephant impact. Previously not browsed trees were selected only in *Capparis tomentosa*, a species seldom eaten by elephants (Makhabu *et al.*, 2006).

This shows that in many species plant responses to browsing or grazing trigger further herbivory, creating a positive feedback loop. In nutrient-rich savanna, repeated grazing of patches with grazing-tolerant grasses has been found to generate short-cropped swards with high aboveground productivity, biomass density, leaf nitrogen concentrations, digestibility, leaf-to-stem ratio and increasing dominance of grazing-tolerant species (Bell, 1971; McNaughton, 1983, 1984; Cromsigt and Olff, 2008). McNaughton (1984) referred to such patches as 'grazing lawns' and described them as resulting from a co-evolution between gregarious, typically migratory, grazing ungulates and grazing-tolerant grasses. In the Chobe ecosystem extensive grazing lawns exist on alluvial soil, mainly on the floodplains, where they probably are created and maintained primarily by hippopotamus and buffalo, and are used by many grazing species. These grazing lawns are dominated by *Cynodon dactylon*, which has expanded greatly with the increase in elephants and grazing ungulates (Chapter 4). *Cynodon dactylon* is also an important component of grazing lawns in the Serengeti, and on the Chobe floodplains shows many of the characteristics described by McNaughton (1984) for grazing-lawn grasses generally (Mathisen, 2005). The enhanced rate of nutrient cycling in grazing lawns, maintained by N-rich urine and faeces from the grazing animals (Ruess *et al.*, 1983; McNaughton *et al.*, 1997), is further enhanced on the Chobe floodplains by the net import of nutrients from elephant dung and urine (Chapter 9). The grazing lawns in Chobe seem to be restricted to the alluvial soils. Kalahari sand is probably too nutrient-poor and too sensitive to heavy trampling by animals for grazing lawns to form, even though both buffalo and elephants graze extensively there in the wet season.

The repeated rebrowsing of certain trees form feedback loops that have been compared to grazing lawns, and described as 'browsing lawns' (McNaughton, 1984; du Toit *et al.*, 1990; Makhabu *et al.*, 2006; Fornara and du Toit, 2007). McNaughton's (1984) concept of the grazing lawn includes a (co-)evolutionary aspect, implying fitness benefits for both plants and herbivores. Whereas the benefit for the herbivores is obvious, a fitness gain for the plant from being eaten is more difficult to envisage (Belsky, 1986). Cromsigt and Kuijper (2011) interpreted the gain for the plant as an increase in relative fitness, as reflected in an increase in the proportion of the browsed species in the vegetation. In the Chobe environment this seems to apply to some intensively browsed woody species, for example, *Combretum mossambicense* and *Combretum elaeagnoides*, which are becoming proportionately more common in the heavily browsed woody vegetation on the alluvium or mixed soil along the Chobe River (Chapter 4), forming 'browsing lawns' *sensu* Cromsigt and Kuijper (2011). For *Combretum mossambicense*, however, its increased abundance could be related both to its low palatability to elephants and to its tolerance of browsing by ruminants. *Combretum elaeagnoides*, in contrast, is browsed mainly by elephants. It is common on alluvium and, in particular, on mixed soils, in many places forming dense stands of plants about 1 m tall, which are browsed down in the dry season and resprout in the wet. The pronounced development of tolerance traits in browsing lawn species requires relatively high resource availability (Cromsigt and Kuijper, 2011), so whether browsing lawns in this sense exist on nutrient-poor sand is doubtful. *Colophospermum mopane* might be a candidate species, however, forming almost monospecific stands, repeatedly browsed by elephants, on compact sand in some areas south of our main study area. Mopane resprouts vigorously following browsing, and elephants preferentially feed on this regrowth (Smallie and O'Connor, 2000). Whether these intensively browsed patches are sustainable and can spread forming a browsing lawn *sensu* Cromsigt and Kuijper (2011) we do not know.

Interactions between elephants, plants and ungulates

Sharing a common browsing resource can lead to resource partitioning based on differences in animal body-size and digestive system, and subsequent resource requirements, as described by the Jarman-Bell principle (Bell, 1971, 1982; Jarman, 1974), or to scramble competition (Illius and Gordon, 1987). In the heavily browsed shrublands on relatively nutrient-rich soil along the Chobe River partial resource partitioning in the dry season is prominent among browsers. Makhabu (2005) demonstrated that elephants largely browsed on different species from those fed on by the ruminant browsers – giraffe, kudu and impala – but they browsed in the same height range as impala and kudu. The three ruminant browsers fed largely on the same tree species, but at different heights, reflecting differences in body size and the existence of scramble competition (Illius and Gordon, 1987; du Toit, 1990; Woolnough and du Toit, 2001; Makhabu, 2005).

Food-mediated interactions between large herbivores, either competition or facilitation, can drive differences in resource use at a range of scales, from height stratification within a tree to habitat segregation. Van de Koppel and Prins (1998) suggest that the interaction between large and small herbivores switches between facilitation at high plant biomass, where large herbivores facilitate the smaller ones, to competition at low plant biomass when small herbivores can outcompete the larger ones, provided nutritive quality is high enough. The two situations might apply to ecosystems with inherently different plant biomass, or to the same system, if plant biomass change over time. Along the Chobe River the strip of raised, nutrient-rich alluvial soil was covered by tree savannah or wood land in the first half of the 20th century (Simpson, 1974). With elephants debarking and killing the old trees, and seedling-eating impala preventing the regeneration of the trees (Moe *et al.*, 2009; Chapter 10), the vegetation has changed to fairly open (about 15% woody cover) shrub vegetation (Chapters 4 and 5). The transition from savanna-woodland to shrub vegetation implies a decrease in above-ground plant biomass of 50–90% (Baccini *et al.*, 2008; Mitchard *et al.*, 2011), potentially promoting a change from predominant facilitation to competition between elephants and ungulates, as suggested by van de Koppel and Prins (1998). We have shown that buffalo on the Chobe floodplain competed with grazing elephants (Taolo, 2003; Chapter 11). Furthermore, the probably elephant-induced alteration of woody species composition on the alluvial flat (Chapters 4 and 10) is unfavourable for elephant foraging, but seemingly provides acceptable forage for ruminant browsers. At present, elephants use the browse on the alluvium to a limited extent (Chapter 13), and do more than 90% of their browsing, measured as the number of branches bitten or broken per unit area, in the woodlands on the Kalahari sand. The nutritive quality of this vegetation is largely too low for ruminant browsers, which subsequently do not control the recruitment of trees in the same way that they do on the richer alluvium.

Thus, the difference in aboveground plant biomass and in historic vegetation dynamics between the alluvium and the sand in the Chobe ecosystem seems to be caused by interactions between elephants and ruminant herbivores, controlled by the difference in soil resources (Wallgren, 2008; Chapters 4, 5, 9, 10 and 13). By extension, this suggests that differences in large herbivore communities and subsequent herbivory regimes might be a key factor in accounting also for the difference in plant biomass generally recorded between rich and poor savannas receiving comparable average precipitation (Bell, 1982; Wallgren, 2008; O'Kane *et al.*, 2011).

References

Agrawal, A.A. (2011) Current trends in the evolutionary ecology of plant defence. *Functional Ecology* 25, 420–432.

Ahmed, O.M.M. & Adam, S.E.I. (1980) The toxicity of *Capparis tomentosa* in goats. *Journal of Comparative Pathology* 90, 187–195.

Alexander, R.M. (1993) The relative merits of foregut and hindgut fermentation. *Journal of Zoology* 231, 391–401.

Augustine, D.J. & McNaughton, S.J. (1998) Ungulate effects on the functional species composition of plant communities: herbivore selectivity and plant tolerance. *Journal of Wildlife Management* 62, 1165–1183.

Baccini, A., Laporte, N., Goetz, S.J., Sun, M. & Dong, H. (2008) A first map of tropical Africa's above-ground biomass derived from satellite imagery. *Environmental Research Letters* 3, 45011. doi:10.1088/1748-9326/3/4/045011.

Bell, R.H.V. (1971) A grazing ecosystem in the Serengeti. *Scientific American* 225, 86–93.

Bell, R.H.V. (1982) The effect of soil nutrient availability on community structure in African ecosystems. In: Huntley, B.J. & Walker, B.H. (eds.) *Ecology of Tropical Savannas.* Springer-Verlag, Berlin, pp. 193–216.

Belsky, A.J. (1986) Does herbivory benefit plants? A review of the evidence. *The American Naturalist* 127, 870–892.

Ben-Shahar, R. (1993) Patterns of elephant damage to vegetation in northern Botswana. *Biological Conservation* 65, 249–256.

Bergström, R. & Danell, K. (1987) Effects of simulated browsing by moose on morphology and biomass of two birch species. *Journal of Ecology* 75, 533–544.

Bergström, R., Skarpe, C. & Danell, K. (2000) Plant responses and herbivory following simulated browsing and stem cutting of *Combretum apiculatum*. *Journal of Vegetation Science* 11, 409–414.

Bond, W.J. & Midgley, J.J. (2001) Ecology of sprouting in woody plants: the persistence niche. *Trends in Ecology and Evolution* 16, 45–51.

Bremner, J., Sengpracha, W., Southwell, I., Bourke, C., Skelton, B. & White, A. (2003) The alkaloids of *Tribulus terrestris*: a revised structure for the alkaloid tribulusterine. *Acta Horticulturae* 677, 11–17.

Briske, D.D. (1996) Strategies of plant survival in grazed systems: a functional interpretation. In: Hodgson, J.I. & Wallingford, A.W. (eds.) *The Ecology and Management of Grazing Ecosystems*. CAB International, Wallingford, pp. 37–67.

Cain, J.W., Owen-Smith, N. & Macandza, V.A. (2012) The cost of drinking: comparative water dependency of sable antelope and zebra. *Journal of Zoology* 286, 58–67.

Coley, P.D., Bryant, J.P. & Chapin, F.S. III, (1985) Resource availability and plant antiherbivore defense. *Science* 230, 895–899.

Cooper, S.M. & Owen-Smith, N. (1986) Effects of plant spinescence on large mammalian herbivores. *Oecologica* 68, 446–455.

Cooper, S.M., Owen-Smith, N. & Bryant, J.P. (1988) Foliage acceptability to browsing ruminants in relation to seasonal changes in leaf chemistry of woody plants in a South African savanna. *Oecologia* 75, 336–342.

Creeper, J.H., Mitchell, A.A., Jubb, T.F. & Colegate, S.M. (1999) Pyrrolizidine alkaloid poisoning of horses grazing a native heliotrope (*Heliotropium ovalifolium*). *Australian Veterinary Journal* 77, 401–402.

Cromsigt, J.P.G.M. & Kuijper, D.P.J. (2011) Revisiting the browsing lawn concept: evolutionary interactions or pruning herbivores? *Perspectives in Plant Ecology, Evolution and Systematics* 13, 207–215.

Cromsigt, J.P.G.M. & Olff, H. (2008) Dynamics of grazing lawn formation: an experimental test of the role of scale-dependent processes. *Oikos* 117, 1444–1452.

Danell, K., Huss-Danell, K. & Bergström, R. (1985) Interactions between browsing moose and two species of birch in Sweden. *Ecology* 66, 1867–1878.

Danell, K., Bergström, R. & Edenius, L. (1994) Effects of large mammalian browsers on architecture, biomass, and nutrients of woody plants. *Journal of Mammalogy* 75, 833–844.

Danell, K., Bergström, R., Edenius, L. & Ericsson, G. (2003) Ungulates as drivers of tree population dynamics at module and genet levels. *Forest Ecology and Management* 181, 67–76.

Díaz, S., Lavorel, S., McIntyre, S., Falczuk, V., Casanoves, F., Milchunas, D.G., Skarpe, C., Rusch, G., Sternberg, M., Noy-Meir, I., Landsberg, J., Zhang, W., Clark, H. & Campbell, B.D. (2007) Plant trait responses to grazing – a global synthesis. *Global Change Biology* 13, 313–341.

du Toit, J.T. (1990) Feeding height stratification among African browsing ruminants. *African Journal of Ecology* 28, 55–61.

du Toit, J.T., Bryant, J.P. & Frisby, K. (1990) Regrowth and palatability of Acacia shoots following pruning by African savanna browsers. *Ecology* 71, 149–154.

Felpin, F.X. & Lebreton, J. (2003) Recent advances in the total synthesis of piperidine and pyrrolidine natural alkaloids with ring-closing metathesis as a key step. *European Journal of Organic Chemistry* 2003, 3693–3712.

Fornara, D.A. & du Toit, J.T. (2007) Browsing lawns? Responses of *Acacia nigrescens* to ungulate browsing in an African savanna. *Ecology* 88, 200–209.

Fryxell, J.M. (1995) Aggregation and migration by grazing ungulates in relation to resources and predators. In: Sinclair, A.R.E. & Arcese, P. (eds.) *Serengeti II: Dynamics, Management, and Conservation of an Ecosystem.* University of Chicago Press, Chicago, pp. 257–273.

Fryxell, J.M., Wilmshurst, J.F., Sinclair, A.R.E., Haydon, D.T., Holt, R.D. & Abrams, P.A. (2005) Landscape scale, heterogeneity, and the viability of Serengeti grazers. *Ecology Letters* 8, 328–335.

Gowda, J.H. (1996) Spines of *Acacia tortilis*: what do they defend and how? *Oikos* 77, 279–284.

Herms, D. & Mattson, W.J. (1992) The dilemma of plants: to grow or defend. *The Quarterly Review of Biology* 67, 283–335.

Hester, A.J., Bergman, M., Iason, G.R. & Moen, J. (2006) Impacts of large herbivores on plant community structure and dynamics. In: Danell, K., Bergström, R., Duncan, P. & Pastor, J. (eds.) *Large Herbivore Ecology, Ecosystem Dynamics and Conservation.* Cambridge University Press, Cambridge, UK, pp. 97–141.

Hobbs, N.T. (1996) Modification of ecosystems by ungulates. *Journal of Wildlife Management* 60, 695–713.

Hrabar, H., Hattas, D. & du Toit, J.T. (2009) Differential effects of defoliation by mopane caterpillars and pruning by African elephants on the regrowth of *Colophospermum mopane* foliage. *Journal of Tropical Ecology* 25, 301–309.

Illius, A.W. & Gordon, I.J. (1987) The allometry of food intake in grazing ruminants. *Journal of Animal Ecology* 56, 989–999.

Jansson, S. & Cronlund, A. (1976) Alkaloids from bark of *Erythrophleum africanum. Acta Pharmaceutica Suecica* 13, 51–54.

Jarman, P.J. (1974) The social organisation of antelope in relation to their ecology. *Behaviour* 48, 215–267.

Löyttyniemi, K. (1985) On repeated browsing of Scots pine saplings by moose (*Alces alces*). *Silva Fennica* 19, 387–391.

Makhabu, S.W. (2005) Resource partitioning within a browsing guild in a key habitat, the Chobe Riverfront, Botswana. *Journal of Tropical Ecology* 21, 641–649.

Makhabu, S.W. & Skarpe, C. (2006) Rebrowsing by elephants three years after simulated browsing on five woody plant species in northern Botswana. *South African Journal of Wildlife Research* 36, 99–102.

Makhabu, S.W., Skarpe, C. & Hytteborn, H. (2006) Elephant impact on shoot distribution on trees and on rebrowsing by smaller browsers. *Acta Oecologica* 30, 136–146.

Mathisen, I.E. (2005) Effects of clipping and nitrogen fertilization on a grazing tolerant grass in Chobe National Park, Botswana. MSc Thesis, Norwegian University of Science and Technology, Trondheim, Norway.

McNaughton, S.J. (1983) Compensatory plant growth as a response to herbivory. *Oikos* 40, 329–336.

McNaughton, S.J. (1984) Grazing lawns: animals in herds, plant form, and coevolution. *The American Naturalist* 124, 863–886.

McNaughton, S.J., Banyikwa, F.F. & McNaughton, M.M. (1997) Promotion of the cycling of diet-enhancing nutrients by African grazers. *Science* 278, 1798–1800.

Midgley, J.J., Lawes, M.J. & Chamaillé-Jammes, S. (2010) Savanna woody plant dynamics: the role of fire and herbivory, separately and synergistically. *Australian Journal of Botany* 58, 1–11.

Mitchard, E.T.A., Saatchi, S.S., Lewis, S.L., Feldpausch, T.R., Gerard, F.F., Woodhouse, I.H. & Meir, P. (2011) Comment on 'A first map of tropical Africa's above-ground biomass derived from satellite imagery'. *Environmental Research Letters* 6, 049001. doi:10.1088/1748-9326/6/4/049001.

Moe, S.R., Rutina, L.P., Hytteborn, H. & du Toit, J.T. (2009) What controls woodland regeneration after elephants have killed the big trees? *Journal of Applied Ecology* 46, 223–230.

O'Kane, C.A., Duffy, K.J., Page, B.R. & Macdonald, D.W. (2011) Are the long-term effects of mesobrowsers on woodland dynamics substitutive or additive to those of elephants? *Acta Oecologica* 37, 393–398.

Omphile, U.J. & Powell, J. (2002) Large ungulate habitat preference in Chobe National Park, Botswana. *Journal of Range Management* 55, 341–349.

Owen-Smith, N. & Chafota, J. (2012) Selective feeding by a megaherbivore, the African elephant (*Loxodonta africana*). *Journal of Mammalogy* 93, 698–705.

Owen-Smith, N. & Novellie, P. (1982) What should a clever ungulate eat? *The American Naturalist* 119, 151–178.

Rooke, T. (2003) *Defences and responses: woody species and large herbivores in African savannas*. PhD Thesis, Swedish University of Agricultural Sciences. Umeå, Sweden.

Rooke, T. & Bergström, R. (2007) Growth, chemical responses and herbivory after simulated leaf browsing in *Combretum apiculatum*. *Plant Ecology* 189, 201–212.

Rooke, T., Bergström, R., Skarpe, C. & Danell, K. (2004) Morphological responses of woody species to simulated twig-browsing in Botswana. *Journal of Tropical Ecology* 20, 281–289.

Ruess, R.W., McNaughton, S.J. & Coughenour, M.B. (1983) The effects of clipping, nitrogen source and nitrogen concentration on the growth responses and nitrogen uptake of an East African sedge. *Oecologia* 59, 253–261.

Scholes, R.J. & Walker, B.H. (1993) *An African Savanna: Synthesis of the Nylsvley Study*. Cambridge University Press, Cambridge, UK.

Scogings, P.F., Hjältén, J. & Skarpe, C. (2011) Secondary metabolites and nutrients of woody plants in relation to browsing intensity in African savannas. *Oecologia*, 167, 1063–1073.

Shrader, A.M., Bell, C., Bertolli, L. & Ward, D. (2012) Forest or the trees: at what scale do elephants make foraging decisions? *Acta Oecologica* 42, 3–10.

Simpson, C.D. (1974) *Ecology of the Zambezi Valley bushbuck Tragelaphus scriptus ornatus Pocock*. PhD Thesis, Texas A&M University, College Station, Texas.

Skarpe, C. & Hester, A.J. (2008) Plant traits, browsing and grazing herbivores, and vegetation dynamics. In: Gordon, I.J. & Prins, H.H.T. (eds.) *The Ecology of Browsing and Grazing*. Springer, Berlin, pp. 217–261.

Skarpe, C., Bergström, R., Bråten, A.-L. & Danell, K. (2000) Browsing in a heterogeneous savanna. *Ecography* 23, 632–640.

Skarpe, C., Bergström, R., Danell, K., Eriksson, H. & Kuntz, C. (2012) Of goats and spines – a feeding experiment. *African Journal of Range & Forage Science* 29, 37–41.

Smallie, J.J. & O'Connor, T.G. (2000) Elephant utilization of *Colophospermum mopane*: possible benefits of hedging. *African Journal of Ecology* 38, 352–359.

Stamp, N. (2003) Out of the quagmire of plant defense hypotheses. *The Quarterly Review of Biology* 78, 23–55.

Stokke, S. & du Toit, J.T. (2000) Sex and size related differences in the dry season feeding patterns of elephants in Chobe National Park, Botswana. *Ecography* 23, 70–80.

Stokke, S. & du Toit, J.T. (2002) Sexual segregation in habitat use by elephants in Chobe National Park, Botswana. *African Journal of Ecology* 40, 360–371.

Strauss, S.Y. & Agrawal, A.A. (1999) The ecology and evolution of plant tolerance to herbivory. *Trends in Ecology and Evolution* 14, 179–185.

Taolo, C.L. (2003) Population ecology, seasonal movement and habitat use of African buffalo (*Syncerus caffer*) in Chobe National Park, Botswana. PhD Thesis, Norwegian University of Science and Technology, Trondheim, Norway.

Vanak, A.T., Shannon, G., Thaker, M., Page, B., Grant, R. & Slotow, R. (2012) Biocomplexity in large tree mortality: interactions between elephant, fire and landscape in an African savanna. *Ecography* 35, 315–321.

van de Koppel, J. & Prins, H.H.T. (1998) The importance of herbivore interactions for the dynamics of African savanna woodlands: an hypothesis. *Journal of Tropical Ecology* 14, 565–576.

Vourc'h, G., Martin, J.-L., Duncan, P., Escarré, J. & Clausen, T.P. (2001) Defensive adaptations of *Thuja plicata* to ungulate browsing: a comparative study between mainland and island populations. *Oecologia* 126, 84–93.

Walker, B.H. & Noy-Meir, I. (1982) Aspects of the stability and resilience of savanna ecosystems. In: Huntley, B.J. & Walker, B.H. (eds.) *Ecology of Tropical Savannas*. Springer-Verlag, Berlin, pp. 556–590.

Wallgren, M. (2008) *Mammal community structure in a world of gradients: effects of resource availability and disturbance across scales and biomes*. PhD Thesis, Swedish University of Agricultural Sciences, Umeå, Sweden.

Wessels, D.C.J., van der Waal, C. & de Boer, W.F. (2007) Induced chemical defences in *Colophospermum mopane* trees. *African Journal of Range & Forage Science* 24, 141–147.

Wilmshurst, J.F., Fryxell, J.M. & Colucci, P.E. (1999) What constrains daily intake in Thomson's gazelles? *Ecology* 80, 2338–2347.

Wink, M. (2000) Interference of alkaloids with neuroreceptors and ion channels. *Studies in Natural Products Chemistry* 21, 3–122.

Woolnough, A.P. & du Toit, J.T. (2001) Vertical zonation of browse quality in tree canopies exposed to a size-structured guild of African browsing ungulates. *Oecologia* 129, 585–590.

Elephants and the Grazing and Browsing Guilds

Christina Skarpe[1], Stein R. Moe[2], Märtha Wallgren[3] and Sigbjørn Stokke[4]

[1] Faculty of Applied Ecology and Agricultural Sciences, Hedmark University College, Norway
[2] Department of Ecology and Natural Resource Management, Norwegian University of Life Sciences, Norway
[3] Forestry Research Institute of Sweden (Skogforsk), Uppsala Science Park, Sweden
[4] Norwegian Institute for Nature Research, Norway

'Are grazers and browsers different beasts' asked Iain Gordon in an article from 2003. The distinction between the two guilds is based on their feeding preferences, browsers feeding on woody and herbaceous dicotyledonous plants and grazers eating graminoids. Exploiting these profoundly different food resources has led to the adaptive evolution of different physiological, morphological, behavioural and life history traits, all of which imply that grazers and browsers are, in many respects, indeed different. Here we ask if the species in these guilds also respond differently to ecosystem changes induced by elephant, *Loxodonta africana*.

Grass and browse differ as food for large herbivores. Browse generally has higher nutrient and lower fibre concentrations than grasses, but more physical and chemical deterrents, which either reduce intake rates or food digestibility, or are toxic. Browse is also more diverse than grass as food for herbivores (Owen-Smith and Novellie, 1982; Clauss *et al.*, 2008; Chapter 12). It occurs in discrete, often widely spaced units – trees, shoots and leaves – each with different nutritive characteristics and growing at a variety

Elephants and Savanna Woodland Ecosystems: A Study from Chobe National Park, Botswana, First Edition. Edited by Christina Skarpe, Johan T. du Toit and Stein R. Moe.
© 2014 John Wiley & Sons, Ltd. Published 2014 by John Wiley & Sons, Ltd.

of heights. Grasses, in contrast, contain silica, which wears down the teeth of grazers. Grass swards often cover large contiguous areas in open landscapes such as savannas, and are perhaps perceived as relatively homogeneous by large and medium-sized grazers such as African buffalo, *Syncerus caffer*, or wildebeest, *Connochaetes taurinus*. The basal and intercalary meristems of grasses ensure that as the plant is cropped from the top, growth continues from the base. As long as water is available, growth replenishes the grazing resource, ensuring that the density of food items persists, even as biomass is periodically reduced. Conversely, when a browser feeds on trees or forbs, it crops whole shoots including the apical meristems, permanently removing the food item, and changing both the total quantity and the density of browse (Jarman, 1974), at least in the short term. Although trees will generally sprout new shoots and leaves from the remaining buds, this depends on plant physiology and time of the year and might not happen until the next growing season.

Following the Jarman-Bell principle (Bell, 1971, 1982; Jarman, 1974), stating that big herbivores tolerate poorer-quality food than do small herbivores, grazing ruminants, utilising a comparatively nutrient-poor but abundant food resource, should be relatively large (Bodmer, 1990; Gagnon and Chew, 2000; Clauss *et al.*, 2008). Browsers, in contrast, use a highly diverse food resource and consequently vary in body mass (Sponheimer *et al.*, 2003; Cerling *et al.*, 2003). Thus, on average browsers should be smaller than grazers (Gagnon and Chew, 2000), although both the largest and the smallest extant ruminants, the giraffe, *Giraffa camelopardalis*, weighing about 1000 kg and the royal antelope, *Neotragus pygmaeus*, weighing less than 3 kg, are browsers.

Dissimilarities in the distribution, structure and growth pattern of grasses and browse also favour different behavioural traits. The homogeneous distribution of grass, even when grazed, allows large and medium-sized grazers, such as buffalo and wildebeest to be markedly gregarious. In contrast, browsers, which selectively crop discrete food items, generally live solitarily or in small groups (Jarman, 1974; Estes, 1991). Differences in the degree of selective feeding by animals with different body size explain much of the variation in behaviour, life history and morphology (Jarman, 1974; Gordon, 2003). For example, the diminutive 15 kg oribi, *Ourebia ourebi*, primarily a grazer, feeds selectively on young grass leaves in much the same way as a similar-sized browser does on dicotyledonous plants, and shows some of the traits generally ascribed to such browsers. These include a relatively pointed incisor arcade and a non-gregarious life style.

In most savanna communities, the difference in distribution and abundance of grass and browse leads to dominance by large or medium sized grazers. Fritz *et al.* (2002), in a review of 31 African savanna ecosystems, found that browsers, grazers and mixed feeders (feeding on both browse and grass) on average contributed about 4%, 81% and 15%, respectively, to the metabolic biomass density of mesoherbivores, those species with a body mass between 4 and 450 kg (Coe *et al.*, 1976). Ungulate populations in African savanna are generally resource limited (Coe *et al.*, 1976; Owen-Smith, 1990;

Fritz and Duncan, 1994; Fritz, 1997; Mduma *et al.*, 1999; Fritz *et al.*, 2002). Hence, any competition and facilitation between sympatric species that influences food availability and quality are likely to be important determinants of community composition and population regulation. The effect that herbivores using the same food resource have on each other is influenced by the difference in feeding mode and body size between the species; by the structure, quantity and quality of the available plant biomass and by the plant responses to herbivory (Bell, 1971, 1982; de Boer and Prins, 1990; van de Koppel and Prins, 1998; Arsenault and Owen-Smith, 2002). Van de Koppel and Prins (1998) suggested that in ecosystems with considerable plant biomass big herbivores should facilitate smaller ones, whereas in systems with little plant biomass small herbivores should compete with larger ones. Large-bodied grazers require substantial quantities of food and can tolerate low-quality forage. They can therefore facilitate smaller species by removing old coarse grass, but cannot outcompete them because the small grazers can sustain themselves by picking small, high-quality food items from the sward. In so doing, small grazers could leave either too little biomass or biomass of inadequate quality for the larger grazers, thereby competing effectively with them (Illius and Gordon, 1987; Murray and Illius, 2000). The same interaction occurs among browsers, but the large species can escape competition from the small ones by browsing higher in the canopy (Illius and Gordon, 1987; Woolnough and du Toit, 2001; Chapter 12). Because browse grows from apical meristems at the shoot tips, large browsers taking big bites can remove the food of smaller species and outcompete them, at least in the short term. But, by reducing apical dominance, a large browser can cause plants to produce more browse lower down, within the reach of smaller species. The nutritional quality of these resprouting shoots can be high, which would favour the smaller species (Bergström *et al.*, 2000; Makhabu *et al.*, 2006). Thus, the interplay between competition and facilitation in browsing systems is more complex than in grazing systems.

In many protected areas in southern Africa, elephant populations are increasing, re-establishing former densities (Lewin, 1986; Skarpe *et al.*, 2004) or in some cases overpopulating confined protected areas (Cumming *et al.*, 1997; Child, 2004). In many of these areas, elephants are important agents of ecosystem change, profoundly influencing plant and animal communities (Herremans, 1995; Cumming *et al.*, 1997; van de Koppel and Prins, 1998; Fritz *et al.*, 2002; Chapters 4, 14 and 15). Reviewing the trophic-guild structure of ungulate communities in southern and eastern Africa, Fritz *et al.* (2002) found a negative correlation between the metabolic biomass density of elephants and that of ungulate browsers and mixed feeders, but no relationship with grazers, and suggested that elephants compete with browsers and mixed feeders but might facilitate grazers (Owen-Smith, 1988).

In previous chapters of this book we have described changes in the Chobe elephant population over the last 150 or so years, from an unknown but presumably high density up to the early 19th century (Campbell, 1990) to virtual extinction at the end of that century, followed by recovery to become, currently, part of the largest and densest elephant population in Africa (Chase, 2011; Chapter 6). Elephant numbers or densities in northern Botswana are not well known, although a number of studies (Junker et al., 2008; Chase, 2011; Chapter 6) suggest a stabilisation at around 130,000 – 150,000 animals, although a recent count by Department of Wildlife and National Parks (2013) provides a much higher estimate (Chapter 14) that still needs to be confirmed by subsequent counts. The large mobility of elephants as well as considerable variation in areas counted contributes to the uncertainty. Nevertheless, the elephant density in the Chobe area is high, and if elephants facilitate grazers, as hypothesised by Owen-Smith (1988), and compete with browsers and mixed feeders, as suggested by Fritz et al. (2002), we expect the proportion of grazers in the metabolic biomass density of herbivores to have decreased, and that of browsing and mixed feeding species to have increased, following the near-extinction of elephants. With the subsequent recovery of elephants, the proportion of grazers is expected to increase whereas that of browsers and mixed feeders should decrease.

How this chapter was compiled

In this chapter we try to trace the population fluctuations of some grazing and browsing species in relation to the dynamics of the elephant population and other factors. The target area is the northernmost part of the Chobe National Park between the Chobe River and the tarmac road between Kasane and Ngoma, but areas included in the different studies referred to varies somewhat. We also discuss food and habitat use by some grazers and browsers in relation to elephant impact and ecosystem type, distinguishing between the relatively nutrient-rich savannas on alluvial soil and the nutrient-poor woodlands on sand (Jarman, 1974; Bell, 1982; Chapter 9). We refer to information available in the form of travel- and hunting accounts, reports, anecdotal accounts, student theses and scientific publications as well as our own studies within the Botswana-Norway Institutional Co-operation and Capacity Building (BONIC) project (Chapter 1; Table 13.1). For historical information before about 1960 only anecdotal information is available. For information on the time before the decline in the elephant population, we rely heavily on Frederick Courteney Selous' accounts of hunting and travelling in the area in the 1870s (Selous, 1881). From the 1960s and 1970s, anecdotal and semi-quantitative information is provided by Graham Child's work primarily from the 1960s (Child, 1968, 1972; Child and von Richter, 1969); Sheppe and Haas's (1976) study in 1971 – 1972 and Simpson's doctoral thesis (Simpson, 1974) and others. Besides his own project, Simpson systematically collected anecdotal information from old residents in the area about status of wildlife before about 1950. With all its shortcomings, from being based on old men's memories, this represents virtually the only information available from that time.

Table 13.1 **Principal mammalian herbivores present in the BONIC study area in Chobe National Park, with qualitative and quantitative estimates of abundance as assessed by different observers over the years: 1870s (Selous, 1881); pre-1950s (Simpson, 1974); 1960–1970 (Child, 1968; Simpson, 1974; Sheppe and Haas, 1976; many unpublished references in these); 2000–2005 (this study, rainy season; Dipotso and Skarpe, 2006; Dipotso et al., 2007; see text for details); 2010 (Chase, 2011 area CNP F, dry season).**

Species	English name	Feeding guild	Status				
			1870s	Pre-1950s	1965–1975	2000–2005 density (km^{-2})*	2010 density (km^{-2})[†]
Loxodonta africana	Elephant	Mixed	Abundant	Scarce	Abundant	5.4	4.0
Equus quagga	Plains zebra	Grazer	Present	Common	Scarce	0.4	1.9
Phacochoerus africanus	Common warthog	Grazer	Present	Common	Abundant	0.2	0.1
Hippopotamus amphibius	Hippopotamus	Grazer	Abundant	Present	Present	Present	0.1
Giraffa camelopardalis	Giraffe	Browser	Present	Common	Scarce	0.2	0.1
Syncerus caffer	African buffalo	Grazer	Abundant	Occasional	Abundant	4.3	1.9
Tragelaphus strepsiceros	Greater kudu	Browser	Present	Common	Common	1.1	0.1
Tragelaphus scriptus	Bushbuck	Browser	Present	Abundant	Common	40 total[‡]	No data
Tragelaphus oryx	Eland	Browser	Present	Common	Scarce	Present	0.2
Connochaetes taurinus	Wildebeest	Grazer	Present	Abundant	Occasional	Present	<0.1
Damaliscus lunatus	Tsessebe	Grazer	Present	Common	Occasional	Present	<0.1
Hippotragus equinus	Roan antelope	Grazer	–	Common	Occasional	Present	0.1
Hippotragus niger	Sable antelope	Grazer	–	Abundant	Present	Present	<0.1
Sylvicapra grimmia	Common duiker	Browser	–	Common	Present	Present	No data
Kobus ellipsiprymnus	Common waterbuck	Grazer	Present	Common	Present	Present	0.0
Kobus leche	Red lechwe	Grazer	Common	Abundant	Present	Present	0.3
Kobus vardonii	Puku	Grazer	Common	Common	Scarce	120 total[‡]	No data
Raphicerus campestris	Steenbok	Browser	–	Occasional	Present	0.1	No data
Aepyceros melampus	Impala	Mixed	Present	Abundant	Occasional	10.2	2.8

*Ground survey data from this study.
[†]Aerial survey data from Chase (2011).
[‡]Estimated number of animals present; not a measure of density.

In the BONIC project, mammals heavier than about 2 kg were counted along transects using Distance sampling (Buckland *et al.*, 2001). Eight transects with a combined length of 67 km were driven repeatedly and in all 3729 km were driven. Methods are described in Stokke *et al.* (2003) and Wallgren (2008). The Distance statistics require large numbers of observations for density estimates (Buckland *et al.*, 2001), meaning that for many species density estimates could not be achieved. In addition we calculated preference indices for the area on alluvial soil close to the river and for the mixed woodland and *Baikiaea* woodland on the sand (Chapter 5) using the design by Manly *et al.* (1993). These include a number of species for which reliable density estimates could not be obtained. Puku, *Kobus vardonii* and bushbuck, *Tragelaphus scriptus*, for which special concern had been expressed by Department of Wildlife and National Parks (DWNP), were the subjects of special studies (Dipotso and Skarpe, 2006; Dipotso *et al.*, 2007), as were buffalo and impala, *Aepyceros melampus* (Halley *et al.*, 2002; Taolo, 2003; Rutina, 2004; Rutina *et al.*, 2005; Moe *et al.*, 2007; Chapters 10 and 11). DWNP has been conducting aerial surveys of wildlife in northern Botswana for about 30 years, and data are summed up in the report of a recent survey by 'Elephants without Boundaries' (Chase, 2011), giving density estimates for relatively large species. A summary of some of the available information is given in Table 13.1.

Grazing and browsing ungulates in the Chobe environment

Among the ungulate species (and elephant) currently assumed to occur in Chobe National Park there are 14 species of grazers, of which 11 are ruminants; 6 species of browsers, all of which are ruminants and 2 of mixed feeders, of which 1 is a ruminant (Table 2.2). In this chapter we refer to preferential browsers as browsers and to preferential grazers as grazers. Only impala and elephant are considered to be mixed feeders. The average body mass for grazers is 254 kg, or 172 kg if only ruminants are included and for browsers, 241 kg (adult female body mass; see Table 2.2). Thus among ruminants, browser species are on average slightly heavier than grazing ones, contrary to the general trend (Gagnon and Chew, 2000). The median mass is slightly larger for grazers, 114 kg, than for browsers, 100 kg (average of bushbuck, *Tragelaphus scriptus*, 42 kg and greater kudu, *Tragelaphus strepsiceros*, 157 kg), and the variation in body-mass between species is much larger in browsers than in grazers standard deviation 294 and 149, respectively (Table 2.2; Sponheimer *et al.*, 2003).

Historical changes in abundance and foraging of some grazing and browsing ungulates

The end of the 19th century and beginning of the 20th century was characterised in the Chobe region by the near-extinction of elephants and by outbreaks of rinderpest. Between 1897 and 1905 rinderpest decimated populations of even-toed ungulates, particularly buffalo, eland, *Taurotragus oryx* and warthog, *Phacochoerus africanus*, as well as any domestic cattle in the area. In contrast, impala were comparatively unaffected (Spinage, 2012; Chapter 6). At the start of the 1900s, human land use, including livestock grazing, widespread burning and logging, intensified. Both logging and livestock grazing ceased in the late 1950s in the area later to become the Chobe National Park and a non-burning policy was adopted (Child, 1968; Simpson, 1974; Chapters 4 and 6). At about the same time, elephants increased and became common in the area (Chapter 6). Concurrent with these impacts, the vegetation on the alluvial flats along the river, and in the sandy areas further south, changed in contrasting ways (Chapter 4). Thus the dynamics of the ungulate community has potentially been influenced by several factors, including the effects of changes in the elephant population. In the following section, we present specific information on some of the species in the area. The data are scanty throughout, and old non-quantitative information in particular should be interpreted with care.

The grazing, the mixed feeding and the browsing guilds

The grazers

Selous (1881) mentioned **hippopotamus**, *Hippopotamus amphibius*, as abundant along the Chobe River. Child (1968) recorded it in most permanent waters in the area, including the river close to Kasane, as well as wandering between seasonal pans further south in the wet season. During the BONIC project we found that hippos fed mainly on the floodplain, but in the wet season, during night-time transect drives, we also encountered them in the shrublands and, occasionally, in the mixed woodland some kilometres from the river. Hippos graze with their lips and, in spite of their large body size, are able to use the short grass on the floodplain, mainly *Cynodon dactylon*. They probably facilitate ruminant grazers by creating grazing lawns – 'hippo lawns' – on the floodplain (Kanga *et al.*, 2013; Chapter 12), and might, like buffalo (Chapter 11), compete with elephants by cropping the grass sward too short for them.

Selous (1881: 156) found the number of **buffalo** along the Chobe River in 1874 'really astonishing'. Between 1897 and 1905 buffalo suffered severely from rinderpest (Spinage, 2012), and were considered widespread but occasional in the first half of the 1900s (Child, 1968; Sheppe and Haas, 1976; Simpson, 1974). Child (1968) found buffalo mainly to be in poor body condition and mentioned a number of die-off events, but did not suggest a reason, other than one instance when the channel at Savuti Marsh ceased flowing in 1965. From the 1940s buffalo apparently increased along the Chobe River close to Kasane concurrently with the growing elephant population (Child, 1968). In the early 2000s the dry-season population there numbered 2500–3000 animals and was still increasing (Taolo, 2003; Chase, 2011; Chapter 11).

Buffalo is the second largest grazer in the area after hippopotamus, and has the second highest population metabolism after elephant. Buffalo show a preference for habitats on alluvium close to the river (Table 13.2). They use the floodplains extensively in the dry season, during which time they feed mostly on *Cynodon dactylon*, but in the wet season they graze mainly on the woodlands further south (Taolo, 2003; Chapter 11). Although primarily grazers, buffalo also browse during the late dry season (Child, 1968), a phenomenon that we occasionally also observed. This small contribution of

Table 13.2 **Habitats preferred and avoided by principal mammalian herbivores in the BONIC study area in Chobe National Park in the wet season. Habitats are alluvium, including shrublands and floodplain, mixed woodland (MW) and *Baikiaea* woodland (BW).**

Scientific name	Common name	Preferred	Avoided
Syncerus caffer	African buffalo	Alluvium	BW
Tragelaphus scriptus	Bushbuck	Alluvium	BW
Sylvicapra grimmia	Common duiker	BW	Alluvium
Loxodonta africana	Elephant	Alluvium	BW
Giraffa camelopardalis	Giraffe	Alluvium	BW
Hippopotamus amphibius	Hippopotamus	Alluvium	BW
Aepyceros melampus	Impala	Alluvium	BW, MW
Tragelaphus strepsiceros	Greater kudu	Alluvium	BW
Kobus leche	Red lechwe	Alluvium	BW
Kobus vardonii	Puku	Alluvium	BW, MW
Hippotragus equinus	Roan antelope	–	Alluvium
Hippotragus niger	Sable antelope	–	–
Raphicerus campestris	Steenbok	BW	Alluvium
Damaliscus lunatus	Tsessebe	–	BW
Phacochoerus africanus	Common warthog	Alluvium	BW, MW
Kobus ellipsiprymnus	Common waterbuck	Alluvium	BW, MW
Equus quagga	Plains zebra	BW	Alluvium, MW

woody species to the diet was not discernible in isotope analyses, however (Halley and Minagawa, 2005). Buffalo and elephant diets overlap, with elephants facilitating buffalo foraging, but buffalo competing with elephants grazing on the floodplain in the dry season (Taolo, 2003; Chapter 11) and potentially also in the woodlands during the wet season.

Selous (1881) encountered **zebra** *Equus quagga* along Chobe River in the 1870s, but did not comment on their abundance. From anecdotal information collected by Simpson (1974), zebra were considered to have been common before 1950 but scarce in 1969–1970. As an odd-toed ungulate, zebra did not suffer from the rinderpest pandemic. The apparent increase in thicket vegetation in the riparian and riverine woodlands on the alluvium during the 1960s (Child, 1968; Simpson, 1974; Chapter 4) obviously implies some degradation of zebra habitat, which could have contributed to their decline in abundance. Another negative factor affecting zebra and other migratory species was the erection of veterinary cordon fences in northern Botswana from the late 1950s, which obstructed their migration routes (Bartlam-Brooks *et al.*, 2011). Child (1968) described zebra in the Chobe area as highly mobile, increasing in density close to permanent water in the dry season and dispersing in the wet season. Both Child and Sheppe and Haas (1976) found zebra in the dry season using the grasslands along the Chobe River, particularly in the west towards Ngoma. During our study in the late 1990s we recorded zebra in the dry season mainly in the woodlands and occasionally in the western part of the alluvial flats and floodplain. As a hind-gut fermenter, zebra can use relatively nutrient-poor grasses on Kalahari sand. We found that zebra preferred habitats on sand during the wet season, when they avoided alluvium, and obviously overlapped in diet with grazing elephants (Table 13.2).

Selous (1881) saw **wildebeest** along the Chobe River, but did not mention their abundance. Simpson (1974) formed the impression from old residents' comments that wildebeest had been abundant in the area before 1950, but he considered the species as only occasional during 1969–1971. Child (1968) indicated that wildebeest density in northern Botswana increased during the first half of the 1900s, but only after the 1920s were there reports of herds 'numbering thousands' and 'blackening out the horizon'. A plausible reason for the increase could have been release from rinderpest after 1905 (Spinage, 2012). This was not mentioned by Child, who attributed the increase to changes in land use. In any case, the population built up along the Chobe River until 1964, when there was a considerable crash, perhaps for the same reasons as those for zebra. Sheppe and Haas (1976) saw just small numbers of wildebeest along the river in the late dry season in 1971. In our study, wildebeest were only occasionally encountered and, according to Chase (2011), are declining throughout northern Botswana.

Selous (1881) described **puku**, *Kobus vardonii*, as common along the Chobe River, where he encountered groups of 50 or more animals. He named the large alluvial flat at Serondela 'Pookoo (Puku) Flats', because he saw so many of these antelope there. A 100 years later the species was recorded as scarce (Child and von Richter, 1969; Sheppe and Haas, 1976; Simpson, 1974). Calculations using data from Child and von Richter (1969) suggest a population then of less than 20 animals (Dipotso and Skarpe, 2006). After the 1970s puku seem to have increased slightly, and Dipotso and Skarpe (2006) and O'Shaughnessy (2011) estimated the population to about 120–130 animals in the early 2000s. Puku is a floodplain species, but when the floodplain is inundated they also use the raised alluvial flats (Selous, 1881; Child and von Richter, 1969; Dipotso and Skarpe, 2006), and the decline and increase in the population might have been influenced by the increasing and decreasing, respectively, woody cover on these flats.

Puku graze primarily on short grasses (about 20 cm tall). Child and von Richter (1969) recorded puku feeding on many species of grasses and a few sedges, varying over the year, with *Brachiaria arrecta*, *Eragrostis rigidior* and *Digitaria eriantha* (modern nomenclature following Chapter 5) among the species contributing most to their diet. At present, *Cynodon dactylon* constitutes a large part of the puku's diet (Dipotso and Skarpe, 2006; O'Shaughnessy, 2011). The change in diet between 1969 and 2010 reflects a turnover in species in the floodplain vegetation between the two studies and implies a possible increase in forage quality (Chapter 4).

Selous (1881) hunted **lechwe**, *Kobus leche*, in the Chobe flood plains in the 1870s and gave the impression that the species was common, being encountered in large groups on both sides of the river, but primarily on the northern side. Lechwe was reported to have been abundant before about 1950 (Simpson, 1974), and even in the early 1960s, large numbers of lechwe apparently used to aggregate on the Chobe flood plains when the river was low, and Child (1968) referred to observations of 1000–2500 lechwe between Kabulabula and Ihaha at that time. He regarded the population as decreasing, a view shared by Sheppe and Haas (1976), who added that the population was based on the northern side of Chobe River, but crossed the river at times to feed on the Botswana side. Around 1970 Simpson recorded lechwe as occasional in the area. In our transect drives it was encountered a few times. As an exclusive floodplain species lechwe might have suffered from the increasing unavailability of large areas of floodplain on the Namibian side of the river, where human landuse now largely excludes wildlife.

Lechwe is a floodplain specialist, and Child and von Richter (1969) showed habitat segregation between the three species of *Kobus* in Chobe, with lechwe using the floodplain closest to the river, puku using the upper floodplain and the lower parts of the shrubland, and common waterbuck, *Kobus ellipsiprymnus*, using the shrublands above the floodplain and the woodlands on sand. O'Shaughnessy (2011) found strong dietary overlap between puku and lechwe, with the same grasses, including *Cynodon dactylon*, constituting the bulk of the diet of both species.

Child and von Richter (1969) recorded **common waterbuck** along Chobe River between Ngoma and Kasane, with a concentration in the eastern part around Serondela, at that time a stronghold for *Cynodon dactylon*. They found that the diet of waterbuck overlapped substantially with that of puku and lechwe, but that the waterbuck diet included more *Cynodon dactylon*. In our study we saw waterbuck occasionally, mainly in the eastern area towards Kasane where, during the wet season, they used the alluvial flats preferentially, avoiding *Baikiaea* woodland (Table 13.2).

The two specialist woodland grazers **sable**, *Hippotragus niger*, and **roan antelope**, *Hippotragus equinus*, were observed coming to the river to drink in the dry season, but were otherwise seen occasionally in the woodlands. Both Simpson (1974) and Child (1968) were told by old residents that both species had declined since before about 1950. Sable and roan are tall grass specialists, selectively feeding on green grass leaves (Macandza *et al.*, 2012), and Child, (1968) suggested that the reduction in such grass caused by trampling and grazing by elephants and other large herbivores might contribute to the decline of the species. During our study, tall grasses were not found within about 4–5 km of the river because of intense grazing and trampling. The shrublands close to the river, with their high concentration of herbivores, also attract predators, providing another reason for predation-sensitive sable and roan to avoid that habitat (Knoop and Owen-Smith, 2006). Sable drink relatively infrequently and can forage far from water (Cain *et al.*, 2012). Our transect counts yielded too few observations to give significant density estimates. Roan was found to avoid habitats on alluvium, and we have the impression that sable do the same, presumably for the reasons given above, although at least roan is reported to select such habitats elsewhere (Heitkönig and Owen-Smith, 1998).

In the Chobe woodlands the diets of both sable and roan in the wet season are likely to overlap with other tall-grass grazers such as elephant, buffalo and zebra, whereas in the dry season elephants switch to browsing (Stokke and du Toit, 2000; Chapter 6) and buffalo largely graze on the floodplains (Taolo, 2003). The density of zebra in the area is low. As a hindgut fermenter, zebra feed less selectively and will more readily accept mature and senescent grass than do sable and roan (Macandza *et al.*, 2012). Hensman *et al.* (2012) found that sable in the Okavango delta browsed substantially on woody species in the dry season. The same might occur in Chobe but was not recorded by us.

The mixed feeders

Selous (1881) recorded **impala** along the Chobe River around 1874, and Simpson (1974) quotes anecdotal information from old residents saying that impala were abundant along the river before 1950, but they were only occasional during his study around 1970. Impala suffered less than many other even-toed ungulates from the rinderpest panzootic between 1896–1897 and 1905 (Spinage, 2012), but apparently declined in numbers 50 years later. A contributing reason might have been the increasing thicket vegetation from about the 1960s (Child, 1968). Child and Sheppe and Haas (1976), counted impala, as well as many other species, regularly along roughly the same 53 km road transect running within about 500 m from the river. The average number counted per transect drive per month was about 10 animals in

1965–1967 (Child, 1968) and 20 in 1971–1972 (Sheppe and Haas, 1976). Both Child (1968) and Sheppe and Haas (1976) implied that the small impala population along the Chobe River was increasing, which they attributed to elephant-induced changes in the vegetation. With ongoing change, impala numbers have continued to increase (Rutina *et al.*, 2005; Chapter 4), so that it is currently the commonest ungulate on the Chobe alluvium. Local densities of up to 50 animals km^{-2} and an overall density in the study area of about 3–6 animals km^{-2} have been recorded (Stokke *et al.*, 2003; Rutina *et al.*, 2005; Wallgren, 2008; Chase, 2011; Table 13.1). We found that impala strongly prefer areas on alluvium and avoid *Baikiaea* woodland and mixed woodland on the sand (Table 13.2).

Impala switch seasonally between feeding guilds, being a browser in the dry season and a grazer in the wet season. The browse diet of impala in our study area overlaps considerably with that of kudu *Tragelaphus strepsiceros* and giraffe. Schoener's index of overlap in plant species utilised by these three species is 0.6–0.8 across both wet and dry seasons; for plant parts it is 0.5–0.7 (Makhabu, 2005). The three different-sized ruminant species partition the browse resource by feeding at different heights. None overlapped significantly with the hind-gut fermenting elephant (Makhabu, 2005; Chapter 12). Schoener's index for tree species and plant parts used by elephant and the three browsing ungulates was around 0.2 and 0.1 respectively. For impala, *Capparis tomentosa* comprised 50% of the browse component of the diet in the dry season and 11% in the wet season, followed by *Combretum mossambicense* (19% and 42% of the dry and wet season browse, respectively) and *Flueggea virosa* (13% and 8%), all based on the number of bites taken (Makhabu, 2005). On most tree species, impala preferred browsing on individuals previously browsed by elephants compared with unbrowsed conspecifics (Makhabu *et al.*, 2006; Chapter 12). At a larger scale, impala along the Chobe River fed preferentially in intensely elephant-impacted habitats (Rutina *et al.*, 2005), as has been found in Hwange National Park in Zimbabwe (Valeix *et al.*, 2011).

The browsers

Selous (1881) mentioned that **giraffe** abounded west of the present National Park, but gives no indication of its abundance in our study area. Simpson (1974) was given anecdotal information from old residents claiming that giraffe was common in the Chobe area before 1950, but he judged it as scarce during his own study around 1970. Child recorded giraffe both along the river and far into the Kalahari sand areas, whereas Sheppe and Haas (1976) saw only three animals. In our study we found giraffe primarily in the shrublands on alluvium. They used this habitat preferentially and avoided the *Baikiaea* woodland (Table 13.2). Child (1968) recorded giraffe feeding primarily on *Boscia albitrunca*, *Terminalia sericea*, *Philenoptera violacea* and *Berchemia discolor*. None of these species was common along the river at the end of the 1990s

(Chapters 4 and 5). Instead, the giraffe in the shrublands mostly browsed *Capparis tomentosa* – which comprised about 46% and 50% of bites taken from woody plants in the wet and dry seasons, respectively – and *Combretum mossambicense*, constituting 20% of bites in the wet season, 16% in the dry (Makhabu, 2005).

Selous (1881) hunted fine trophies of **greater kudu** at the Chobe River in the mid-1870s, but did not comment on the species' abundance. Old residents thought kudu to have been common before 1950, and Simpson (1974) classified it as such around 1970. Our transect data suggest that kudu preferentially use the *Capparis* and *Combretum* shrublands on alluvium and select against the *Baikiaea* woodlands on sand. This accords with other observations (Omphile and Powell, 2002; Kazonganga, 2011; Kandume, 2012). At the end of the dry season, kudu also use mixed woodlands where much *Combretum apiculatum* is eaten. In the shrublands on alluvial soil, kudu mostly eat *Capparis tomentosa*, constituting 20–30% of bites taken, and *Combretum mossambicense*, 15–20%, with other important species being *Psydrax livida* (formerly *Canthium huillense*), *Erythroxylum zambesiacum* and *Markhamia zanzibarica* (Makhabu, 2005; Kazonganga, 2011; Kandume, 2012). Kudu in the study area exhibit sexual segregation in foraging, as found in other ecosystems (du Toit, 1995; Ginnett and Demment, 1997), but showed little difference between sexes in selectivity among and within tree species (Kandume, 2012; Kazonganga, 2011). Both sexes preferred habitats on alluvium, but groups of females and juveniles tended to forage further from the river than males, possibly trading forage quality for a reduced risk of predation (Kandume, 2012) as suggested by du Toit (1995).

The decline since the 1960s of the Chobe **bushbuck** has long concerned the DWNP. Selous (1881) mentioned bushbuck, but gave no information on its abundance. Child (1968) suggested that the spread of thicket vegetation on the alluvial soil along the Chobe River around 1960 was the likely reason for an 'eruptive' increase in the bushbuck population, which he proposed peaked around 1965. Simpson (1974) judged the species as common and estimated a population of more than 500 animals along a roughly 50 km stretch of the Chobe River between Kasane and Ihaha (Figure 2.2). He hypothesised that bushbuck in the early 1900s might have been limited by the then-common impala until increasing thicket vegetation favoured bushbuck and disadvantaged impala. With the subsequent increase in elephant and buffalo populations opening up the thickets bushbuck numbers seem to have dropped and the width of range used by them along the Chobe River to have narrowed from about 3 miles (4.8 km; Child, 1968) to about 0.75 miles (1.2 km; Simpson, 1974). Although the estimates of the bushbuck population are uncertain, there is evidence that the population declined substantially between the studies in the 1960s (Child, 1968; Simpson, 1974) and that in 1991 (Addy, 1993), but probably not between 1991 and 2002–2003 (Dipotso *et al.*, 2007). Using individually recognised animals and capture-mark-recapture methodology (Dipotso *et al.*, 2007), we estimated the population in 2002–2003 along the same section of the river as studied by Child

(1968) and Simpson (1974) to be about 40 adult bushbuck. Currently, the fairly open vegetation along the river likely provides suboptimal habitat for bushbuck (Simpson, 1974; Addy, 1993; Dipotso *et al.*, 2007). In addition, the increase in the impala population could be leading to competition for food between these two fairly equally sized antelopes with similar browse preferences, as well as to apparent – that is, predator-mediated – competition (Omphile, 1997; Dipotso *et al.*, 2007).

The bushbuck is a shrub and thicket-specialist, and is almost exclusively found in the denser parts of the *Capparis* shrublands and fragments of riparian forest along the river (Dipotso *et al.*, 2007). Bushbuck diet in 1969–1971 (Simpson, 1974) overlapped with that of impala, with *Flueggea virosa* as the most important species, followed by *Combretum mossambicense*, *Capparis tomentosa* and *Dichrostachys cinerea*. As these species are common in the habitat presently used by bushbuck (Dipotso *et al.*, 2007; Chapter 5), there is no reason to believe that diet has changed to any great extent.

Contrary to what might be expected from the Bell's hypothesis (Bell, 1982), the smallest browsers, **steenbok**, *Raphicerus campestris* and **common duiker**, *Sylvicapra grimmia*, preferentially used the *Baikiaea* woodlands on nutrient-poor Kalahari sand (Table 13.2). Preference for seemingly low-quality habitats have been observed elsewhere in these species (du Toit, 1993; Lunt, 2011). Steenbok, at least, feed mostly on forbs (du Toit, 1993), and *Baikiaea* woodlands have the greatest abundance and species richness of this important food resource (Aarrestad *et al.*, 2011, Chapter 5). This might partly explain their habitat preference, although minimising the risk of predation should also be considered. The shrublands on the alluvium have high densities of a variety of predators, many of which could prey on these small antelopes (Sinclair *et al.*, 2003).

Elephants and the grazing and the browsing guild

The fall and rise of the elephant population in northern Botswana makes an ideal 'experiment' for studying the impact of changes in elephant abundance on ungulate communities (Fritz *et al.*, 2002). Historical data on herbivore population sizes are anecdotal at best, however, and even modern population estimates can be far from accurate. Moreover, many factors other than elephant density have directly and indirectly influenced herbivore populations in Chobe. What is reasonably certain is that buffalo and impala populations increased, and bushbuck decreased, concurrently with the increase in the elephant population (Taolo, 2003; Rutina *et al.*, 2005, Dipotso *et al.*, 2007). There has also likely been a decline in 'plains game' – wildebeest and zebra – since about the 1950s (Child, 1968; Simpson, 1974). Of these species, buffalo, zebra and wildebeest are grazers, impala is a mixed feeder and bushbuck a browser. Thus, no clear pattern of population change emerges in relation to different feeding guilds.

Using female body mass (Table 2.2) raised to 0.75, and densities from Chase (2011: only dry season data are available), we can calculate the metabolic biomass of the meso-herbivore (*sensu* Coe *et al.*, 1976) community in the whole of Chobe National Park. The browsing, grazing and mixed-feeding species there contribute 14%, 73% and 13%, respectively, to the total. Compared with the average from the meta-analysis by Fritz *et al.* (2002) – 4% browsers, 81% grazers and 15% mixed feeders – this means a considerably larger proportion of browsers than would be expected in this elephant-dense environment. These calculations do not include giraffe, a browser, which is classified as a megaherbivore (Coe *et al.*, 1976; Fritz *et al.*, 2002). In contrast to Fritz *et al.* (2002),who classify eland as a mixed feeder, we count it as a browser (Sponheimer *et al.*, 2003; Cerling *et al.*, 2003). Defining eland instead as a mixed feeder would obviously reduce the proportion of browsers and correspondingly increase that of mixed feeders, but this would still be at variance with the suggestion by Fritz *et al.* (2002) that high elephant densities negatively affect the metabolic biomass densities of both mesobrowsers and mesomixed feeders. If we look only at Chase's sampling block Chobe National Park (CNP) F (Chase, 2011), roughly corresponding with the BONIC study area in the north-eastern corner of Chobe National Park, the proportion of grazers is larger, about 84%, reducing the relative contribution by browsers and mixed feeders. This large proportion of grazers is a result of the migration in the dry season of many large, primarily grazing species, such as buffalo and zebra, to the permanent water in the river. These species tend to leave the proximity of the river when the rains commence. Data from the Distance sampling in the BONIC project (Table 13.1) suggest that grazers in the wet season contribute roughly 67% of the metabolic mass of mesoherbivores in the study area, and mixed feeders about 25%. This data set has fewer species than that based on Chase (2011) as many species, including eland, had too few observations for us to use Distance statistics. Even if density estimates should be interpreted with great caution, they indicate a surprisingly large proportion of browsers and mixed feeders in the mesoherbivore community in this environment with high elephant density.

We have recorded and inferred considerable interaction between elephants and smaller herbivores at different spatial and temporal scales. At a small scale we have seen grazing elephant on the floodplain facilitate grazing by buffalo, and buffalo competing with elephants (Taolo, 2003; Chapter 11), similar to results obtained by Vesey-Fitzgerald (1969) and de Boer and Prins (1990). Furthermore, also at a small scale, browsing elephants facilitate browsers and mixed feeders, both mega- and mesoherbivores, with giraffe, kudu and impala in the shrublands alongside the river feeding preferentially from trees previously browsed by elephants (Makhabu *et al.*, 2006; Chapter 12). In the woodlands on sand, twig-biting, presumably mainly by giraffe and kudu (because the small impala does not use that environment: Wallgren, 2008), was positively correlated with the accumulated impact of elephant browsing, at the scale of both individual trees and vegetation patches (Chapter 12). This is caused by morphological and chemical response of trees to browsing, potentially leading to increased accessibility and palatability of browse (Skarpe *et al.*, 2000; Makhabu *et al.*, 2006; Makhabu and Skarpe, 2006). There is evidence that trees respond differently to defoliation by small herbivores, such as insects, and to pruning by, for example, elephant, with responses to the latter often including coppicing and increased shoot- and leaf size and palatability (du Toit *et al.*, 1990; Danell *et al.*, 1994; Hrabar *et al.*, 2009; Scogings *et al.*, 2011; Chapter 12).

At a larger scale, intensive browsing can change vegetation structure and species composition, with preferentially browsed species in many cases being replaced by those less utilised by the principal browser (du Toit *et al.*, 1990; Danell *et al.*, 2003; Holdo, 2003). We infer this to have taken place in the nutrient-poor woodland on sand, where *Baikiaea plurijuga*, which is not eaten by either elephants or ungulates, seems to have become more dominant in the canopy-tree community at the expense of, for example, *Pterocarpus angolensis* (Chapter 4). On the relatively nutrient-rich alluvium, species such as *Acacia* spp., *Faidherbia albida* and *Dichrostachys cinerea*, palatable to both megabrowsers and mesobrowsers, have declined in abundance, a result of elephants killing the large trees and impala, browsing on the seedlings, preventing regeneration (Moe *et al.*, 2009; Chapters 4 and 10). Increasing species such as *Capparis tomentosa* and *Combretum mossambicense* are not palatable to elephants but are frequently eaten by ruminant browsers. Their seedlings, however, seem to be less palatable to or more tolerant of browsing by impala than those of the decreasing species (Makhabu, 2005; Moe *et al.*, 2009; Chapters 4, 10 and 12). Thus, in this environment, elephants do not seem to compete with ruminant browsers or mixed feeders but instead change woody vegetation composition in a way that makes it unfavourable as food for elephants but acceptable for ruminant browsers.

In our study area elephant density in the vegetation on nutrient-rich alluvium close to the river is high, particularly in the dry season, because the elephants spend time there drinking, bathing, utilising clay and mineral licks and socialising, although they forage mainly in the woodlands on sand (Chapters 9 and 12). The herbaceous vegetation on the alluvium is short-cropped by grazing mesoherbivores and hippo and offers little grazing to elephants, and the woody vegetation largely is not palatable to them. Hence, were it not for the vicinity to the permanent water in the river and short distance to foraging areas on the Kalahari sand, this area would support few elephants. It is, however, preferentially used by the majority of mesoherbivore species (Table 13.2). Conversely, in the woodlands on nutrient-poor sand, the density of ungulate grazers, browsers and mixed feeders is low, about 20% of that on alluvium (Wallgren, 2008), because the nutritive quality of the plants there is largely insufficient to sustain these relatively small-bodied herbivores (Jarman, 1974; Bell, 1982). But elephants, because of their large body size and hind-gut fermentation, can tolerate low-quality forage, and so do most of their foraging in these woodlands (Chapter 12). Our observations from Chobe therefore suggest that in the patches of relatively nutrient-rich savanna, mesoherbivores can competitively exclude elephant foraging by direct competition in the same way as described for mesograzers (Vesey-Fitzgerald, 1969; de Boer and Prins, 1990; Taolo, 2003; Arsenault and Owen-Smith, 2002) and by browsing seedlings of woody species preferred by elephants (Prins and van de Jeugd, 1993; Moe *et al.*, 2009). The subsequent change in woody vegetation structure and species composition disadvantaged elephants more than ruminant browsers (Makhabu, 2005). In contrast, in the nutrient poor woodlands foraging by elephants dominates. Thus, there is a negative correlation between density of mesoherbivores of all three feeding guilds and elephant foraging. Were it not for the position of the patches of rich savanna adjacent to the only source of permanent water in the area, this would likely translate into corresponding differences in densities. Similarly, Fritz *et al.* (2002) found a negative correlation between density of elephants and that of mesobrowsers and mesomixed-feeders, but for opposite reasons. While Fritz *et al.* (2002) attributed the negative correlation to elephants competing with smaller browsers and mixed feeders, we found in our area that the smaller herbivores likely competitively excluded foraging by elephants in vegetation of high enough nutritive quality. The discrepancy testifies to the complexity of the relationship between elephants and the grazing and the browsing guilds, and to the need for improved understanding of the significance of interactions between herbivores in savanna ecology.

References

Aarrestad, P.A., Masunga, G.S., Hytteborn, H., Pitlagano, M.L., Marokane, W. & Skarpe, C. (2011) Influence of soil, tree cover and large herbivores on field layer vegetation along a savanna landscape gradient in northern Botswana. *Journal of Arid Environment* 75, 290–297.

Addy, J.E. (1993) Impact on elephant induced vegetation change on the status of the bushbuck (*Tragelaphus scriptus ornatus*) along the Chobe River in northern Botswana. MSc Thesis, University of the Witwatersrand, Johannesburg, South Africa.

Arsenault, R. & Owen-Smith, N. (2002) Facilitation versus competition in grazing herbivore assemblages. *Oikos* 97, 313–318.

Bartlam-Brooks, H.L.A., Bonyongo, M.C. & Harris, S. (2011) Will reconnecting ecosystems allow long-distance mammal migrations to resume? A case study of a zebra *Equus burchelli* migration in Botswana. *Oryx* 45, 210–216.

Bell, R.H.V. (1971) A grazing ecosystem in the Serengeti. *Scientific American* 225, 86–93.

Bell, R.H.V. (1982) The effect of soil nutrient availability on community structure in African ecosystems. In: Huntley, B.J. & Walker, B.H. (eds.) *Ecology of Tropical Savannas.* Springer-Verlag, Berlin, pp. 193–216.

Bergström, R., Skarpe, C. & Danell, K. (2000) Plant responses and herbivory following simulated browsing and stem cutting of *Combretum apiculatum*. *Journal of Vegetation Science* 11, 409–414.

Bodmer, R.E. (1990) Ungulate frugivores and the browser-grazer continuum. *Oikos*, 57, 319–325.

Buckland, S.T., Anderson, D.R., Burnham, K.P., Laake, J.L., Borchers, D.L. & Thomas, L. (2001) *Introduction to Distance Sampling – Estimating Abundance of Biological Populations.* Oxford University Press, Oxford.

Cain, J.W., Owen-Smith, N. & Macandza, V.A. (2012) The costs of drinking: comparative water dependency of sable antelope and zebra. *Journal of Zoology* 286, 58–67.

Campbell, A.C. (1990) History of elephants in Botswana. In: Hancock, P., Cantrell, M. & Hughes, S. (eds.) *The Future of Botswana's Elephants*; Publisher: Kalahari Conservation Society in conjunction with the Department of Wildlife and National Parks Editor: Hancock, P. *Proceedings of the Kalahari Conservation Society Symposium, Gaborone, Botswana*, pp. 5–15.

Cerling, T.E., Harris, J.M. & Passey, B.H. (2003) Diets of East African Bovidae based on stable isotope analysis. *Journal of Mammalogy* 84, 456–470.

Chase, M. (2011) *Dry Season Fixed-wing Aerial Survey of Elephants and Wildlife in Northern Botswana*. Elephants Without Borders, Kasane, Botswana; Department of Wildlife and National Parks, Botswana; and Zoological Society of San Diego, USA. [online] http://www.elephantdatabase.org/population_submission_attachments/102.

Child, G. (1968) *An Ecological Survey of North-eastern Botswana*. Food and Agriculture Organization of the United Nations, Rome.

Child, G. (1972) Observations on a wildebeest die-off in Botswana. *Arnoldia* 31, 1–13.

Child, G. (2004) Growth of modern nature conservation in southern Africa. In: Child, B. (ed.) *Parks in Transition: Biodiversity, Rural Development and the Bottom Line*. Earthscan, London, UK, pp. 7–27.

Child, G. & von Richter, W. (1969) Observations on ecology and behaviour of lechwe, puku and waterbuck along the Chobe River, Botswana. *Zeitschrift für Sägetierkunde* 34, 275–295.

Clauss, M., Kaiser, T. & Hummel, J. (2008) The morphophysiological adaptations of browsing and grazing mammals. In: Gordon, I.J. & Prins, H.H.T. (eds.) *The Ecology of Browsing and Grazing*. Springer, Berlin, pp. 47–88.

Coe, M.J., Cumming, D.H.M. & Phillipson, J. (1976) Biomass and production of large African herbivores in relation to rainfall and primary production. *Oecologia* 22, 341–354.

Cumming, D.H.M., Fenton, M.B., Rautenbach, I.L., Taylor, R.D., Cumming, G.S., Cumming, M.S., Dunlop, J.M., Ford, G.S., Hovorka, M.D., Johnston, D.S., Kalcounis, M.C., Mahlanga, Z. & Portfors, C.V. (1997) Elephants, woodlands and biodiversity in southern Africa. *South African Journal of Science* 93, 231–236.

Danell, K., Bergström, R. & Edenius, L. (1994) Effects of large mammalian browsers on architecture, biomass, and nutrients of woody plants. *Journal of Mammalogy* 75, 833–844.

Danell, K., Bergström, R., Edenius, L. & Ericsson, G. (2003) Ungulates as drivers of tree population dynamics at module and genet levels. *Forest Ecology and Management* 181, 67–76.

de Boer, W.F. & Prins, H.H.T. (1990) Large herbivores that strive mightily but eat and drink as friends. *Oecologia* 82, 264–274.

Department of Wildlife and National Parks (2013) *Wildlife Aerial Surveys of 2012 Report*. Department of Wildlife and National Parks, Ministry of Environment, Wildlife and Tourism, Government Printer, Gaborone, Botswana.

Dipotso, F.M. & Skarpe, C. (2006) Population status and distribution of puku in a changing riverfront habitat in northern Botswana. *South African Journal of Wildlife Research*, 36, 89–97.

Dipotso, F.M., Skarpe, C., Kelaeditse, L. & Ramotadima, M. (2007) Chobe bushbuck in an elephant-impacted habitat along the Chobe River. *African Zoology* 42, 261–267.

du Toit, J.T. (1993) The feeding ecology of a very small ruminant, the steenbok (*Raphicerus campestris*). *African Journal of Ecology* 341, 35–48.

du Toit J.T. (1995) Sexual segregation in kudu: sex differences in competitive ability, predation risk or nutritional needs? *South African Journal of Wildlife Research* 25, 127–132.

du Toit, J.T., Bryant, J.P. & Frisby, K. (1990) Regrowth and palatability of Acacia shoots following pruning by African savanna browsers. *Ecology* 71, 149–154.

Estes, R.D. (1991) *The Behavior Guide to African Mammals: Including Hoofed Mammals, Carnivores, Primates.* University of California Press, Berkeley, USA.

Fritz, H. (1997) Low ungulate biomass in west African savannas: primary production or missing megaherbivores or large predator species. *Ecography* 20, 417–421.

Fritz, H. & Duncan, P. (1994) On the carrying capacity for large ungulates of African savanna ecosystems. *Proceedings of the Royal Society of London B* 256, 77–82.

Fritz, H., Duncan, P., Gordon, I.J. & Illius A.W. (2002) Megaherbivores influence trophic guilds structure in African ungulate communities. *Oecologia* 131, 620–625.

Gagnon, M. & Chew, A.E. (2000) Dietary preferences in extant African Bovidae. *Journal of Mammalogy* 81, 490–511.

Ginnett, F.T. & Demment, W.M. (1997) Sex differences in giraffe foraging behavior at two spatial scales. *Oecologia* 110, 291–300.

Gordon, I.J. (2003) Browsing and grazing ruminants: are they different beasts? *Forest Ecology and Management* 181, 13–21.

Halley, D.J. & Minagawa, M. (2005) African buffalo diet in a woodland and bush dominated biome, as determined by stable isotope analysis. *African Zoology* 40, 160–163.

Halley, D.J., Vandewalle, M., Mari, M. & Taolo, C. (2002) Herd-switching and long-distance dispersal in female African buffalo *Syncerus caffer*. *African Journal of Ecology* 40, 97–99.

Heitkönig, I.M.A. & Owen-Smith, N. (1998) Seasonal selection of soil types and grass swards by roan antelope in a South African savanna. *African Journal of Ecology* 36, 57–70.

Hensman, M.C., Owen-Smith, N., Parrini, F. & Erasmus, B.F.N. (2012) Dry season browsing by sable antelope in northern Botswana. *African Journal of Ecology* 50, 513–516.

Herremans, M. (1995) Effects of woodland modification by African elephant *Loxodonta africana* on bird diversity in northern Botswana. *Ecography* 18, 440–454.

Holdo, R. (2003) Woody plant damage by African elephants in relation to leaf nutrients in western Zimbabwe. *Journal of Tropical Ecology* 19, 189–196.

Hrabar, H., Hattas, D. & du Toit, J.T. (2009) Differential effects of defoliation by mopane caterpillars and pruning by African elephants on the regrowth of *Colophospermum mopane* foliage. *Journal of Tropical Ecology* 25, 301–309.

Illius, A.W. & Gordon, I.J. (1987) The allometry of food intake in grazing ruminants. *Journal of Animal Ecology* 56, 989–999.

Jarman, P.J. (1974) The social organisation of antelope in relation to their ecology. *Behaviour* 48, 215–267.

Junker, J., van Aarde, R.J. & Ferreira, S.M. (2008) Temporal trends in elephant *Loxodonta africana* numbers and densities in northern Botswana: is the population really increasing? *Oryx* 42, 58–65.

Kandume, J.N. (2012) Sexual segregation in foraging of greater kudu (*Tragelaphus strepsiceros*) in a heterogeneous savanna in Chobe National Park, Botswana. MSc Thesis, Hedmark University College, Evenstad, Norway.

Kanga, E.M., Ogutu, J.O., Piepho, H.-P. & Olff, H. (2013) Hippopotamus and livestock grazing: influences on riparian vegetation and facilitation of other herbivores in the Mara Region of Kenya. *Landscape and Ecological Engineering* 9, 47–58.

Kazonganga, B. (2011) Sexual segregation in greater kudu (*Tragelaphus strepsiceros*) in a heterogeneous savanna in Chobe National Park, Botswana. MSc Thesis, Hedmark University College, Evenstad, Norway.

Knoop, M.-C. & Owen-Smith, N. (2006) Foraging ecology of roan antelope: key resources during critical periods. *African Journal of Ecology* 44, 228–236.

Lewin, R. (1986) In ecology, change brings stability. *Science* 234, 1071–1073.

Lunt, N. (2011) Role of small antelope in ecosystem functioning in the Matobo Hills, Zimbabwe. PhD Thesis, Rhodes University, Grahamstown, South Africa.

Macandza, V.A., Owen-Smith, N. & Cain, J.W. (2012) Dynamic spatial partitioning and coexistence among tall grass grazers in an African savannah. *Oikos* 121, 891–898.

Makhabu, S.W. (2005) Resource partitioning within a browsing guild in a key habitat, the Chobe Riverfront, Botswana. *Journal of Tropical Ecology* 21, 641–649.

Makhabu, S.W., Skarpe, C. & Hytteborn, H. (2006) Elephant impact on shoot distribution on trees and on rebrowsing by smaller browsers. *Acta Oecologica* 30, 136–146.

Manly, B.F.J., McDonald, L.L. & Thomas D.L. (1993) *Resource Selection by Animals.* Chapman & Hall, London, UK.

Mduma, S.A.R., Sinclair, A.R.E. & Hilborn, R. (1999) Food regulates the Serengeti wildebeest: a 40-year record. *Journal of Animal Ecology* 68, 1101–1122.

Moe, S.R., Rutina, L.P. & du Toit, J.T. (2007) Trade-off between resource seasonality and predation risk explains reproductive chronology in impala. *Journal of Zoology* 273, 237–243.

Moe, S.R., Rutina, L.P., Hytteborn, H. & du Toit, J.T. (2009) What controls woodland regeneration after elephants have killed the big trees? *Journal of Applied Ecology* 46, 223–230.

Murray, M.G. & Illius, A. (2000) Vegetation modification and resource competition in grazing ungulates. *Oikos* 89, 501–508.

Omphile, U.J. (1997) Seasonal diet selection and quality of large savanna ungulates in Chobe National Park, Botswana. PhD Thesis, University of Wyoming, Laramie, Wyoming, USA.

Omphile, U.J. & Powell, J. (2002) Large ungulate habitat preference in Chobe National Park, Botswana. *Journal of Range Management* 55, 341–349.

O'Shaughnessy, R. (2011) Comparative diet and habitat selection of puku (*Kobus vardonii*) and lechwe (*Kobus leche*) on the Chobe River floodplain, Botswana. MSc Thesis, University of the Witwatersrand, Johannesburg, South Africa.

Owen-Smith, N. (1990) Demography of a large herbivore, the greater kudu *Tragelaphus strepsiceros*, in relation to rainfall. *Journal of Animal Ecology* 59, 893–913.

Owen-Smith, N. & Novellie, P. (1982) What should a clever ungulate eat? *The American Naturalist* 119, 151–178.

Owen-Smith, R.N. (1988) *Megaherbivores: The Influence of Very Large Body Size on Ecology.* Cambridge University Press, Cambridge, UK.

Prins, H.H.T. & van de Jeugd, H. (1993) Herbivore population crashes and woodland structure in East Africa. *Journal of Ecology* 81, 305–314.

Rutina, L.P. (2004) Impalas in an elephant-impacted woodland: browser-driven dynamics of the Chobe riparian zone, Northern Botswana. PhD Thesis, Agricultural University of Norway, Ås, Norway.

Rutina, L.P., Moe, S.R. & Swenson, E.E. (2005) Elephant *Loxodonta africana* driven woodland conversion to shrubland improves dry-season browse availability for impalas *Aepyceros melampus*. *Wildlife Biology* 11, 207–213.

Scogings, P.F., Hjältén, J. & Skarpe, C. (2011) Secondary metabolites and nutrients of woody plants in relation to browsing intensity in African savannas. *Oecologia* 167, 1063–1073.

Selous, F.C. (1881) *A Hunter's Wanderings in Africa*. Richard Bentley & Son, London, UK.

Sheppe, W. & Haas, P. (1976) Large mammal populations of the lower Chobe River, Botswana. *Mammalia* 40, 223–244.

Simpson, C.D. (1974) Ecology of the Zambezi Valley bushbuck *Tragelaphus scriptus ornatus* Pocock. PhD Thesis, Texas A&M University, College Station, Texas.

Sinclair, A.R.E., Mduma, S. & Brashares, J.S. (2003) Patterns of predation in a diverse predator–prey system. *Nature* 425, 288–290.

Skarpe, C., Bergström, R., Bråten, A.-L. & Danell, K. (2000) Browsing in a heterogeneous savanna. *Ecography* 23, 632–640.

Skarpe, C., Aarrestad, P.A., Andreassen, H.P., Dhillion, S., Dimakatso, T., du Toit, J.T., Halley, D.J., Hytteborn, H., Makhabu, S., Mari, M., Marokane, W., Masunga, G., Modise, D., Moe, S.R., Mojaphoko, R., Mosugelo, D., Motsumi, S., Neo-Mahupeleng, G., Ramotadima, M., Rutina, L., Sechele, L., Sejoe, T.B., Stokke, S., Swenson, J.E., Taolo, C., Vandewalle, M., & Wegge, P. (2004) The return of the giants: ecological effects of an increasing elephant population. *Ambio* 33, 276–282.

Spinage, C.A. (2012) *African Ecology – Benchmarks and Historical Perspectives*. Springer-Verlag, Berlin.

Sponheimer, M., Lee-Thorp, J.A., DeRuiter, D.J., Smith, J.M., van der Merwe, N.J., Reed, K., Grant, C.C., Ayliffe, L.K., Robinson, T.F. Heilelberger, C. & Marcus, W. (2003) Diets of southern African Bovidae: stable isotope evidence. *Journal of Mammalogy* 84, 471–479.

Stokke, S. & du Toit, J.T. (2000) Sex and size related differences in the dry season feeding patterns of elephants in Chobe National Park, Botswana. *Ecography* 23, 70–80.

Stokke, S., Motsumi, S., Skarpe, C. & Swenson, J.E. (2003) Ungulate population densities and composition in relation to habitat types and browsing pressure from elephant and other browsers in the Chobe River area, northern Botswana. In: Vandewalle, M. (ed.) *Effects of Fire, Elephants and Other Herbivores on the Chobe Riverfront Ecosystem*. Government Printer, Gaborone, Botswana, pp. 57–58.

Taolo, C.L. (2003) Population ecology, seasonal movement and habitat use of African buffalo (*Syncerus caffer*) in Chobe National Park, Botswana. PhD Thesis, Norwegian University of Science and Technology, Trondheim, Norway.

Valeix, M., Fritz, H., Sabatier, R., Murindagomo, F., Cumming, D. & Duncan, P. (2011) Elephant-induced structural changes in the vegetation and habitat selection by large herbivores in an African savanna. *Biological Conservation* 144, 902–912.

van de Koppel, J. & Prins, H.H.T. (1998) The importance of herbivore interactions for the dynamics of African savanna woodlands: an hypothesis. *Journal of Tropical Ecology* 14, 565–576.

Vesey-Fitzgerald, D.F. (1969) Utilisation of the habitat by buffalo in Lake Manyara National Park. *East African Wildlife Journal* 7, 131–145.

Wallgren, M. (2008) Mammal community structure in a world of gradients: effects of resource availability and disturbance across scales and biomes. PhD Thesis, Swedish University of Agricultural Sciences, Umeå, Sweden.

Woolnough, A.P. & du Toit, J.T. (2001) Vertical zonation of browse quality in tree canopies exposed to a size-structured guild of African browsing ungulates. *Oecologica* 129, 585–590.

14

Cascading Effects on Smaller Mammals and Gallinaceous Birds of Elephant Impacts on Vegetation Structure

Sigbjørn Stokke[1], Sekgowa S. Motsumi[2,3], Thato B. Sejoe[4,5] and Jon E. Swenson[6,7]

[1]Norwegian Institute for Nature Research, Norway
[2]Department of International Environment and Development Studies, Norwegian University of Life Sciences, Norway
[3]Department of Environmental Affairs, Botswana
[4]Department of International Environment and Development Studies, Norwegian University of Life Sciences, Norway
[5]PO Box 1826, Gaborone, Botswana
[6]Department of Ecology and Natural Resource Management, Norwegian University of Life Sciences, Norway
[7]Norwegian Institute for Nature Research, Norway

The elephant *Loxodonta africana* population in Botswana, currently estimated at 207,273 (Department of Wildlife and National Parks, 2013), has been increasing steadily since the 1930s as a result of protection and the conservation of habitats. The impact of such a large number of megaherbivores on the vegetation can have a dominant influence on the composition and abundance of both plant and animal communities (Skarpe *et al.*, 2004; Asner and Levick, 2012). The effects of this impact on other animal species can be either positive, for example, by creating mosaic habitat diversity that promotes the coexistence of a number of animals (Owen-Smith, 1988; van de Koppel and Prins, 1998; Valeix *et al.*, 2011), or negative, by causing changes that lower species richness (Cumming *et al.*, 1997; Fritz *et al.*, 2002). Elephants have

Elephants and Savanna Woodland Ecosystems: A Study from Chobe National Park, Botswana,
First Edition. Edited by Christina Skarpe, Johan T. du Toit and Stein R. Moe.
© 2014 John Wiley & Sons, Ltd. Published 2014 by John Wiley & Sons, Ltd.

thus been termed a keystone species in savannas, because of their pronounced impact on vegetation (Owen-Smith, 1988).

Studies summarised in this book show that there is a strong gradient of elephant impact, with increasing current and long-term browsing pressure from the interior *Baikiaea* woodland through mixed woodland to the shrublands near the Chobe River (Mosugelo *et al.*, 2002). Current elephant browsing was highest in the mixed wood-land and in the *Combretum* shrubland, whereas current ungulate browsing peaked in the *Capparis* shrubland. The shrubland habitats near the Chobe River have been subjected to intensive long-term herbivore impacts (Child, 1968; Simpson, 1975; Som-merlatte, 1976; Melton, 1985), resulting in vegetation types that are partly a product of long-term cumulative elephant impact. The high current elephant impact in the mixed woodland is largely the result of an abundance of palatable species, such as *Combre-tum elaeagnoides and Friesodielsia obovata*. In the shrublands, however, high long-term utilization has resulted in a mosaic of vegetation dominated by species that are unpalat-able to elephants, such as *Capparis tomentosa* and *Croton megalobotrys*, which therefore show little observable impact. As a result, the long-term cumulative impact of elephants is highest in the shrubland habitats. The increasing population of elephants since the 1930s has been a major agent in creating the present vegetation pattern in the study area (Mosugelo *et al.*, 2002; Skarpe *et al.*, 2004).

The main objective of our study was to determine habitat selection by medium-sized and small mammals, and gallinaceous birds (families Phasianidae and Numididae), in northern Chobe National Park, especially in relation to the long-term impacts that elephants have had on the vegetation structure. The working hypothesis of the Botswana–Norway Institutional Cooperation Project (BONIC), of which this research was part, was that the elephant population facilitates nutrient transfer from a slow and deep tree-mediated nutrient cycle into a fast and shallow cycle that supports herbaceous vegetation, palatable woody plants, and ultimately a higher biomass and diversity of other animal species (Chapter 3). The specific goal of our study was to infer potential cascading effects of elephants on gallinaceous birds and small and medium-sized mammals, based on estimates of their preference and avoidance of habitats subject to long-term, elephant-induced impacts on the vegetation; large herbivores are covered in Chapter 13. Thus, following the model of Pickett *et al.* (2003), we consider these species to be responders, through their habitat selection, with elephants being the agents, through their modification of the habitats. Soil productivity and other site factors are the controllers in this system, although these are not covered here.

We worked on different scales for the different vertebrate groups. Differences in scale are important, because they influence the questions asked, how they are answered, the data obtained, and interpretation of the results (Wiens, 1989). We studied medium-sized mammals and gallinaceous birds at the habitat scale. We also used a more local scale for the gallinaceous birds, comparing horizontal and vertical cover at sites where we observed birds and random sites within high bird-density habitat. Our study of small mammals was the most local, as we only live-trapped small mammals at three sites in *Baikiaea* woodland, which has been substantially reduced during the last 35 years, because of elephant browsing and fire outbreaks (Mosugelo *et al.*, 2002). Although the most numerous tree species in the woodlands, *Baikiaea plurijuga*, is not used much by elephants (Ben-Shahar, 1996; Stokke, 1999), many of the other species in this woodland type are.

Our study area was about 403 km² between the Kasane-Ngoma tarmac road to the south and the Chobe River to the north. All transects from which we estimated animal habitat selection were located between Kaluwizi Valley near Chobe Game Lodge in the east and route 16 further west (Figure 14.1). The Chobe River is a permanent source of water for wildlife in the study area. The substrate supporting the vegetation in Chobe National Park and its surroundings is Kalahari Sand, deep, well-drained, nutrient-poor sands of largely aeolian origin (Sommerlatte, 1976; Chapter 2). The vegetation of the study area is described in Chapter 5. We based our study on the vegetation types identified and mapped by Skarpe *et al.* (2004): seasonally flooded floodplain (called 'floodplain'), *Capparis* shrubland, *Combretum* shrubland, mixed *Baikiaea* woodland, and *Baikiaea* woodland. A few of the transects close to the river included a mixture of several habitats, such as riparian woodland and riverine shrublands dominated by *Combretum* spp. and *Capparis tomentosa* which we included in floodplain, making it, for our purposes, a mosaic of shrublands, riparian woodland and floodplain. The vegetation types were generally not clearly separated, but had gradual and patchy boundaries. For example, pockets of mixed *Baikiaea* woodland occurred within the *Baikiaea* woodland, and relics of riparian woodland existed in the shrublands near the river. The only relatively distinct vegetation type was the seasonally flooded floodplain (Figure 14.1).

Medium-sized mammals and gallinaceous birds: large-scale habitat selection and diversity

We estimated habitat selection and species diversity indices from observations made while driving transects over a 4 year period, October 1998 to February 2002, covering four dry seasons and three wet seasons. These transects comprised five relatively straight tracks, formerly used as firebreaks, running perpendicular to the river, and three tourist routes that ran parallel to the river (Figure 14.1). Most of the firebreaks were rarely used by unofficial vehicles and one of them was hard to follow, because it had not been used for many years. The tourist routes, on the other hand, were heavily used during daytime. We drove along the transects at $15-20\,\mathrm{km\,hr^{-1}}$, with two observers, using binoculars, standing in the back of the vehicle. To avoid autocorrelation, we waited at least one day before repeating a transect. We drove the transects in the morning (07h00–11h00) and afternoon (15h00–18h00) to target peak activity

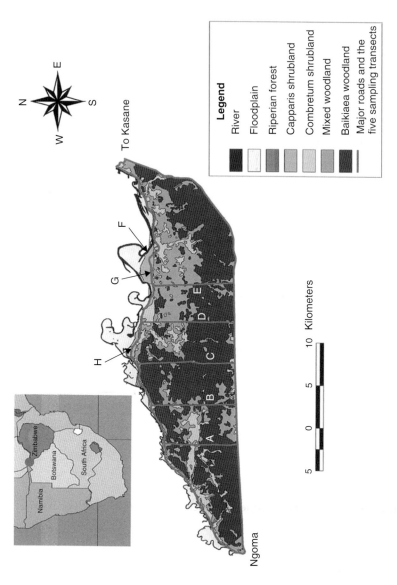

Figure 14.1 Habitat map of the study area in northern Chobe National Park, showing the habitat types and transects perpendicular (A – E) to and parallel (F – H) with the Chobe River. (Source: C. Skarpe 2004 in AMBIO 33, 276–282; Copyright Royal Swedish Academy of Sciences.)

periods, and at night (21h00–01h00), using a spotlight on each side of the road to locate nocturnal mammals. We used a rangefinder to measure the perpendicular distances of observed animals from the transects. For groups of animals of a species, distances were measured to the centre of the group. The longest transect was 12 km and was perpendicular to the river; the shortest was 4.7 km, parallel to the river. Distances travelled per habitat type were: *Baikiaea* woodland, 2024 km; floodplain, 897 km; mixed *Baikiaea* woodland, 604 km; *Capparis* shrubland, 105 km; and *Combretum* shrubland, 99 km.

The proportions of sightings of mammals and birds in each habitat were compared with the proportions of total road distance driven along transects in each habitat. This sampling design means that a standardised selection ratio, B_i, of habitat preferences could be calculated as follows (Manly *et al.*, 1993):

$$B_i = \frac{w_i}{\sum\limits_{i=1}^{n} w_i} \tag{14.1}$$

where $w_i = u_i/a_i$, with u_i being the proportion of animal or bird sightings in habitat type i (i.e. proportional 'use') and a_i being the proportion of total transect distance surveyed in habitat type i (i.e. proportional 'availability'). The individual habitat preferences sum to one. This selection index measures the estimated probability that a category i resource unit would be the next one selected if all resource units were equally available (Manly *et al.*, 1993). This is a convenient way of presenting selection ratios, because removing one or more categories from the analysis leaves the indices largely unaffected (Manly *et al.*, 1993). The null hypothesis was that animals select habitats in proportion to availability. To test for statistically significant departures from the null hypothesis of no habitat selection, we applied a log-likelihood chi-square statistic (Manly *et al.*, 1993). If the test value was significantly large, we computed simultaneous confidence intervals for the population proportions of the different types of resources used and compared these with the available proportions, using Bonferroni-adjusted confidence intervals (Manly *et al.*, 1993). For these tests to be valid, the expected frequencies should be five or more. If they are not, the test could still be valid, but obviously the outcome should be treated with some reservation. Selection indices were calculated separately for each habitat type for combined dry and wet seasons and presented as either against (selected less than expected), for (selected more than expected) or none (selected as expected). These calculations were executed in a customised Visual FoxPro (9.0 SP 2) programme.

In addition, we calculated the diversity index and evenness for animal species per habitat type and season (dry, wet and combined) using the Shannon-Wiener diversity index H:

$$H = -\sum p_{u_i} \log_e p_{u_i} \tag{14.2}$$

where u_i is the proportion of species u in habitat i (Shannon, 1948). Evenness J was calculated as $H/\log_e k$, where k is the number of species (Pielou, 1977).

Baikiaea woodland and mixed Baikiaea woodland were most often avoided by the species we studied and the shrublands and floodplain were the most often selected (Tables 14.1 and 14.2). This pattern was consistent for gallinaceous birds, carnivores, and other mammals (Table 14.2). We found 28 cases of selection for and 11 cases of selection against the three habitats with high long-term elephant impact, compared with 2 and 30 cases respectively, for the two woodland habitats, which had low and moderate elephant impact. This difference was statistically significant (χ^2 with Yate's correction $= 30.9$, df $= 1$, $p < 0.0001$). The result would be similar even if we disregarded floodplain, with 11 cases of selection for and no cases of selection against the two shrubland types (Fisher's exact test, $p < 0.0001$).

Species richness was independent of season (Table 14.3, $\chi^2 = 0.875$, df $= 4$, $p = 0.928$). Nevertheless, the number of species differed among habitats within seasons and were always largest in floodplain and lowest in Combretum shrubland (Table 14.3, wet: $\chi^2 = 10.82$, df $= 4$, $p = 0.028$; dry: $\chi^2 = 10.04$, df $= 4$, $p = 0.040$; combined: $\chi^2 = 9.211$, df $= 4$, $p = 0.056$). In addition, Baikiaea woodland always had the second largest species richness (Table 14.3). The diversity index was always largest for Baikiaea woodland and smallest for Combretum shrubland (Table 14.3). Similarly, evenness was always largest for Baikiaea woodland, except during the wet season, when mixed Baikiaea woodland had a marginally greater evenness. This suggests that, even though there were always more species in floodplains, the species there were less evenly distributed; some were abundant whereas others were rare (see list of species in Table 14.1).

Gallinaceous birds: selection of cover within high-density habitats

Whenever we sighted a gallinaceous bird (helmeted guineafowl, Numida meleagris, red-billed francolin, Francolinus adspersus, Swainson's francolin, Francolinus swainsonii, and crested francolin, Francolinus sephaena) along the transects in the shrubland habitats in August 2001 and February 2002 (one dry season and one wet season), we established a plot of 10-m radius centred on the position where each bird or group had been sighted. This plot was termed a 'bird plot' and the following attributes were measured: horizontal visual obscurity, vertical cover, GPS position, number of elephant dung piles (dry and moist), current elephant impact (recorded on a scale of 0–2, where 0 = no observable impact, 1 = recognisable impact but with no change of growth form, and 2 = browsing has changed growth form), bird species, number of adults and juveniles, and behaviour (still in sun, still in shade, feeding on ground, feeding on elephant dung, moving, and alert).

Table 14.1 Selection for (+) and against (−) the major habitats in the Chobe National Park by medium-sized mammals and gallinaceous birds based on the standardised selection index.

Group and species	Scientific name	BW (0.543) Observed	Index	Selection	MW (0.162) Observed	Index	Selection	COM (0.026) Observed	Index	Selection	CAP (0.028) Observed	Index	Selection	FP (0.241) Observed	Index	Selection	Log-likelihood χ^2 value
Carnivores																	
African wild cat	*Felis sylvestris*	17	0.050	−	7	0.070	0	1	0.062	0	10	0.577	+	36	0.241	+	**49.84**
Caracal	*Felis caracal*	1	1	−	0	0	−	0	0	−	0	0	−	0	0	−	1.22
Leopard	*Panthera pardus*	0	0	−	0	0	−	0	0	−	0	0	−	1	1	−	2.84
Lion	*Panthera leo*	0	0	−	0	0	−	0	0	−	3	0.618	0	16	0.382	+	**50.42**
African civet	*Civettictis civetta*	2	0.077	−	0	0	−	0	0	−	1	0.749	−	2	0.174	−	4.74
Honey badger	*Mellivora capensis*	2	0.228	−	0	0	−	0	0	−	0	0	−	3	0.772	−	4.26
Brown hyaena	*Hyaena brunnea*	0	0	−	0	0	−	0	0	−	0	0	−	1	1	−	2.84
Spotted hyaena	*Crocuta crocuta*	4	0.062	−	0	0	−	0	0	−	0	0	−	27	0.938	+	**57.88**
African wild dog	*Lycaon pictus*	2	0.166	−	3	0.834	−	0	0	−	0	0	−	0	0	−	6.64
Black-backed jackal	*Canis mesomelas*	0	0	−	0	0	−	0	0	−	2	0.896	0	2	0.104	0	**14.44**
Side-striped jackal	*Canis adustus*	0	0	−	0	0	−	0	0	−	0	0	−	5	1	+	**14.22**
Bat-eared fox	*Otocyon megalotis*	7	0.092	−	3	0.132	−	0	0	−	2	0.509	−	9	0.267	−	7.66
Large-spotted genet	*Genetta tigrina*	44	0.123	−	16	0.150	0	5	0.292	0	3	0.163	0	43	0.271	+	**14.66**
Small-spotted genet	*Genetta genetta*	4	0.263	−	2	0.441	−	0	0	−	0	0	−	2	0.296	−	1.22
Banded mongoose	*Mungos mungo*	3	0.024	−	2	0.054	0	0	0	−	3	0.468	0	25	0.453	+	**49.66**
Meller's mongoose	*Rhynchogale melleri*	1	0.069	−	2	0.463	−	0	0	−	0	0	−	3	0.467	−	4.90
Selous' mongoose	*Paracynictis selousi*	2	0.055	−	2	0.184	−	1	0.575	−	0	0	−	3	0.186	−	4.44
Slender mongoose	*Herpestes sanguineus*	8	0.044	−	3	0.056	0	3	0.347	0	2	0.217	0	27	0.336	+	**37.48**
White-tailed mongoose	*Ichneumia albicauda*	0	0	−	0	0	−	0	0	−	1	0.896	0	1	0.104	−	7.22
Yellow mongoose	*Cynictis penicillata*	0	0	−	0	0	−	0	0	−	0	0	−	2	1	−	5.70
White-naped weasel	*Poecilogale albinucha*	1	1	−	0	0	−	0	0	−	0	0	−	0	0	−	1.22

(continued overleaf)

Table 14.1 (*Continued*)

Group and species	Scientific name	BW (0.543)			MW (0.162)			COM (0.026)			CAP (0.028)			FP (0.241)			Log-likelihood χ^2 value
		Observed	Index	Selection	Observed	Index	Selection	Observed	Index	Selection	Observed	Index	Selection	Observed	Index	Selection	
Primates																	
Baboon	*Papio hamadryas*	4	0.004	–	2	0.007	–	28	0.602	+	15	0.299	+	38	0.088	+	**213.06**
Bush-baby sp.	*Galago sp*	16	0.315	–	4	0.264	–	0	0	–	1	0.378	–	1	0.044	–	7.90
Lesser Bush-baby	*Galago senegalensis*	17	0.465	+	0	0	–	0	0	–	1	0.535	0	0	0	–	**20.20**
Green monkey	*Cercopithecus pygerythrus*	0	0	–	0	0	–	0	0	–	3	0.838	0	5	0.162	0	**25.08**
Rodents, Lagomorphs and Springhares																	
Porcupine	*Hystrix africaeaustralis*	2	1	–	0	0	–	0	0	–	0	0	–	0	0	–	2.44
Scrub hare	*Lepus saxatilis*	4	0.051	–	0	0	–	1	0.262	0	0	0	–	24	0.687	+	**48.82**
Springhare	*Pedetes capensis*	21	0.011	–	5	0.009	–	7	0.079	0	69	0.719	+	150	0.181	+	**487.12**
Tree squirrel	*Paraxerus cepapi*	9	0.017	–	37	0.231	+	8	0.314	0	4	0.144	0	70	0.294	+	**135.68**
Pigs																	
Warthog	*Phacochoerus africanus*	8	0.019	–	1	0.008	–	1	0.052	0	13	0.610	+	57	0.310	+	**135.64**
Aardvark																	
Aardvark	*Orycteropus afer*	2	0.307	–	0	0	–	0	0	–	0	0	–	2	0.693	–	2.58
Gallinaceous birds																	
Coqui francolin	*Francolinus coqui*	2	1	–	0	0	–	0	0	–	0	0	–	0	0	–	2.44
Crested francolin	*Francolinus sephaena*	2	0.005	–	7	0.057	0	7	0.356	0	6	0.284	0	54	0.297	+	**126.92**
Red-billed francolin	*Francolinus adspersus*	7	0.002	–	24	0.021	–	32	0.176	+	98	0.501	+	507	0.300	+	**1399.40**
Swainson's francolin	*Pternistis swainsoni*	11	0.005	–	23	0.034	–	28	0.259	+	60	0.516	+	186	0.186	+	**549.10**
Helmeted guineafowl	*Numida meleagris*	73	0.015	–	105	0.074	0	73	0.319	+	101	0.407	+	394	0.185	+	**850.40**

No selection either way is also shown (0). The indices are based on combined dry and wet season data. Log-likelihood χ^2 test values are also given, with significant results shown in bold. The habitats are: *Baikiaea* woodland (BW), mixed woodland (MW), *Combretum* shrubland (COM), *Capparis* shrubland (CAP) and floodplain (FP). The proportions of each habitat are given in brackets.

Table 14.2   **The number of species in three vertebrate categories from Table 14.1 selecting significantly for or against the five habitat types on the Chobe National Park study area.**

Habitat type	Long-term elephant impact	Selection	Gallinaceous birds (5)	Carnivores (21)	Other mammals (9)	Total (35)
Baikiaea woodland	Low	For	–	–	1	1
		Against	4	8	6	18
Mixed woodland	Moderate	For	–	–	1	1
		Against	2	4	6	12
Combretum shrubland	High	For	3	–	1	4
		Against	–	5	2	7
Capparis shrubland	High	For	3	2	3	8
		Against	–	2	1	3
Floodplain	High	For	4	7	5	16
		Against	–	–	1	1

The total number of species in each category is shown in parenthesis.

We estimated horizontal visual obscurity from the ground level up to 1 m using a 7-cm wide, 1-m long modified vegetation profile board (Nudds, 1977). Estimates were made for each 10 cm from a height of about 30 cm and a distance of 10 m from each of the four cardinal directions. For each 10 cm, we recorded the obscured portion of the board, based on four categories (0 = totally visible, 1 = more than 50% visible, 2 = less than 50% visible, and 3 = totally obscured), with a minimum score of 0 and a maximum score of 12. We grouped the scores into three categories; low (0–0.20 m), medium (0.21–0.50 m), and high (0.51–1.00 m). We estimated vertical cover within a 10-m radius from the centre of each plot at different height classes (0 < 0.5 m, 0.5–2 m, 2–10 m and > 10 m), using an 11-point scale from 0 to 100% cover, (0 = 0%, points 1–10 at successive 10% intervals). For consistency, only one observer (SSM) took the measurements.

Control plots were sampled in the same habitats in the same way, except for bird data. In transects parallel to the river, control plots were placed systematically every 250 m on alternating sides of the transects, with the first plot being selected randomly from the start of the transect. Along the perpendicular transects, the control plots were 500 m apart, because these transects were about twice as long as those parallel to the river. To avoid road effects, the control plots were placed 20 m from the road. GPS positions were taken at each plot to locate them on the habitat map. We did not establish any

Table 14.3 **Richness, Shannon–Wiener diversity indices (*H*), and derived evenness indices (*J*) of vertebrate species from Table 14.1 per habitat and on the Chobe National Park study area.**

Season	Habitat	Richness	Diversity *H*	Evenness *J*
Wet	BW	17	2.472	0.87
	MW	9	1.942	0.88
	CAP	13	2.045	0.80
	COM	7	1.662	0.85
	FP	22	2.124	0.69
Dry	BW	25	2.569	0.80
	MW	16	1.835	0.66
	CAP	15	1.912	0.71
	COM	11	1.792	0.75
	FP	27	2.208	0.67
Combined	BW	28	2.621	0.79
	MW	18	1.995	0.69
	CAP	20	2.006	0.67
	COM	13	1.857	0.72
	FP	30	2.225	0.65

The habitats are: *Baikiaea* woodland (BW), mixed woodland (MW), *Combretum* shrubland (COM), *Capparis* shrubland (CAP), and floodplain (FP).

plots in *Baikiaea* woodland or mixed *Baikiaea* woodland habitats, because none of the gallinaceous species selected these woodland habitats (Table 14.2).

We used a multivariate General Linear Model (GLM) within the Statistical Package for Social Sciences (SPSS) to analyse how the gallinaceous bird species responded to the explanatory variables across habitats in the dry and wet seasons. The following explanatory variables were examined: horizontal visual obscurity, vertical cover, total number of dry and moist elephant dung piles, and estimates of elephant impact in the high bird-density shrubland habitats. We built one model per bird species and merged the remaining species with the control plots to contrast the target species to both the other species and the controls. In so doing, we assumed that the interspecific competition was low. This approach allowed us to investigate the effect of explanatory variables on each species. In addition, we could compare the species by comparing the species-specific GLM coefficients.

The explanatory variables differed among the plots between the seasons for all bird species (Wilks' lambda = 0.000 for all species). As expected, we counted fewer elephant dung and registered more horizontal visual obscurity and vertical cover in the wet season (Table 14.4). Elephant impact, however, did not vary between seasons (Table 14.4).

Table 14.4 **Comparisons among species of gallinaceous birds regarding the habitat-related explanatory variables between seasons (dry = D and wet = W) and plot types (bird plots = plot and control plots = control) on the Chobe National Park study area.**

Source	Explanatory variables	Sum of squares	df	F	Significance
Crested francolin (n = 38)					
Season (D vs. W)	Elephant dung	1,741.515	1	200.314	0.000
	Browsing impact	1.017	1	1.984	0.159
	Horizontal < 0.2	10,066.470	1	249.503	0.000
	Horizontal 0.2–0.5	9,224.938	1	103.109	0.000
	Horizontal > 0.5	14,561.012	1	81.735	0.000
	Vertical < 0.5	655.377	1	402.202	0.000
	Vertical 0.5–2	35.177	1	41.272	0.000
	Vertical 2–10	0.004	1	0.001	0.973
	Vertical > 10	0.402	1	1.657	0.198
Category (plot vs. control)	Elephant dung	23.308	1	2.681	0.102
	Browsing impact	0.100	1	0.196	0.658
	Horizontal < 0.2	3.566	1	0.088	0.766
	Horizontal 0.2–0.5	2.680	1	0.030	0.863
	Horizontal > 0.5	452.861	1	2.542	0.111
	Vertical < 0.5	0.664	1	0.407	0.523
	Vertical 0.5–2	2.036	1	2.388	0.123
	Vertical 2–10	3.341	1	1.109	0.293
	Vertical > 10	0.013	1	0.055	0.815
Red-billed francolin (n = 143)					
Season (D vs. W)	Elephant dung	1,765.582	1	202.499	0.000
	Browsing impact	0.813	1	1.598	0.207
	Horizontal < 0.2	10,217.599	1	254.919	0.000
	Horizontal 0.2–0.5	9,337.222	1	104.646	0.000
	Horizontal > 0.5	14,099.362	1	79.071	0.000
	Vertical < 0.5	661.416	1	406.997	0.000
	Vertical 0.5–2	34.535	1	40.381	0.000
	Vertical 2–10	0.000	1	0.000	0.991
	Vertical > 10	0.351	1	1.452	0.229
Category (plot vs. control)	Elephant dung	5.865	1	0.673	0.412
	Browsing impact	2.897	1	5.697	0.017
	Horizontal <0.2	187.762	1	4.684	0.031
	Horizontal 0.2-0.5	170.809	1	1.914	0.167

(*continued overleaf*)

Table 14.4 **(*Continued*)**

Source	Explanatory variables	Sum of squares	df	F	Significance
	Horizontal >0.5	338.296	1	1.897	0.169
	Vertical <0.5	3.704	1	2.279	0.132
	Vertical 0.5-2	0.009	1	0.011	0.917
	Vertical 2-10	0.316	1	0.105	0.746
	Vertical >10	0.435	1	1.799	0.180
Helmeted guineafowl (n = 105)					
Season (D vs. W)	Elephant dung	1499.290	1	177.739	0.000
	Browsing impact	0.396	1	0.781	0.377
	Horizontal < 0.2	8051.121	1	219.904	0.000
	Horizontal 0.2 – 0.5	7117.438	1	83.873	0.000
	Horizontal > 0.5	11693.821	1	67.038	0.000
	Vertical < 0.5	602.191	1	374.358	0.000
	Vertical 0.5 – 2	27.565	1	32.723	0.000
	Vertical 2 – 10	0.442	1	0.147	0.701
	Vertical > 10	0.567	1	2.345	0.126
Category (plot vs. control)	Elephant dung	203.565	1	24.132	0.000
	Browsing impact	3.862	1	7.614	0.006
	Horizontal < 0.2	2,606.203	1	71.184	0.000
	Horizontal 0.2 – 0.5	3,214.522	1	37.880	0.000
	Horizontal > 0.5	3,041.782	1	17.438	0.000
	Vertical < 0.5	15.215	1	9.458	0.002
	Vertical 0.5 – 2	8.954	1	10.630	0.001
	Vertical 2-10	13.356	1	4.456	0.035
	Vertical >10	0.553	1	2.290	0.131
Swainson's francolin (n = 102)					
Season (D vs. W)	Elephant dung	1,779.389	1	204.724	0.000
	Browsing impact	0.272	1	0.541	0.462
	Horizontal < 0.2	9,195.030	1	230.537	0.000
	Horizontal 0.2 – 0.5	7,961.980	1	90.420	0.000
	Horizontal > 0.5	12,592.890	1	71.112	0.000
	Vertical < 0.5	618.093	1	380.992	0.000
	Vertical 0.5 – 2	27.232	1	32.450	0.000
	Vertical 2 – 10	0.407	1	0.136	0.713
	Vertical > 10	0.690	1	2.873	0.091

(*continued overleaf*)

Table 14.4   (*Continued*)

Source	Explanatory variables	Sum of squares	df	F	Significance
Category (plot vs. control)	Elephant dung	24.902	1	2.865	0.091
	Browsing impact	7.195	1	14.322	0.000
	Horizontal < 0.2	324.771	1	8.143	0.004
	Horizontal 0.2 – 0.5	987.381	1	11.213	0.001
	Horizontal > 0.5	1,194.386	1	6.745	0.010
	Vertical < 0.5	5.648	1	3.481	0.062
	Vertical 0.5 – 2	11.187	1	13.330	0.000
	Vertical 2 – 10	13.326	1	4.446	0.035
	Vertical > 10	1.500	1	6.243	0.013

The explanatory variables are: dung = number of elephant dung piles, browsing impact = current elephant browsing impact, horizontal = horizontal visual obscurity in three heights, <0.2 m, 0.2–0.5 m and 0.5–1.0 m and vertical = vertical cover in four height classes, <0.5 m, 0.5–2.0 m, 2–10 m and >10 m.

For all species, except the crested francolin, we found significant differences between bird and control plots (Wilks' lambda = 0.231). Red-billed francolins were seen in plots with less cover below 20 cm than observed in the control plots, even though elephant impact in the bird plots was lower than in the control plots (Table 14.4). Helmeted guineafowl were observed in plots with higher elephant impact and more dung compared to control plots. It was thus not surprising that horizontal visibility was higher around the birds than in the control plots (Table 14.4). A high level of elephant impact was also associated with patches used by Swainson's francolins. These plots were also associated with a higher horizontal visibility than observed in the controls (Table 14.4). Both guineafowl and Swainson's francolins preferred plots with less vertical cover than observed in the control plots (Table 14.4). Thus we suggest that the species can be classified to three groups regarding cover and choice of plot sites. Crested francolins apparently used areas that did not differ from the surrounding areas. Red-billed francolins seemed to occupy an intermediate position, as they chose areas with increased horizontal visibility below 20 cm. Helmeted guineafowl and Swainson's francolins seemed to prefer more open areas regarding both horizontal and vertical cover.

When we compared bird plots among the species studied, we found that guineafowl apparently selected plots containing more dung than the other species, except crested francolin, which did not differ from guineafowl (Table 14.5). Overall, elephant impact was highest in areas used by Swainson's francolins, but this did not differ significantly from the other species, except for red-billed francolins, which used areas with the lowest impact (Table 14.5). Helmeted guineafowl selected areas with the overall highest

Table 14.5 Comparison of estimated marginal means among plots where gallinaceous birds were sighted with regard to habitat-related explanatory variables on the Chobe National Park study area.

Explanatory variables	Crested francolin 95% confidence interval		Red-billed francolin 95% confidence interval		Helmeted guineafowl 95% confidence interval		Swainson's francolin 95% confidence interval	
	Lower	Upper	Lower	Upper	Lower	Upper	Lower	Upper
Elephant dung	2.340	4.228	2.213	3.186	3.275	4.429	1.454	2.641
Browsing impact	1.149	1.607	1.185	1.419	1.470	1.753	1.540	1.825
Horizontal <0.2	12.203	16.270	12.477	14.562	8.567	10.973	11.557	14.100
Horizontal 0.2–0.5	12.006	18.062	12.251	15.361	7.652	11.314	9.910	13.687
Horizontal >0.5	17.405	25.951	17.474	21.872	10.536	15.787	12.358	17.714
Vertical <0.5	1.935	2.753	2.120	2.540	1.857	2.361	1.992	2.504
Vertical 0.5–2	1.513	2.104	1.423	1.728	1.121	1.486	1.082	1.450
Vertical 2–10	0.866	1.978	0.805	1.377	0.447	1.136	0.439	1.135
Vertical >10	0.016	0.331	0.062	0.224	0.025	0.220	-0.023	0.175

The explanatory variables are: dung = number of elephant dung piles, browsing impact = current elephant browsing impact, horizontal = horizontal visual obscurity in three heights, <0.2 m, 0.2–0.5 m and 0.5–1.0 m and vertical = vertical cover in four height classes, <0.5 m, 0.5–2.0 m, 2–10 m and >10 m.

horizontal visibility, particularly compared to crested and red-billed francolins, but less so compared to Swainson's francolins (Table 14.5). However, vertical cover did not vary much among the plots selected by these species, although guineafowl and Swanson's francolins tended to choose less vertical cover than the other two species (Table 14.5).

The frequency of behavioural activities was similar among species and between seasons. The dominant activities, aggregated across the survey period, were 'feeding on ground' and 'still in shade' ($N = 405$ observations). The proportions of each activity were: feeding on ground 52%, still in shade 32%, moving 8%, feeding on dung 6%, still in sun 1%, and alert 1%. Some of the activities were difficult to distinguish, as a bird could be engaged in more than one activity concurrently.

Small mammals

Small mammals were studied using mark-capture-recapture live trapping on three study sites in the mixed *Baikiaea* woodland from November to December 1998, thus encompassing the last portion of the dry and beginning of the wet season. The mixed *Baikiaea* woodland was dominated by trees and shrubs usually less than 5 m high. For the trees more than 5 m high, more than 10% of the canopy cover was mature *Baikiaea plurijuga*. The total canopy cover of woody species was typically about 50–100%, of which *Baikiaea plurijuga* contributed less than 50%. Important species were *Combretum elaeagnoides, Combretum mossambicense, Combretum apiculatum, Baphia massaiensis, Lonchocarpus nelsii, Markhamia acuminata*, and *Burkea africana* (Chapter 5). The three trapping sites consisted of a large site with modest current elephant browsing impact 17 km west of Kasane and 3.7 km south of the Chobe River, a small site of modest current browsing impact 9 km west of Kasane and 3.4 km south of the river, and a small site dominated by shrubs, a result of substantial current browsing impact 21 km west of Kasane and 3.6 km west of the river. The mean field layer coverage of the site with substantial current browsing impact was 56% grass and 40% forbs, compared with 35% and 20% for the large site with moderate impact and 47% and 35% for the small site with moderate impact, respectively.

The large site was 100×100 m, divided into a 10×10 grid with three traps, one of each size, set within each grid square. The small sites were 50×50 m subdivided into a 5×5 m grid, also with one trap of each size within each square. We used three sizes of Sherman live traps ($5 \times 6 \times 16$ cm, $8 \times 9 \times 23$ cm, and $8 \times 8 \times 30$ cm; Sejoe *et al.*, 2002). All the traps were ventilated, to prevent overheating of the animals. The traps were baited with a mix of rolled oats, peanut butter, peanuts, sunflower seeds, dry fruits, and maize meal (Smithers, 1971; Barnett and Dutton 1995). The traps were set in the afternoon and checked and disarmed the following morning while it was still cool. The large site was trapped for seven sessions and the small sites were trapped for three sessions. Each session comprised five trapping nights followed by a two-night interval, except between sessions six and seven, which were one night apart. At the end of the last session, all captured animals were killed in the field by euthanizing them in a sealed plastic bag with a chloroform-soaked ball of cotton (Barnett and Dutton, 1995). The total number of trap-nights per site (1 trap set for 1 night = 1 trap-night) ranged from 1125 (small plots) to 10,500 (large plot), a total 12,750 trap-nights overall. Immediately prior to this (6–13 November 1998), we trapped for 1050 trap-nights 200 m from the large site to collect specimens for species identification. All animals were identified using Skinner and Smithers (1990) and Stuart and Stuart (1988). We also recorded the grid coordinates for the trap station, the trapping session and night concerned, and the size of the trap. Each animal was individually marked by ear notching and toe clipping and the animal was released at the point of capture

Not all individuals are equally trappable (Krebs and Boonstra, 1984), so we estimated the probability of capture (Yoccoz et al., 1993) for population estimation using seven trapping sessions. Otis et al. (1978) recommended at least five trapping sessions. We calculated both measured and corrected densities; in the latter we increased the site size with a border area equal to one-half of the mean movement distance recorded for all recaptured mice in that session (distance from previous capture). The individuals captured in the last five-night session in each site were not included in the density estimates, as they were killed on capture and could not be included in estimates of probability of capture, except for session three on the small site with substantial impact, because the animals there were only killed during the last day of trapping.

We used chi-square goodness of fit tests to test for differences in species occurrence among the three sites. We analysed the trapping data with the programme CAPTURE (Otis et al., 1978), using the programme's model selection procedure to choose the most appropriate model for each trapping period, which was the model of heterogeneity of trapping probabilities (M_h) (Pollock et al., 1990). CAPTURE gave population estimates and confidence intervals for each trapping session. We used these estimates to calculate mean population density for each site and then tested for differences among sites using a one-way ANOVA (analysis of variance). We applied the least significant difference test to determine which site(s) were different when a global ANOVA showed a significant difference among them. All statistical tests were performed using SPSS with $\alpha = 5\%$ level of significance.

We trapped 69 individuals of six species of small mammals. *Steatomys pratensis* dominated in all sites (Table 14.6). Species composition varied by site, when comparing the frequency of *S. pratensis* against that of the other species ($\chi^2 = 10.15$, df $= 2$, $p = 0.006$). The frequency of occurrence of *S. pratensis* was not statistically different between the large site with moderate browsing impact (89%) and the corresponding small site (77%) ($p > 0.05$ with a Bonferroni adjustment for significance levels), but it was significantly less in the site with substantial current ungulate browsing impact (44%, $p < 0.001$ with a Bonferroni adjustment for significance levels).

Table 14.6 **Rodent species trapped at three sites in Chobe National Park, differing in size, vegetation cover, and extent of browsing impact, mostly by elephants (see footnotes).**

Species	Site		
	LMI	SGI	SMI
Steatomys pratensis	42	4	10
Saccostomys campestris	2	3	3
Steatomys parvus	1	0	0
Steatomys krebsii	1	0	0
Mus indutus	1	0	0
Pellomys fallax	0	2	0
Trapping sessions	7*	3	3*
Capture rate (new ind.100 trap-nights^{-1})	0.67	1.96	1.42
Uncorrected density Mean ($\pm$SD) (ind.ha^{-1})	11.5 (9.3)	57.3 (38.9)	12.0 (0.0)
Corrected density Mean ($\pm$SD) (ind.ha^{-1})	9.4 (6.2)	47.9 (33.8)	8.6 (0.4)

Standard deviations are given in parentheses.
LMI = large site with moderate browsing impact.
SGI = small site dominated by shrubs, substantial browsing impact.
SMI = small site, moderate browsing impact.
*Captures during the last session were not used for the population estimate.

The uncorrected mean densities of all mammals were significantly different among sites (ANOVA, $F = 5.29$, df $= 2$, $p = 0.03$, Table 14.6). The density in the site with substantial impact was significantly higher than both the large ($p = 0.01$, least significant difference test) and small sites with moderate impact ($p = 0.04$), but the latter two sites did not differ ($p = 0.98$). Similar results were obtained when using corrected densities, although the density values were lower (Table 14.6). These densities also differed (ANOVA, $F = 5.29$, $d = 2$, $p = 0.03$); the least significant difference test showed that the site with substantial impact was still significantly higher than the large ($p = 0.02$) and small sites with moderate impact ($p = 0.04$), but these latter two did not differ significantly ($p = 0.96$).

Discussion

During the 1960s the Chobe Riverfront was famous for its large riverine *Acacia* forests, which probably were derived from a cohort of seedlings established when populations of megaherbivores and ungulates were low, as a result of heavy exploitation of ivory and the rinderpest epizootic. Following the recovery of the ungulates from rinderpest, and with the protection of elephants, herbivore numbers increased and the animals became major agents influencing the present composition of vegetation (Mosugelo *et al.*, 2002; Skarpe *et al.*, 2004). We have no information about the distribution and density of gallinaceous birds and medium-sized mammals in the study area when large herbivores were rare, but our results suggest a positive correlation between habitat selection of gallinaceous birds and smaller mammals and the browsing and grazing impacts of elephants and other large mammals.

There are two critical factors in our methodological approach that might affect the calculated preference indices. First, comparisons among species hinge on the assumption that all species have an equal detection probability. This assumption was most likely not met because of differences in size, behaviour and conspicuousness. For instance, a large guineafowl will not normally hide in cover, whereas a smaller crested francolin typically will, thereby avoiding detection. As a result, the easily detectable species in open habitats tended to be overestimated and less detectable ones in closed habitats underestimated. Second, we counted all individuals within each species and habitat type to maximize sample sizes for calculating habitat selection indices. This might have increased the number of observations in more open habitats relative to more closed ones. Nevertheless, the average documented distance of observed animals from the transects showed little intraspecific variation among habitat types, suggesting that our results were comparable among habitat types (results not shown). Although this approach is susceptible to error, we feel that the large differences in observations among habitat types for many species cannot be explained by method biases alone. Thus, we feel confident that our approach reflects the true distribution of species in the study area.

At the scale of the study area, we found a great difference in habitat selection by medium-sized mammals and gallinaceous birds between combined *Baikiaea* woodland and mixed *Baikiaea* woodland, with the least and moderate long-term elephant impact, respectively (few species selecting for these habitats), and the shrublands, with the greatest long-term impact and the highest number of species. Although other factors, including soil fertility and distance to water, complicate the above relationship, Mosugelo *et al.* (2002) showed that the area of shrublands on our study area had increased from 5% to 33% from 1962 to 1998, coincident with a dramatic increase in the elephant population. They concluded that elephant browsing was an important factor in the conversion of woodland to shrublands near the Chobe River, along with fire (Mosugelo *et al.*, 2002).

Our findings within shrubland habitats also showed a positive correlation between gallinaceous bird occurrence and long-term elephant impact. This suggests a beneficial effect of elephant impact on gallinaceous bird communities, by creating a mosaic diversity of habitats, with open areas that have a short grass layer, interspersed with shrub thickets of *Combretum* and *Croton*, in contrast to the closed thickets of mixed *Baikiaea* woodland and the generally homogeneous *Baikiaea* woodland. This mosaic might be important for gallinaceous birds, which generally respond positively to a mosaic of vegetation structure with a high degree of habitat edge (Wolff *et al.*, 2002; Dörgeloh, 2000). Concentrations of large herbivores along the river front, especially during the dry season, could also be important for providing food, such as partially and undigested seeds in their dung, or insects attracted to the dung. These are important food sources for gallinaceous birds (Mentis *et al.*, 1975; Little *et al.*, 2000; Barnes, 2001; van Niekerk, 2001). We observed all four species picking at elephant dung, in aggregate 6% of their time, suggesting that elephant dung is an important food source for these birds. However, helmeted guineafowl and crested francolins were most associated with elephant dung.

All the small mammal species caught in this study were rodents that had been recorded earlier in this area by Smithers (1971) and Sheppe and Haas (1981), except for *Mus indutus*. Smithers (1971) reported that small-mammal species richness was low in the mixed *Baikiaea* woodland, compared to the entire Chobe National Park. Within this woodland, we found that the density and species composition of the small mammal communities were similar in two sites with moderate elephant impact on the vegetation, but that density was statistically greater in the site with a substantial impact. This difference could be a result of vegetation structure. Although the major species of plants were similar in all sites, those with moderate impact had a more woodland character, with a greater coverage of trees more than 1 m high. The site with substantial impact was primarily a shrubland, with most trees less than 1 m high, and the highest field layer coverage of grasses and forbs. This suggests that the density and species structure of rodent communities can be sensitive to vegetation structure, even within a habitat type. The impact of elephants in this habitat could be a major reason for the observed difference, as elephants can create heterogeneity in the vegetation, as well as concentrating their activities in nutritional 'hot spots' (Nellemann *et al.*, 2002), which could also be an important factor.

There is not enough space to discuss our habitat-level results at the species level. Nevertheless, the overall pattern of more species selecting floodplain and shrubland habitats is also evident at that level. Large carnivores, such as the predatory lion and the scavenging spotted hyaena and side-striped jackal, selected the floodplain habitat and a similar pattern was found among many of the herbivores, on which they prey (Chapter 13). Medium-sized predators, such as the slender and banded mongoose, large-spotted genet, and wild cat are predators on large invertebrates, reptiles, small mammals, and birds and also showed selection for the floodplain. We found that large rodents, lagomorphs, springhares and gallinaceous birds also selected the floodplain (and often associated shrublands) and our small-mammal results suggested that densities were higher in more browsing-impacted vegetation. Thus, the prey of these medium-sized predators probably also is more abundant in the floodplain habitat.

Conclusion

Elephants have been termed keystone species in savannas, because of their pronounced impact on vegetation (Owen-Smith, 1988). Their effects on biodiversity are controversial, however, and have been widely debated (Owen-Smith *et al.*, 2006). Despite examples where elephants are the probable cause of species loss (Cumming *et al.*, 1997; Fritz *et al.*, 2002; Skarpe *et al.*, 2004), many of their reported negative ecological effects seem to be exaggerated (Owen-Smith, 1988; van de Koppel and Prins, 1998; Owen-Smith *et al.*, 2006). Our studies suggest that small and medium-sized mammals and gallinaceous birds generally may have benefitted from the impacts of elephants and other large herbivores on the vegetation, principally through increasing the shrubland habitats that most of these species prefer.

References

Asner, G.P. & Levick S.R. (2012) Landscape-scale effects of herbivores on treefall in African savannas. *Ecology Letters* 15, 1211–1217.

Barnes, M.E. (2001) Seed predation, germination and seedling establishment of *Acacia erioloba* in Northern Botswana. *Journal of Arid Environments* 49, 541–554.

Barnett, A. & Dutton, J. (1995) *Expedition Field Techniques. Small Mammals (excluding bats).* Royal Geographical Society with IBG, London, UK.

Ben-Shahar, R. (1996) Woodland dynamics under the influence of elephants and fire in Northern Botswana, *Vegetatio* 123, 153–163.

Child, G. (1968) *An Ecological Survey of North-eastern Botswana.* Food and Agriculture Organization of the United Nations, Rome.

Cumming, D.H.M., Fenton, M.B., Rautenbach, I.L., Taylor, R.D., Cumming, G.S., Cumming, M.S., Dunlop, J.M., Ford, G.S., Hovorka, M.D., Johnston, D.S., Kalcounis, M.C., Mahlanga, Z. & Portfors, C.V. (1997) Elephants, woodlands and biodiversity in southern Africa. *South African Journal of Science* 93, 231–236.

Department of Wildlife and National Parks (2013) *Wildlife Aerial Surveys of 2012.* Department of Wildlife and National Parks, Ministry of Environment, Wildlife and Tourism, Gaborone, Botswana.

Dörgeloh, W.G. (2000) Relative densities and habitat utilisation of non-utilised, terrestrial game bird populations in a natural savanna, South Africa. *African Journal of Ecology* 38, 31–37.

Fritz, H., Duncan, P., Gordon, I.J. & Illius A.W. (2002) Megaherbivores influence trophic guilds structure in African ungulate communities. *Oecologia* 131, 620–625.

Krebs, C.J. & Boonstra, R. (1984) Trappability estimates for mark-recapture data. *Canadian Journal of Zoology* 62, 2440–2444.

Little, R., Crowe, T. & Barlow, S. (2000) *Gamebirds of Southern Africa.* Struik Publishers, Cape Town, South Africa.

Manly, B.F.J., McDonald, L.L. & Thomas D.L. (1993) *Resource Selection by Animals.* Chapman & Hall, London, UK.

Melton, D.A. (1985) The status of elephants in northern Botswana. *Biological Conservation* 31, 317–334.

Mentis, M.T., Poggenpoel, B., & Maguire, R.R.K. (1975) Food of helmeted guineafowl in highland Natal. *Journal of Southern African Wildlife Management Association* 5, 23–25.

Mosugelo, D. K., Moe, S. R., Ringrose, S. & Nellemann C. 2002. Vegetation changes during a 36-year period in northern Chobe National Park, Botswana. *African Journal of Ecology* 40, 232–240.

Nellemann, C., Moe, S.R., & Rutina, L.P. (2002) Links between terrain characteristics and forage patterns of elephants (*Loxodonta africana*) in northern Botswana. *Journal of Tropical Ecology* 18, 835–844.

Nudds, T.D (1977) Quantifying the vegetative structure of wildlife cover. *Wildlife Society Bulletin* 5, 113–117.

Otis, D.L., Burnham, K.P., White, G.C. & Anderson, D.R. (1978) Statistical inference for capture data on closed animal populations. *Wildlife Monographs* 62, 1 – 135.

Owen-Smith, N., Kerley, G.I.H., Page, B., Slotow, R. & van Aarde, R.J. (2006) A scientific perspective on the management of elephants in the Kruger National Park and elsewhere. *South African Journal of Science* 102, 389–394.

Owen-Smith, R.N. (1988) *Megaherbivores: The Influence of Very Large Body Size on Ecology*. Cambridge University Press. Cambridge, UK.

Pickett, S.T., Cadenasso, M.L. & Benning, T.L. (2003) Biotic and abiotic variability as key determinants of savanna heterogeneity at multiple spatiotemporal scales. In: du Toit, J.T., Rogers, K.H. & Biggs, H.C. (eds.) *The Kruger Experience. Ecology and Management of Savanna Heterogeneity*. Island Press, Washington DC, pp. 22–40.

Pielou, E.C. (1977) *Mathematical Ecology*. John Wiley & Sons, Inc., New York.

Pollock, K.K., Nichols, J.D., Brownie, C. & Hines, J.E. (1990) Statistical inference for capture-recapture experiments. *Wildlife Monographs* 107, 1 – 97.

Sejoe, T.B., Swenson, J.E. & Moe, S.R. (2002) Relative capture efficiency of small, medium and large Sherman live traps for small mammals in northern Botswana. *Game and Wildlife Science* 19, 247–254.

Shannon, C.E. (1948) A mathematical theory of communication. *Bell System Technical Journal* 27, 379–423.

Sheppe, W. & Haas, P. (1981) The annual cycle of small mammal populations along the Chobe River, Botswana. *Mammalia* 45, 157 – 176.

Simpson, C.D. (1975) A detailed vegetation study on the Chobe river in north-east Botswana. *Kirkia* 10, 185–227.

Skarpe, C., Aarrestad, P.A., Andreassen, H.P., Dhillion, S., Dimakatso, T., du Toit, J.T., Halley, D.J., Hytteborn, H., Makhabu, S., Mari, M., Marokane, W., Masunga, G., Modise, D., Moe, S.R., Mojaphoko, R., Mosugelo, D., Motsumi, S., Neo-Mahupeleng, G., Ramotadima, M., Rutina, L., Sechele, L., Sejoe, T.B., Stokke, S., Swenson, J.E., Taolo, C., Vandewalle, M., & Wegge, P. (2004) The return of the giants: ecological effects of an increasing elephant population. *Ambio* 33, 276–282.

Skinner, J.D. & Smithers, R.H.N. (1990) *The Mammals of the Southern African Subregion*. University of Pretoria, Pretoria, South Africa.

Smithers, R.H.N. (1971) *The Mammals of Botswana*. Museum Memoir, 4. The Trustees of the National Museums of Rhodesia, Salisbury [Harare, Zimbabwe].

Sommerlatte, M.W. (1976) *A Survey of Elephant Populations in North-eastern Botswana.* Department of Wildlife and National Parks, Gaborone, Botswana.

Stokke, S. (1999) Sex differences in feeding-patch choice in a megaherbivore: elephants in Chobe National Park, Botswana. *Canadian Journal of Zoology* 77, 1723–1732.

Stuart, C. & Stuart, T. (1988) *Field Guide to Mammals of Southern Africa.* Struik Publishers (Pty) Ltd, Cape Town.

Valeix, M., Fritz, H., Sabatier, R., Murindagomo, F., Cumming, D. & Duncan, P. (2011) Elephant-induced structural changes in the vegetation and habitat selection by large herbivores in an African savanna. *Biological Conservation* 144, 902–912.

van de Koppel, J. & Prins, H.H.T. (1998) The importance of herbivore interactions for the dynamics of African savanna woodlands: an hypothesis. *Journal of Tropical Ecology* 14, 565–576.

van Niekerk, J.H. (2001) Notes on the winter diet of the crested francolin in South Africa. *South African Journal of Wildlife Research* 31, 66–67.

Wiens, J.A. (1989) *The Ecology of Bird Communities. Foundations and Patterns*, Vol. 1. Cambridge University Press, Cambridge, UK.

Wolff, S., Bothma J. Du P. & Viljoen, P.J. (2002) Gamebirds: general management. In: Bothma, J. du P. (ed.) *Game Ranch Management.* Van Schaik, Pretoria, pp. 226–241.

Yoccoz, N.G., Steen, H., Ims, R.A. & Stenseth, N.C. (1993) Estimating demographic parameters and the population size: an updated methodological survey. In: Stenseth, N.C. & Ims, R.A. (eds.) *The Biology of Lemmings.* Academic Press, San Diego, California, pp. 565–587.

15

The Chobe Riverfront Lion Population: A Large Predator as Responder to Elephant-Induced Habitat Heterogeneity

Harry P. Andreassen[1], Gosiame Neo-Mahupeleng[2], Øystein Flagstad[3] and Per Wegge[4]

[1] Faculty of Applied Ecology and Agricultural Sciences, Hedmark University College, Norway
[2] Poso House, Gaborone, Botswana
[3] Norwegian Institute for Nature Research, Norway
[4] Department of Ecology and Natural Resource Management, Norwegian University of Life Sciences, Norway

Introduction

Carnivores are the top trophic level of food webs, eventually only superseded by man. Being at the top, they often require large spaces from which to obtain enough prey, a requirement that can create severe management complications when they roam outside protected areas or wildlife management zones. This can cause human-wildlife conflicts, usually resulting in the interface becoming a population sink for large predators (Woodroffe and Ginsberg, 1998). Human-carnivore relations are therefore often the proximate cause of the distribution and abundance of these predators today. But it is not only the carnivores' relationship with man that limits their abundance. In spite of being at the top of the food web, carnivore populations can be regulated by other trophic levels, most commonly through food availability.

Carnivores can also be regulated by intraguild competition, whereby the composition of a particular carnivore guild and their interspecific interactions determine the abundance of a given carnivore species within it. Furthermore, carnivores are potentially

Elephants and Savanna Woodland Ecosystems: A Study from Chobe National Park, Botswana,
First Edition. Edited by Christina Skarpe, Johan T. du Toit and Stein R. Moe.

regulated not only by other species in their guild, but also by conspecifics. Carnivores, as a taxonomic group and irrespective of size, tend to be territorial, with at least one sex maintaining a territory. Most species are solitary, but a number, such as some canids and lions *Panthera leo*, form territorial groups. This individual or group territoriality forms the basis for strong self-regulation of carnivore populations (Wolff, 1997).

Therefore, in addition to interactions with people, populations of large predators are generally thought to be regulated by (i) food availability, (ii) intraguild competition and (iii) self-regulation. All these factors can be affected by elephant-induced habitat heterogeneity along the Chobe riverfront, where elephants *Loxodonta africana* accentuate the difference between woodland and shrubland, creating variation in the shrubland, and even among individual shrubs within the shrubland (Chapters 5 and 17). In the following section we summarise how habitat heterogeneity can affect the regulation of carnivore populations, before focussing on the lion population along the Chobe riverfront as a responder to the changes induced there by elephants (Chapter 4).

Habitat heterogeneity and population regulation in carnivores

Food availability

For most carnivores the lack of prey or lack of hunting success is a major threat to the recruitment of young into the population (Schaller, 1972; van Orsdol *et al.*, 1985). Even in areas with abundant prey, the proportion of available prey can be limited or the hunt can fail. Available prey and hunting success are a result of the continuous arms race in predator-prey systems resulting in a variety of predator and anti-predator adaptations (Stephens and Peterson, 1984; Matter and Mannan, 2005). Habitat heterogeneity affects this arms race by modifying prey availability at all points along the predation process, from searching, encountering and killing prey, through to its consumption (Endler, 1986). Habitat change is also likely to affect the composition of prey communities, the sex and age structure of the populations, and the behaviour of prey (Gorini *et al.*, 2012).

Because of this tight arms race, habitat selection by predators and prey are interlinked. Whereas predators select habitat that will maximize their hunting success (Hopcraft *et al.*, 2005), prey seek to minimize predation by avoiding habitat patches with high predator density, substantial perceived risk (Lima and Dill, 1990), or during certain phases of their life cycle (Olsson *et al.*, 2008). Spatial or temporal refugia for prey within the landscape mosaic is expected to make predator-prey systems more enduring over time (Fryxell *et al.*, 1988). For instance, prey can detect predators more easily in open landscapes (Mills *et al.*, 2004), but open landscapes also improve visibility for the predators. Dense vegetation provides refuge for prey, but can also function to reduce the probability of detecting approaching predators. For pursuit predators the terrain through which they chase their prey is more important for hunting success than vegetation cover, whereas for an ambush or stalking predator the cover between the predator and its prey is more significant (Husseman *et al.*, 2003; Gorini *et al.*, 2012). For example, where cover is available, hunting lions usually stalk from one patch of cover to another until they are close enough to rush out at their prey, or they wait hidden until their prey is sufficiently close to attack.

The outcome of the arms race between predators and their prey is therefore not easily predicted because the effects of habitat heterogeneity can either benefit or burden both prey and predator. For lions, elephant-induced habitat heterogeneity in the shrubland along the Chobe riverfront, in which open grasslands with abundant prey are situated close to shrubland and woodland, should therefore be favoured landscapes, with the open savannas and shrubland being preferred habitat for hunting, and the shrubs and trees providing shade and shelter for resting.

Intraguild competition

Competition between carnivore species can result in one species killing another. This interspecific killing is called intraguild predation (Polis *et al.*, 1989) and is fairly common in mammalian carnivores (Palomares and Caro, 1999), particularly among canids and felids. The phenomenon has received considerable attention (Linnell and Strand, 2000) because removing a top predator releases controls on other predator populations, which in turn can affect lower trophic levels (i.e. mesopredator release: Meffe *et al.*, 1994). In theory, intraguild predation could release food resources that would otherwise be consumed by the victim, or remove a threat of mortality to the killer or its offspring (Polis *et al.*, 1989). It is less likely that the killing actually confers any energetic benefits on the killer as their victims are seldom eaten.

In African savannas, African wild dog *Lycaon pictus* and cheetah *Acinonyx jubatus* have often been observed to be victims of interspecific killing, with *Panthera* species often being the killer (Palomares and Caro, 1999). This interspecific mortality can be substantial (Creel and Creel, 1996; Kelly *et al.*, 1998). Lions and spotted hyenas, *Crocuta crocuta*, caused 13–50% of all mortality among African wild dogs (Mills and Biggs, 1993; Creel and Creel, 1996, 1998). In cheetahs, 68% of mortality among cubs and young animals resulted from killings by lions, spotted hyenas and leopards *Panthera pardus* (Laurenson, 1994). Leopards and hyenas reportedly killed 8% of lion cubs and youngsters (Schaller, 1972). In the Serengeti, Laurenson (1995) found an inverse relation between the population densities of lions and cheetahs, with interspecific killing being suggested as a potential factor limiting the cheetah population.

We expect an arms race among sympatric mammalian carnivores to result in increased kill rates and improved measures being taken to avoid being killed. Quarry species most often either avoid using areas within the killer species' home range, or change their habitat use. For instance do African wild dogs and cheetahs avoid suitable habitat in the presence of lions (Mills and Gorman, 1997; Durant, 1998), avoid proximity to lions (Kelly *et al.*, 1998), or adjust their activity rhythm to avoid meeting lions (Mills and Biggs, 1993). All three are possible and none is exclusive, but it is unclear under what circumstances each is the best response.

Just as elephant-modified shrublands along the riverfront are advantageous to lions in their normal predator-prey relations, they also serve as cover for lions in their interactions with other predators. This cover can hamper attempts by cheetah and African wild dogs to avoid lions, putting them at a disadvantage in intraguild competition with these larger predators.

Self-regulation

Social organisation in carnivores ranges from solitary living, with individuals only associating briefly for breeding purposes, through long-standing pairs and larger social groups of various kinds, such as packs in many canids and prides in lions, to colonial breeding aggregations (e.g. eared seals, Family Otariidae). Even within a species, social organisation can vary with both environmental and social circumstances, including habitat availability; the density, diversity and biomass of prey; population density; group size and composition; and the extent of intra- and interspecific competition and predation (Bekoff *et al.*, 1984; Meena, 2009). Whereas most carnivores are solitary (Sandell, 1989), with large overlapping home ranges and a smaller included defended area, usually around a breeding site, group-living carnivores more often maintain large, vigorously defended, intergroup territories (Mosser and Packer, 2009). Even then, these territories are not necessarily wholly exclusive and actively defended. There is a large variation in the degree of territoriality and exclusivity within species and even within populations (Grant *et al.*, 1992). Territoriality therefore seems to be flexible and continuous rather than a stringent entity with some form of spatial dominance.

Territorial behaviour, in which individuals attempt to appropriate space for their exclusive use, is a prerequisite for intrinsic population regulation. Intrinsic population regulation, or self-regulation, occurs when increasing population densities induce behavioural or physiological changes that in turn reduce population growth rates (Wolff, 1997). Infanticide is such a mechanism that has a direct numeric effect on population growth, as adult individuals kill neonates or cubs to improve their own reproductive success. Infanticide committed by males is difficult to observe, but has often been documented in primates, terrestrial carnivores, and some rodents. It occurs when a male takes over the territory of another dominant male, for instance in lions during the takeover of a pride by a new coalition of males (Bertram, 1975; Packer and Pusey, 1984).

Sex-selected infanticide occurs when a male increases his reproductive success by killing dependent young he has not sired himself (Swenson, 2003). This can reduce the interbirth period and gives the infanticidal male the opportunity to sire the next litter. Carnivores usually have a polygynous mating, in which a male mates with several females. Hence, males tend to roam over larger areas than females to overlap several female home ranges (Nilsen *et al.*, 2005). The turnover of one male can thus reduce the recruitment of young into the population from several females. The effect of infanticide caused by high mortality of males can be detrimental to mammal populations (Swenson *et al.*, 1997; Andreassen and Gundersen, 2006).

The effect of habitat heterogeneity on infanticidal behaviour is not known, however. Habitats such as woodland or other dense vegetation could presumably help cubs hide from infanticidal males, but whether complex habitat structure actually improves cub survival has not yet been documented. Given elephant-induced heterogeneity of the habitat we would expect females with cubs to avoid open shrubland.

The lion riverfront population

Number of animals

We studied the Chobe riverfront lion population for four years from April 1999 to April 2003 by direct observations and radio-telemetry. In August 2000 we were confident that we had observed all lions belonging to the riverfront population and had employed at least one radio collar on each separate lion group. From August 2000 until the end of the study the riverfront lion population size varied from 18 to 46 individuals: two adult males; 8 – 13 adult females organized in five female groups each consisting of 2 – 3 reproducing females; and a varying number of subadults and cubs (Figure 15.1, Table 15.1).

Unknown or foreign lions were only noted twice, once in January 2002, when an adult male appeared for a couple of months (Figure 15.1), and again in January 2003, when an adult male was observed for a short period associating with female groups close to Kasane. This male managed to establish himself as a pride male of two female groups. The two original pride males disappeared soon after the end of the study in April 2003.

In September 2001, the two female groups nearest Kasane joined permanently, forming a pride of five adult females and six subadults and cubs, which then left the riverfront and established itself towards Leshomo and the Zimbabwe border. The original areas of these two groups along the riverfront were left abandoned. This emigration could have been a response to the two old males who contracted their territory towards Ngoma Bridge, abandoning the female groups closest to Kasane before these left the area. As the riverfront lion population along Chobe was small and vulnerable to both stochastic demographic and environmental events, such relocations can cause large variation in the number of animals. In this case, the emigration of the two female

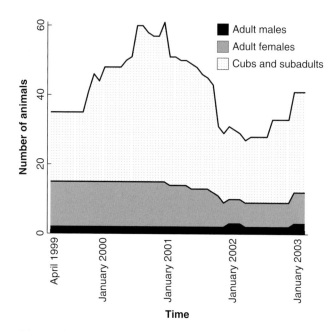

Figure 15.1 The population trajectory of the Chobe lion riverfront population. Since August 2000 we are confident that we regularly observed all animals in the population because all female groups were equipped with radio collars. The figure show cumulative numbers of adult males (black), adult females (grey), and cubs and subadults (white with dots).

Table 15.1 Composition of the various female lion groups along the Chobe riverfront in August 2000.

Animal category	Number of individuals in each group				
	Sedudu	Moselesele	Kabulabula	Mokweto	Ihaha
Adult females	2	2	2	3	3
Subadults	0	1	6	0	2
Cubs	5	3	0	3	7
Total	7	6	8	6	12

groups from the riverfront substantially reduced the size of the riverfront population (Figure 15.1).

Social structure and spacing behaviour

The five female groups had extremely small home ranges both during the dry season (June–November 2000: mean 95% kernel utilization distribution calculated in Arcview GIS 3.2 = 20 km^2; minimum convex polygon range = 18.4–68.9 km^2) and the wet season (middle of November 2000–January 2001: mean 95% kernel utilization distribution = 28 km^2; minimum convex polygon range = 11.5–105.0 km^2; Neo, 2001). Lion home ranges vary considerably from 20–45 km^2 in Manyara National Park and Ngorongoro Crater (Schaller, 1972) to more than 2000 km^2 in Etosha (Nowell and Jackson, 1996). Our observed home ranges are thus among the smallest observed.

Although the population was small, population density was high because of the small home ranges. Across the whole study area the number of lions varied between 8 and 19 animals 100 km^{-2} (4–16 adults 100 km^{-2}). If we estimate the density only for the areas used by the lions, it was as high as 29–66 animals 100 km^{-2} (14–22 adults 100 km^{-2}). Such densities are comparable with the highest known density of 30 lions 100 km^{-2} observed in Maasai Mara National Reserve (Nowell and Jackson, 1996).

The home ranges of female groups were linearly arranged along the riverfront with some movement north into Caprivi Strip of Namibia (a subsistence mixed agricultural area) and some minor excursions south into the *Baikiaea* woodlands during the wet season. In voles, such a linear (narrow rectangular) arrangement results in small home ranges, probably because there are only two short borders, the ends of each rectangle, where intruders are expected (Fauske *et al.*, 1997). For the lions in Chobe, however, if the linear arrangement of the home ranges is contributing to their small size, it requires there to be no neighbouring lion prides to the south of the riverfront population because these would require a higher degree of vigilance southwards as well, yielding a more quadratic or semi-circular area. The lack of adjacent prides was confirmed by searching for spoor southwards in areas adjacent to the riverfront. No spoor was found except in Nogatsaa, 20 km south of the riverfront. We did not search for spoor north of the Chobe River in the Namibian Caprivi Strip.

The two pride males patrolled the whole riverfront, their range covering all five female groups located in the shrubland along the river (Figure 15.2). They were less often seen towards Kasane, spending only 1% of their time there with the Sedudu female group. They shared their time more evenly among the other female groups (9–14% of the time), but were most often observed separate from the females (53% of male observations).

Except for the Sedudu and Moselesel female groups located furthest east (towards Kasane), which we saw together 68% of the time, the female groups seldom met and

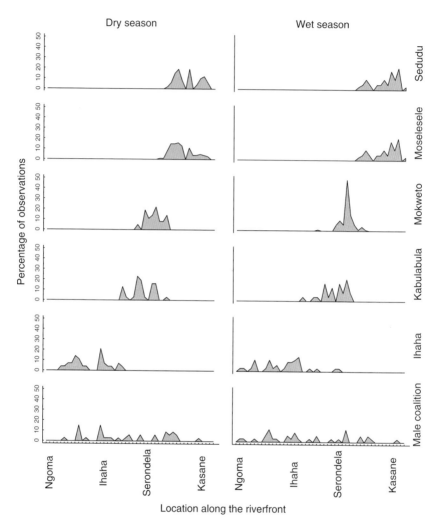

Figure 15.2   The utilization distribution of the home ranges along the west east direction of the riverfront for the five female groups (Sedudu, Moselesele, Mokweto, Kabulabula, Ihaha) and the male coalition. The X-axis shows the west–east direction along the riverfront from Ngoma to Kasane, and the Y-axis the percentage of observations for each lion group observed from radio tracking in the dry (June–mid November 2000; left panel) and wet (mid November 2000–January 2001; right panel) season.

some were never observed together. In addition to the occasional merging of the Sedudu and Moselesel groups, the Kabulabula and Mokwet groups were observed together 3% of the time, and Mokwet and Ihaha groups 1% of the time.

Kin relations

We were able to collect tissue samples from all adult females, the two adult males and some of the subadults in 2002. Genetic analyses from tissue samples indicated that all adults (of both sexes) were somewhat related, suggesting a low rate of turnover of individuals along the riverfront. All adult females along the riverfront were probably related, and the two adult males originally in the population were probably brothers. Local knowledge suggests that these two adult males had been patrolling the riverfront from the beginning of the 1990s, that is, for approximately 10 years. One of the adult males had mated with at least one of his sisters and their common progeny. Furthermore, up to 2002, before the intrusion of the new adult male, the two old males had sired all the cubs.

Habitat and food

The lions along the Chobe riverfront preferred the shrubland close to the river, and avoided the *Baikiaea* woodland (Figure 15.3). The shrubland is the area with the highest abundance of prey and cover for hunting (Chapter 13), and the one most heavily impacted by elephant. To this extent, lions are benefitting from and responding to elephant-induced changes in habitat structure.

We recorded a total of 109 lion kills rather opportunistically, both while radio-tracking lions and through information from tourist guides. We adjusted the frequency with which each prey was recorded according to the probability of detecting a kill (i.e. the expected time spent feeding on a carcass), which was assumed to be proportional to the species' live weight. We further estimated the yearly consumption of prey by assuming the food intake of an adult female lion to be 2500 kg (Schaller, 1972) and converting the riverfront lion population to 30 adult female lion equivalents (Table 15.1).

Approximately 10% of the kills were large prey such as elephant, giraffe, *Giraffa camelopardalis*, and African buffalo, *Syncerus caffer*, which together comprised 61% of the consumed biomass (Table 15.2). This disproportionate number of large prey fits with the finding that larger carnivores take proportionately larger prey relative to their body size than do smaller carnivores (Vézina, 1985). Even though prey availability and composition vary considerably between seasons (Chapter 13) the lion population did not seem to respond to this heterogeneity either through altered behaviour

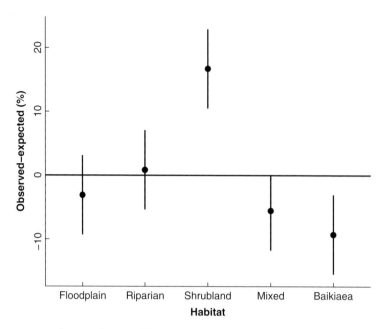

Figure 15.3 **Habitat preference of lions along the Chobe riverfront. The horizontal line at point 0 on the vertical axis indicates neither preference nor avoidance relative to habitat availability within the study area. Positive values show a preference while negative values show an avoidance. For each habitat we present the mean and 95% confidence interval.**

(i.e. changes in habitat preference) or through changes in the composition of kills (Figure 15.4).

Breeding and survivorship

We observed a total of 21 litters. The average litter size when cubs were first observed on leaving the den was 2.5 (range: 1–6), in line with other lion populations (Bertram, 1975; Smuts *et al.*, 1978; Packer and Pusey, 1997). The interbirth interval was 20–26 months on the four occasions that we observed the same female giving birth to a second litter during the study. This is similar to that found in the Serengeti (Pusey and Packer, 1987). The litter was highly male-biased (67% male cubs, 95% CL = 50–80%), which has often been noted in disturbed lion populations (e.g. where males are hunted: Smuts 1978; Creel and Creel, 1997; Milner *et al.*, 2007). It is also a feature of large cohorts, which usually result when females breed synchronously following a takeover of a pride by

Table 15.2 **The distribution of observed kills.**

Species	Observed	Percentage	Prey body mass (kg)	Yearly no. of prey per LEQ	Yearly no. of prey per 30 LEQ
Large					
Elephant	8	7.5	400	0.47	14.1
Giraffe	3	2.8	600	0.12	3.6
Buffalo	56	52.3	420	3.11	93.3
Medium					
Kudu	3	2.8	173	0.40	12.0
Zebra	2	1.9	164	0.29	8.7
Sable	3	2.8	160	0.44	13.2
Small					
Puku	1	0.9	50	0.45	13.5
Warthog	6	5.6	40	3.50	105.0
Impala	11	10.3	32	8.05	241.5
Others	17	15.9	20	19.88	596.4

Prey body mass was taken to be 75% of adult female body mass (from Schaller, 1972) and own estimates of young elephants. One lion equivalent (1 LEQ) assumed to be an adult female (and subadults > 2 years old), consuming on average 2500 kg yearly (Schaller, 1972); one adult male = 1.5 LEQ, and 1 cub or subadult less than 2 years old = 0.5 LEQ. The lion population along the Chobe riverfront thus comprises a total of 30 LEQ. 'Others' include cattle, crocodile and other species.

one or more new males and the cubs from the previous male(s) are killed (Packer and Pusey, 1997).

Cub survival to one year of age was high (75%), in the upper end of the range found by van Orsdol *et al.* (1985) across a range of habitats (27–86%). The subadult stage seemed to be risky, however: out of the 10 female cubs that survived at least one year, only two survived to reproduce in the riverfront population. Of the others, two were found shot and six were lost – either emigrated or dead (four left the national park with the two female groups and were never observed again). The probability of subadult females remaining on the riverfront as reproducing females was therefore low (20%). Our results for male subadults are similar: out of the 17 male cubs that had survived at least 1 year, one settled and became sexually mature in the riverfront; two were still resident subadults about 3 years old at the end of the study; one was killed in a car accident; two were killed by the intruding male; and 11 either dispersed or died (four left the national park with the two female groups and were never observed again).

The two original males appear to have persisted as pride males along the riverfront for at least a decade, which is somewhat longer than for male lions in the Serengeti, the best studied population (mean 2.8 years, range 0.4–10.8 years: Packer *et al.*, 1988).

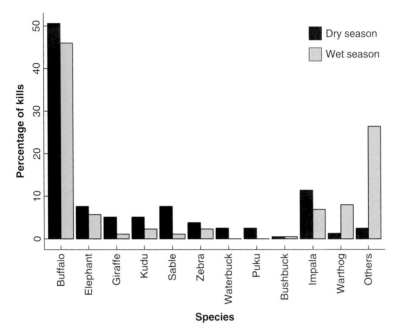

Figure 15.4 **Seasonal diet composition of the lions along the Chobe riverfront.**

The yearly survival of adult females was 92% (range 77–100%). Except for the two female groups that left the riverfront during the last 18 months of the study, when we lost contact with four of the five adult females in these groups, we only lost contact with one of the eight remaining adult females.

Elephant impact and the lion riverfront population

The lion population along the riverfront seems to have been successful. The shrub-lands favoured by the lions comprise a mosaic of open areas with abundant prey, along with bush and remnant forest patches that give shelter and facilitate the lions' stalk-ing method of hunting. Proximity to the Chobe River and its floodplains provide easy access to water and to additional prey. This heterogeneity in vegetation composition and structure is created largely by elephants. We expect that the change in habitat induced by elephants is probably the main reason for the dense lion population along the riverfront. Abundant prey enables successful breeding and high cub survival. The low number of immigrant males and the resulting low number of pride takeovers by foreign males, with its associated infanticide, almost certainly facilitated high cub survival. The only observed incidence of lethal intraspecific interaction was in 2002 when an intruding male killed two subadult males.

Nevertheless, the riverfront lion population is small and isolated, with no immigration and substantial inbreeding. The coalition of the two adult males along the Chobe riverfront could have been stable for more than a decade, during which time they appear to have fathered all the cubs. With pride males elsewhere only holding their position for about 2.8 years on average (Packer *et al.*, 1988), and with a mean inter-birth interval among females of around two years (Pusey and Packer, 1987), male lions usually only foster one cohort before being displaced. Genetic analysis confirmed this stable and unusual situation of the two old males having been present without competitors for an unusually long time. These genetic results, the long and stable male coalition, the lack of observations of foreign lions and the lack of lion spoors in the adjacent areas suggest that the riverfront lion population was largely isolated from other lion groups. The occasional successful immigration of a foreign adult male, as happened towards the end of our study, should be sufficient to avoid genetic erosion, but should be monitored. The acute challenge for the population is its small size, making it highly vulnerable to stochastic events. If the immigration of adult males is as low as one per 10 years, the death of a resident adult male could have severe consequences for the future viability of the population. Conversely, the intrusion of a foreign male could cause substantial disturbance in such a small population, with recruitment being temporarily interrupted because of infanticide.

The changes observed in the size of the lion population over four years resulted mainly from variations in the survival of cubs and subadults. A sharp decline was noted around the time that an adult foreign male was seen in the area, which initiated a period of low recruitment. In addition, dispersing subadults were confronted with dangers caused by human activities, such as traffic and hunting, making them the age category with the highest mortality. This risk to human-induced mortality adds to the vulnerability of this small population.

Conclusion

Habitat heterogeneity probably affects all major aspects of population regulation in large carnivores the same way. The lion population seems to respond to the spatial heterogeneity in vegetation by accessing abundant prey on the floodplain, and sheltering in the nearby shrubland and woodland. Shrubland also facilitates stalking and ambush hunting and, in the vicinity of dense *Baikiaea* woodland, provides good hiding places for cubs. Much of the heterogeneity in the shrublands apparently results from elephant activity (Chapter 5), something to which the riverfront lion population appears to respond and benefit.

Whereas the turnover of males appears to have only an occasional influence on population regulation, intraguild competition does not seem to contribute at all. Lions are the top predator and the killer rather than the victim in intraguild competition. Although hyenas are known to scavenge on lion kills (Cooper, 1991), there is little reason to believe that this affects the lion population negatively in our study area. Instead, the dense lion population could be the reason for the limited number of cheetahs and African wild dogs in the area. If such intraguild competition is taking place we might expect it to be enhanced by elephant-induced habitat heterogeneity in the shrubland, making it difficult for competitors to observe and avoid lions.

In conclusion, the elephant-modified habitats benefit the lion riverfront population. The main risk for the population seems to be connected to large-scale landscape factors such as the high risk of mortality that subadults experience when they start roaming over large areas and encounter human activity, and the long distance to other lion populations. This isolation could be detrimental if adverse stochastic events, acting on the small population, cannot be outweighed by immigration.

References

Andreassen, H.P. & Gundersen, G. (2006) Male turnover reduces population growth: an enclosure experiment on voles. *Ecology* 87, 88–94.

Bekoff, M., Daniels, T.J. & Gittleman, J. L. (1984) Life history patterns and the comparative social ecology of carnivores. *Annual Review of Ecology and Systematics* 15, 191–232.

Bertram, B.C.R. (1975) Social factors influencing reproduction in wild lions. *Journal of Zoology* 177, 463–482.

Cooper, S.M. (1991) Optimal hunting group size: the need for lions to defend their kills against loss to spotted hyaenas. *African Journal of Ecology* 29, 130–136.

Creel, S. & Creel, N.M. (1996) Limitation of African wild dogs by competition with larger carnivores. *Conservation Biology* 10, 526–538.

Creel, S. & Creel, N.M. (1997) Lion density and population structure in the Selous Game Reserve: evaluation of hunting quotas and offtake. *African Journal of Ecology* 35, 83–93.

Creel, S. & Creel, N.M. (1998) Six ecological factors that may limit African dogs, *Lycaon pictus*. *Animal Conservation* 1, 1–9.

Durant, S.M. (1998) Competition refuges and coexistence: an example from Serengeti carnivores. *Journal of Animal Ecology* 67, 370–386.

Endler, J.A. (1986) Defense against predators. In: Federand M.E. & Lauder G.V. (eds.) *Predator-Prey Relationships: Perspective and Approaches from the Study of Lower Vertebrates*. University of Chicago Press, Chicago, Illinois, USA, pp. 109–134.

Fauske, J., Andreassen, H.P. & Ims, R.A. (1997) Spatial organization in a small population of the root vole *Microtus oeconomus* in a linear habitat. *Acta Theriologica* 42, 79–90.

Fryxell, J.M., Greever, J., & Sinclair, A.R.E. (1988) Why are migratory ungulates so abundant? *The American Naturalist* 131, 781–798.

Gorini, L., Linnell, J.D.C., May, R., Panzacchi, M., Boitani, L., Odden, M. & Nilsen, E.B. (2012) Habitat heterogeneity and mammalian predator-prey interactions. *Mammal Review* 42, 55–77.

Grant, J.W.A., Chapman, C.A. & Richardson, K.S. (1992). Defended versus undefended home range size of carnivores, ungulates and primates. *Behavioral Ecology and Sociobiology* 31, 149–161.

Hopcraft, J.G.C, Sinclair, A.R.E. & Packer, C. (2005) Planning for success: Serengeti lions seek prey accessibility rather than abundance. *Journal of Animal Ecology* 74, 559–566.

Husseman, J.S., Murray, D.L., Power, G., Mack, C., Wenger, C.R. & Quigley, H. (2003) Assessing differential prey selection patterns between two sympatric large carnivores. *Oikos* 101, 591–601.

Kelly, M.J., Laurenson, M.K., FitzGibbon, C.D., Collins, D.A., Durant, S.M., Frame, G.W., Bertram, B.C.R. & Caro, T.M. (1998) Demography of the Serengeti cheetah population: the first twenty-five years. *Journal of Zoology* 244, 473–488.

Laurenson, M.K. (1994) High juvenile mortality in cheetahs (*Acinonyx jubatus*) and its consequences for maternal care. *Journal of Zoology* 234, 387–408.

Laurenson, M.K. (1995) Implications of high offspring mortality for cheetah population dynamics. In: Sinclair, A.R.E. & Arcese, P. (eds.) *Serengeti II: Dynamics, Management and Conservation of an Ecosystem*. University of Chicago, Chicago, pp. 385–399.

Lima, S.L. & Dill, L.M. (1990) Behavioral decisions made under the risk of predation: a review and prospectus. *Canadian Journal of Zoology* 68, 619–640.

Linnell, J.D.C. & Strand, O. (2000) Conservation implications of aggressive intra-guild interactions among mammalian carnivores. *Diversity and Distributions* 6, 169–176.

Matter, W.J. & Mannan, R.W. (2005) How do prey persist? *Journal of Wildlife Management* 69, 1315–1320.

Meena, V. (2009) Variation in social organisation of lions with particular reference to the Asiatic Lions *Panthera leo persica* (Carnivora:Felidae) of the Gir forest, India. *Journal of Threatened Taxa* 1, 158–165.

Meffe, G.K., Carroll, C.R. & Pimm, S.L. (1994) Community-level conservation: species interaction, disturbance regimes, and invading species. In: Meffe, G.K. & Carroll, C.R. (eds.) *Principles of Conservation Biology*. Sinauer Associates, Sunderland, Massachusetts, USA, pp. 209–236.

Mills, M.G.L. & Biggs, H.C. (1993) Prey apportionment and related ecological relationships between large carnivores in Kruger National Park. *Symposium of the Zoological Society of London* 65, 253–268.

Mills, M.G.L., & Gorman, M.L. (1997) Factors affecting the density and distribution of wild dogs in the Kruger National Park. *Conservation Biology* 11, 1397–1406.

Mills, M.G.L, Broomhall, L.S. & du Toit, J.T. (2004) Cheetah *Acinonyx jubatus* feeding ecology in the Kruger National Park and a comparison across African savanna habitats: is the cheetah only a successful hunter on open grassland plains? *Wildlife Biology* 10, 177–186.

Milner, J.M., Nilsen, E.B. & Andreassen, H.P. (2007) Demographic side effects of selective hunting in ungulates and carnivores. *Conservation Biology* 21, 36–47.

Mosser, A. & Packer, C. (2009) Group territoriality and the benefits of sociality in the African lion, *Panthera leo*. *Animal Behaviour* 78, 359–370.

Neo, G. (2001) Population structure, group dynamics, home ranges and habitat use of lions (*Panthera leo*) in north-east Chobe National Park, Botswana. MSc Thesis, Agricultural University of Norway, Ås, Norway.

Nilsen, E.B., Herfindal, I. & Linnell, J.D.C. (2005) Can intra-specific variation in carnivore home-range size be explained using remote sensing estimates of environmental productivity? *Ecoscience* 12, 68–75.

Nowell, K., & Jackson, P. (1996) *Wild Cats: Status Survey and Conservation Action Plan*. IUCN, Gland, Switzerland.

Olsson, O., Brown, J.S. & Helf, K.L. (2008) A guide to central place effects in foraging. *Theoretical Population Biology* 74, 22–33.

Packer, C. & Pusey, A.E. (1984) Infanticide in carnivores. In: Hausfater, G. & Hrdy, S.B. (eds.) *Infanticide: Comparative and Evolutionary Perspectives*. Aldine Publishing Company, New York, pp. 31-42.

Packer, C. & Pusey, A.E. (1997) Intrasexual cooperation and the sex ratio in African lions. *American Naturalist* 130, 636–642.

Packer, C., Herbst, L., Pusey, A.E., Bygott, J.D., Hanby, J.P., Cairns, S.J. & Mulder, M.B. (1988) Reproductive success of lions. In: Clutton-Brock, T.H. (ed.) *Reproductive Success*. University of Chicago Press, Chicago, pp. 363–383.

Palomares, F. & Caro, T.M. (1999) Interspecific killing among mammalian carnivores. *The American Naturalist* 153, 492–508.

Polis, G.A., Myers, C.A. & Holt, R.D. (1989) The ecology and evolution of intraguild predation: potential competitors that eat each other. *Annual Review of Ecology and Systematics* 20, 297–330.

Pusey, A.E. & Packer, C. (1987) The evolution of sex biased dispersal in lions. *Behaviour* 101, 275–310.

Sandell, M. (1989) The mating tactics and spacing patterns of solitary carnivores. In: Gittleman, J.L. (ed.) *Carnivore Behavior, Ecology, and Evolution*. Springer, New York, pp. 164–182.

Schaller, G.B. (1972) *The Serengeti Lion: A Study of Predator-Prey Relations*. University of Chicago Press, Chicago, USA.

Smuts, G.L. (1978) Effects of population reduction on the travels and reproduction of lions in Kruger National Park. *Carnivore* 1, 61–72.

Smuts, G.L., Hanks, J. & Whyte, I.J. (1978). Reproduction and social organization of lions from the Kruger National Park. *Carnivore* 1, 17-28.

Stephens, P.W. & Peterson, R.O. (1984) Wolf-avoidance strategies of moose. *Holarctic Ecology* 7, 239–244.

Swenson, J.E. (2003) Implications of sexually selected infanticide for the hunting of large carnivores. In: *Festa-Bianchet, M.* & Apollonio, M. (eds.) *Animal Behavior and Wildlife Conservation*. Island Press, Washington, DC, pp. 171–189.

Swenson, J.E., Sandegren, F., Söderberg, A., Bjärvall, A., Franzén, R. & Wabakken, P. (1997) Infanticide caused by hunting of male bears. *Nature* 386, 450–451.

Van Orsdol, P.J.A., Hanby, J.P. & Bygott, J.D. (1985) Ecological correlates of lion social organization (*Panthera leo*). *Journal of Zoology* 206, 97–112.

Vézina, A.F. (1985) Empirical relationships between predator and prey size among terrestrial vertebrate predators. *Oecologia* 67, 555–565.

Wolff, J.O. (1997) Population regulation in mammals: an evolutionary perspective. *Journal of Animal Ecology* 66, 1–13.

Woodroffe, R. & Ginsberg, J. R. (1998) Edge effects and the extinction of populations inside protected areas. *Science* 280, 2126–2128.

Part VI
Elephants in Social-Ecological Systems

Elephants and Savanna Woodland Ecosystems: A Study from Chobe National Park, Botswana,
First Edition. Edited by Christina Skarpe, Johan T. du Toit and Stein R. Moe.
© 2014 John Wiley & Sons, Ltd. Published 2014 by John Wiley & Sons, Ltd.

16

Human Dimensions of Elephant Ecology

Eivin Røskaft[1], Thor Larsen[2], Rapelang Mojaphoko[3],
A. H. M. Raihan Sarker[1] and Craig Jackson[1]

[1] Department of Biology, Norwegian University of Science and
Technology, Norway
[2] Noragric, Norwegian University of Life Sciences, Norway
[3] Ministry for Environment, Wildlife and Tourism, Botswana

African elephants, *Loxodonta africana*, inhabit about 85,000 km² of northern Botswana, across which the relatively sparse human population occurs in scattered communities relying predominantly on subsistence farming. Whereas the region's human population has increased 2- to 3-fold since 1981, the elephant population has increased about 16-fold since 1960 (Chase and Griffin, 2003). Estimated at 8000 in 1960, the current elephant population is just under 130,000 animals (Chase, 2011). Concurrent increases both in the rural communities and in elephant populations result in heightened competition for space and resources. Botswana depends highly on wildlife; ecotourism generates the largest amount of foreign revenue after the mining industry. From a Western perspective, most interactions with wildlife in this region are positive, taking the form of non-consumptive ecotourism and, to a far lesser degree, trophy hunting. This sentiment, however, is not necessarily shared by local villagers.

The Chobe region is a world-renowned wildlife safari destination. The Chobe National Park and surrounding region supports some of the highest concentrations of Botswana's large elephant population. The protected area is large and unfenced, facilitating elephant movement into surrounding areas. The region's agro-pastoralist communities depend largely on subsistence farming. Fertile soils and access to water are consequently preferred by farmers, but are equally important and attractive to elephants. This sets the stage for a battle between humans and elephants, hereafter referred to as human-elephant conflict (HEC).

Elephants and Savanna Woodland Ecosystems: A Study from Chobe National Park, Botswana,
First Edition. Edited by Christina Skarpe, Johan T. du Toit and Stein R. Moe.
© 2014 John Wiley & Sons, Ltd. Published 2014 by John Wiley & Sons, Ltd.

HEC is prevalent in much of northern Botswana, being associated mainly with crop raiding and risk to human life (Jackson *et al.*, 2008). This phenomenon is indeed common throughout the ranges of both the African and the Asian elephant, *Elephas maximus*, but in this chapter we review the history of human-elephant interactions and complexities of HEC in Africa. Understanding the dynamics of HEC is vital for identifying management strategies to protect both humans and elephants, with traditional mitigation measures being diverse yet often ineffective because they address the symptoms of HEC rather than its causes. Sustainable solutions to HEC require that spatial and temporal patterns of elephant movement be incorporated into land-use planning. Similarly, the needs of local communities must be recognised in mitigation schemes, to ensure that the costs of HEC directly outweigh the elephant-associated benefits.

History of human-elephant conflict (HEC) and coexistence

Human-elephant interactions date back to the early hominids and proboscids, spanning several million years and several continents, but with human-elephant coexistence originating in Africa. Historical evidence from well-studied early human cultures (700,000–70,000 BP) suggests that large prey species were often exploited (Jerozolimski and Peres, 2003), and Stone Age rock art indicates that elephants have been hunted by humans for millennia (Carrington, 1958). Although traditional societies frequently over-harvested their resources (Diamond, 1988), the population-level effects of human predation on elephants are not known for anywhere in Africa prior to the influences of European exploration and settlement. Nevertheless, the advent of agriculture saw human settlements expanding across fertile lands favourable for cultivation, thus encroaching on elephant habitats and elevating the level of HEC. Elephant distribution in pre-colonial Africa was probably a major hindrance to the development of agriculture and farmers prospered only when residing in larger villages (Laws *et al.*, 1975; Parker and Graham, 1989a,b). In this way elephants directly influenced human ecology and to some degree shaped the development of traditional societies in sub-Saharan Africa.

The influx of Europeans and Arabs into Africa during the 19th and 20th centuries marked a major change in the history of human-elephant interactions. Ivory has always been a prized commodity but the introduction of firearms led to sport and ivory hunters decimating elephant populations across the African savannas, with Botswana being no exception (Chapter 6).

Colonial governments in Africa began to realise the need to protect selected wildlife areas, with several game reserves being created in the early 20th century and the idea of national parks emerging in 1928. The Convention Relative to the Preservation of Fauna and Flora in their Natural State, enacted in London in 1933, made provision for

Contracting Governments to control settlement in national parks. In consequence, local people were often evicted without either consultation or compensation. This process was later referred to as the 'fences and fines' policy: 'Identify rich wildlife areas, declare them protected, evict the people and punish them if they attempt to come back' (Kideghesho, 2006).

The demand for ivory peaked during the 1970s and early 1980s. The African elephant population is estimated to have more than halved during the decade 1979–1989, a decline from 1.3 million to about 0.6 million elephants (Douglas-Hamilton, 1987; Douglas-Hamilton et al., 1992). This precipitated a total ban on the sale of ivory by the Convention on the International Trade in Endangered Species of Wild Fauna and Flora (CITES) (Bonner, 1993). The ivory trade ban therefore came into effect when elephant populations were at an all-time low (Tchamba, 1996).

Following the protection provided by national park authorities in some colonial and post-colonial African countries, and perhaps assisted by the CITES trade restrictions in recent years, elephant populations have slowly recovered in many areas (Said et al., 1995; Skarpe et al., 2004; Blanc et al., 2007; Loarie et al., 2009). In southern Africa in particular, the issue of whether or not to reduce elephant densities inside certain large protected areas, such as Chobe, Hwange and Kruger, has for several decades now been a focus of public debate (Chapter 17).

The continuing growth of human populations and the associated fragmentation of habitats around recovering elephant populations limit dispersal possibilities (Loarie et al., 2009; van Aarde and Jackson, 2007). Restricting elephant ranges increases densities within protected areas and has led to a concern that elephants might have a detrimental effect on vegetation structure and biodiversity (Waitaka, 1993; Van de Vijver et al., 1999; Pamo and Tchamba, 2001; Chira and Kinyamario, 2009). Furthermore, expanding human populations increase the extent of HEC around many protected areas in southern and eastern Africa.

Whereas elephants, along with other wildlife such as primates and large carnivores, frequently pose problems for agriculture and livestock and even threaten human lives (Løe and Røskaft, 2004; Packer et al., 2005), the economic value of wildlife is increasing rapidly with expanding tourism, particularly in Africa. Pressure from Western countries to preserve biodiversity is also increasing despite local African communities having to bear the social and economic costs of wildlife but receiving little or no economic benefit.

During the latter part of the 20th century the importance of community participation in conservation was recognised and various community-based conservation (CBC) programmes were initiated. The CBC ideology spread rapidly, particularly in southern Africa (Kideghesho, 2010b; Nyahongo, 2010), although since the 1990s there has been a growing debate between proponents of the CBC concept and those advocating more conventional conservation methods (Hutton et al., 2005).

Elephants as threats to humans and their crops

Cultivation alongside elephant-occupied habitats results in elephants being attracted to a concentrated source of nutritious forage. With subsistence farming having transformed most of the mesic savanna landscapes of Africa, crop depredation represents the dominant cause of HEC, with affected communities incurring both direct and indirect costs.

Direct costs include the destruction of crops, infrastructure, water supplies and injury or loss of human life (Tchamba, 1996). Crop damage can have severe implications, particularly for small-scale farmers, but crop raiding is generally sporadic and unpredictable, usually not resulting in the loss of all crops. When compared with other factors affecting agricultural production in Botswana and elsewhere, crop loss to livestock, primates, bush pigs, antelopes, rodents, birds and insects normally exceed losses attributed to elephants (Cumming *et al.*, 1990; Gadd, 2005; Warren *et al.*, 2007). Despite this, farmers are intolerant of elephants because of their large size and obvious danger to human life.

Indirect costs are incurred when time and productivity are lost to chores such as guarding fields, as well as safety concerns that disrupt daily routines such as firewood collection or children walking to school. Furthermore, depredation by elephants can prevent certain profitable crops from being grown, thereby impeding local economic development. In comparison with direct costs, indirect costs – sometimes referred to as 'opportunity costs' – are more difficult to quantify yet can have a more serious impact on rural communities.

The probability of elephants ranging into human-dominated areas is influenced by spatial and temporal factors. Protected areas are intended to minimise interactions between humans and wildlife, but the edges of these areas are the conflict zones (Hart and O'Connell, 1998; Stokes *et al.*, 2010). These pose the greatest challenges to conservation management (Bell, 1984; Thouless, 1994; Hoare, 1999; Chiyo *et al.*, 2005). Despite various smaller species actually causing greater economic losses, affected communities typically report elephants as the primary problem associated with living alongside wildlife reserves (Hill, 1997; Gillingham and Lee, 1999; Sarker and Røskaft, 2010, 2011b).

Proximity to water resources increases the likelihood of HEC during the dry months. Elephants need to drink regularly and can be attracted to cultivated lands when water is scarce elsewhere, because water is generally more available in such areas (Hoare, 1999; Chamaille-Jammes *et al.*, 2007). During the wet months, elephants have ample food in their natural habitats but crop-raiding increases towards the end of the growing season when crops are ready for harvesting and thus at their most nutritious (Barnes, 1996; Osborn and Parker, 2002; Osborn, 2004). Elephants can turn to crop-raiding even when natural foods are plentiful, presumably attracted by the crops' greater palatability and nutritional value (Barnes, 1991). Crop-raiding is thus not necessarily caused by either a shortage of natural forage or a surplus of crops (Barnes, 2002). For example, in the Caprivi region of Namibia, crop damage was found to fluctuate between seasons depending on elephant movements, rainfall patterns and crop quality, but the highest levels of HEC always occurred in villages bordering protected areas with high elephant densities (O'Connell-Rodwell *et al.*, 2000).

Patterns of HEC also vary depending on the age and sex structure of the elephant groups involved. In Botswana, as elsewhere in the African elephant's range, adult males were found to be more frequent and destructive crop raiders than adult females and calves (Marchais, 2008). Even though bull elephants commonly forage alone they probably have substantially greater impact on crops than do family units (Chiyo *et al.*, 2011).

Elephants usually raid crops under the cover of darkness to minimise detection and retaliation by villagers (Marchais, 2008), with crop raiding being lowest during the full moon (Barnes *et al.*, 2007). With humans being active mainly during daylight hours, however, that is when most attacks on people occur (Dunham *et al.*, 2010).

Management and mitigation of HEC

Managing HEC is complex and no single management strategy will resolve the core and associated issues. In many African countries, centralised interventions suffer from logistical problems whereas traditional methods are generally ineffective. In addition, management interventions, including reducing crop loss, generally stem from organisations outside the affected community. Such organisations include government wildlife departments, external development organisations and international non-governmental organisations (NGOs), many of which are interested in elephant conservation (e.g. International Union for Conservation of Nature (IUCN), World wildlife Fund (WWF), International Elephant Foundation, African Elephant Conservation Trust, African Wildlife Foundation, Born Free, among others). Many of these NGOs attempt to pressure African governments, sometimes using questionable arguments.

Humans have tried many different measures to reduce crop raiding by elephants. Some, such as placing a watchman in the fields and making noise by shooting, singing, beating drums or tin cans, are effective in the short-term but habitual raiders are undeterred. Additional measures can include placing scarecrows in the fields, cracking whips, throwing stones and burning sticks at the elephants, keeping fires burning at night, burning bamboo that makes explosive sounds like gunshots, placing sharp objects on elephant pathways, erecting low-cost barriers with cowbells around the fields, planting thorny plants on the perimeter as 'live fencing', using dogs, and building pitfalls, among others (Sarker, 2010).

Chilli-based mitigation measures have been introduced to deter crop-raiding elephants (Osborn and Parker, 2002), with chilli-grease being used as a repellent in several African countries where it is applied on simple traditional fences (Hoare, 2001). The toxic smoke from a mixture of ground chilli and elephant or cow dung is also effective in deterring elephants from crops (Osborn and Rasmussen, 1995; Hoare, 2001). The advantages are that such techniques use simple, locally produced materials, and although they have short-term noxious effects to elephants, these do not accumulate in the environment. Such unpalatable deterrents (e.g. chilli peppers) could be cultivated peripherally to buffer crops (Hoare, 2000), although chilli peppers are more often recommended for use in powder form, as an intense olfactory irritant (Osborn and Rasmussen, 1995; Osborn and Parker, 2002). Farmers have also used poisoned decoys which can either kill the elephants or make them sick (Thouless, 1994). The use of fire around crop fields has proved fairly effective (Hoare, 2000; Osborn and Rasmussen, 1995), as have kerosene lamps on wooden poles, flaming torches and powerful flashlights (Damiba and Ables, 1993; Nyhus et al., 2000). Discharging firearms near approaching or crop-raiding elephants – shooting at them or above their heads with rubber bullets – can be effective (Hoare, 2000; Sitati et al., 2003) and broadcasting elephant alarm calls has also been attempted (Hoare, 2001) but the success of all such measures varies greatly in space and time.

Various forms of fencing (conventional and electric), trenches, moats, stonewalls, and barriers of sharp objects (stones or sharpened wooden stakes) can act as buffers, although elephants are effective in seeking out the weak points of such installations, so that they are only temporarily effective or need continuous maintenance (Davies et al., 2011; Hoare, 2000; Nyhus et al., 2000; O'Connell-Rodwell et al., 2000). The most drastic measure is obviously to kill each 'problem animal', although in most range states an elephant may be killed only if the animal in question has injured or killed a person (Kiiru, 1995). Hence, shooting might not reduce the problem of crop damage but will usually placate local people by satisfying their sense of revenge and providing meat and skin from the carcass. In short, it is more of a political measure than an elephant management tool (Tchamba and Elkan, 1995). Translocation of problem elephants is also an option but is prohibitively costly for most African countries. The costs are not well documented but can exceed US $100,000 per translocated animal (Njumbi et al., 1996). As with killing, it can be difficult to ensure that the correct animal is identified for translocation. There also has to be a suitable and sufficiently distant site for relocation where local authorities and communities will accept the problem animal. It could become a problem in its new area or – unless the release site is fenced – return to its home area to be a problem again (Nyhus et al., 2000; Fernando et al., 2012).

More recently beehives have been suggested as elephant repellents based on elephants avoidance behaviour towards African honeybees (O'Brien, 2002; Vollrath and Douglas-Hamilton, 2002; King et al., 2009), including alarm calls (King et al., 2010). The concept of a beehive fence is not only popular and desired by local communities but the cost of constructing beehives can be recovered through the sale of honey and other bee products (King *et al.*, 2009). Nevertheless, the considerable media attention given to beehives as a way of mitigating HEC is based on initial findings from limited studies. Further research is needed before any conclusions can be made (Hoare, 2012).

Although most HEC mitigation measures have proved ineffective under some conditions, a combination of methods – adapted to local conditions – probably provides the best way forward. People experiencing crop losses and other HEC problems need solutions and, in the absence of effective intervention from the responsible agencies, will understandably become disenchanted and might tolerate or even become involved in illegal killing of elephants. This is compounded by the recent increase in the black market value of ivory, making commercial poaching more profitable, creating greater incentives and leading to more poaching. As long as local communities have no influence over elephant management and no formal user rights, the illegal killing of elephants will solve their immediate problem. Donor-funded technical solutions, such as electric fences, are seldom sustainable because of high recurrent maintenance costs. Long-term solutions thus have to be custom-made for the socio-economic level and technological capacity of each affected local community (Kangwana, 1996). Range-wide elephant conservation depends on some level of coexistence of people and elephants outside protected areas, which can occur only when and where costs imposed on local people by elephants are adequately offset by direct elephant-related benefits. This in turn requires effective governance that recognises the role of local communities in wildlife stewardship and implements proven strategies to mitigate HEC.

Economic compensation for damage by wildlife is common in Western countries that can afford it, relieving farmers of the economic burden of living with wildlife, but implementing such policies is problematic in Africa. For example, Kenya had a national policy of paying compensation for wildlife damage but suspended it in 1989 because of widespread false claims, corruption and high administration costs (Thouless, 1994). An alternative and more realistic option for paying for crop losses is limited compensation scheme, a concept developed in Namibia, whereby farmers are paid fixed amounts for losses caused by certain wildlife species from the local authorities that have a collective value for conservation. Payments are made through the community to local farmers within a framework of clear rules and conditions developed by it. A new form of reduced compensation, termed 'consolation', is being trialled in Tanzania and aims to provide at least some financial payments when wildlife-induced hardships are incurred (Hoare, 2012). It can be counter-productive to directly compensate for elephant depredation as this can reduce incentives to protect crops (Hoare, 2000). Another option for offsetting the costs of crop raiding by elephants is through economic benefits to local communities from legal elephant hunting managed within a CBC plan (e.g. Communal Areas Management Programme for Indigenous Resources, CAMPFIRE, in Zimbabwe). When local people have user rights and realise that elephants have a value in the form of safari license fees and meat, their attitudes towards living with elephants can change, as has been found with human-predator conflicts (Romañach *et al.*, 2007). Furthermore, locally empowered stakeholders in the elephant resource can become hostile towards poachers who are effectively stealing local assets.

Humans as threats to elephants

The focus of discussion on HEC is the impacts elephants might have on human life and well-being, but the conflict obviously works both ways with humans having an over-riding influence on the life and well-being of elephants. In the Serengeti region, Tingvold *et al.* (2013) found the general level of stress hormones in elephants to be elevated just outside the park in the vicinity of human settlements, suggesting that elephants find human-dominated landscapes stressful. These findings complement those of Ahlering *et al.* (2011) who found that crop-raiding elephants had significantly higher metabolite levels of glucocorticoid stress hormones than elephants residing in nearby protected areas in northern Kenya. High risk of hunting – whether legal or illegal – is another factor causing stress in elephants, with faecal stress hormones remaining elevated in survivors for several days after experiencing the shooting of a group member (Burke *et al.*, 2008).

Elephants tend to be more common where human densities are relatively low (Barnes, 1983; Hoare and du Toit, 1999; Blake *et al.*, 2007) as found for example in Dzanga-Sangha Reserve in the Central African Republic, where the abundance of elephants, duikers *Cephalophus* spp., and Western lowland gorillas, *Gorilla gorilla*, was highest farthest away from human settlements (Remis and Kpanou, 2011). The corollary is that rapidly increasing elephant densities within protected areas are caused by immigration from places where human densities are increasing, with the immigrants adding to the growth of the resident population. Awareness of being in unprotected versus protected areas is evident from elephant movements being significantly faster when travelling through unprotected areas (Douglas-Hamilton *et al.*, 2005). This response to humans calls for further research on the elephant 'landscape of fear' and its possible use in controlling elephant movements and crop-raiding behaviour.

Elephant macroeconomics – ivory and tourism

Elephants have provided ivory, meat, hides and draft power to humans throughout recorded history, although demands for ivory caused the extinction of elephants in North Africa about 1500 years ago (Spinage, 1994). Profitable trade with the Middle East, China, India and subsequently Europe facilitated extensive settlements of Swahili ivory and slave traders along the East African coast around 600 years ago (Hakansson,

2004). During those times Europe was importing 100–200 tonnes of ivory annually, rising to 700 tonnes per year during the late 19th century (Spinage, 1994), the product of about 60,000 elephants (Blanc *et al.*, 2003).

The ivory market peaked again during the 'ivory crisis' in the 1970s and 1980s. In 1989 CITES took action by banning the trade of ivory and other elephant products. Following its implementation, international demand for ivory fell, although illicit trading has increased dramatically during the past decade, with the associated risks driving prices to as much as US $1500 kg^{-1} in the Far East. Individual poachers receive only a small fraction of the profits retained by middle-men in ivory-smuggling syndicates. There is thus no benefit to the national economies of countries in which elephants are killed to supply this illicit trade.

National economic benefits arising from wildlife-based tourism can be lucrative and sustainable with elephants, other charismatic species, and wilderness experiences being strong attractants for increasing number of foreign tourists visiting sub-Saharan Africa. For example, since 1994, tourism revenues in Botswana have grown by 10.1% per annum, accounting for 3.7% of GNP by 2010, and contributing more than 2 million bed nights to the hospitality industry (Department of Tourism, 2010). Foreign income from tourism is now second highest after the mining industry (Mbaiwa, 2003; Gereta, 2010). Countries with high numbers of elephants, such as Botswana, South Africa and Tanzania, are all experiencing growth in tourism to their conservation areas (Gereta, 2010).

Elephant microeconomics – community-based conservation initiatives

The importance of including local communities in conservation programmes has become recognised in recent decades, with the mainstream tourism industry generating some jobs but distributing negligible revenues to local communities in or around wildlife areas. CBC initiatives have been implemented in attempts to reduce unsustainable exploitation of wildlife, including elephants and their habitats, whilst recognising and incorporating the needs of local people by providing wildlife-derived incentives for conservation (Nyahongo, 2010). Some CBC initiatives such as CAMPFIRE in Zimbabwe, Namibia's Communal Wildlife Conservancies, and Community-based Natural Resources Management (CBNRM) in Botswana have assigned utilisation rights and concessions to local communities so that the bulk of revenues are retained within the local economy. Theoretically, the economic potential of sustainably managed safari hunting in such programmes can exceed income derived from non-consumptive tourism.

Initially lauded as a promising approach, several studies are now critical of the CBC approach. Factors affecting the success of CBC programmes include perceptions that the CBC idea was imported from the West (Rozemeijer, 2002), intra-community conflicts of interests (Adams and McShane, 1996), economic constraints (Holmern et al., 2002), poor governance by politicians and managers (Mfunda, 2010; Nyahongo, 2010) and several other factors in conflict with more conventional approaches to wildlife conservation (Hutton et al., 2005; Holmern, 2010; Kideghesho, 2010b). The unwillingness of central governments to delegate power to local institutions is a tradition inherited from colonial powers (see Kideghesho, 2010a for some examples and discussion of inconsistent applications of legislation and regulations). Moreover, the effects of wildlife-related benefits on human behaviour remain poorly tested, although studies report that incentives can change conservation attitudes (Nyahongo, 2010). Providing communities with wildlife-related benefits might not, however, reduce unwanted behaviour. Local people could instead use the new benefits to complement existing income rather than replace it. For instance, modelling suggests that CBCs relying on money transfers do not always act to conserve wildlife, especially when benefits are low compared with wildlife-induced damage or the potential income gained from illegal hunting (Hofer et al., 2000). Several studies implicate the disproportionately low level of benefits or uneven distribution thereof as key elements in the continued persistence of illegal activities (reviewed by Nyahongo, 2010). Careful consideration is therefore required before the CBC concept is promoted as a mitigating option for HEC. The political economy of a community has to be reconstructed to create an organisation that is transparent, highly participatory, equitable, functional, and supported by legislation and regulations that truly empower local people and secure their access to income derived from using wildlife. Only if revenues are distributed in a participatory and democratic manner will conservation outcomes be improved (Child, 2006).

Attitudes and conflicts

The attitudes of local inhabitants reflect their experiences. In rural sub-Saharan Africa, negative attitudes towards elephants indicate the existence of HEC, which most commonly arises from elephant crop-raiding and its associated problems (Damiba and Ables, 1993; Hoare, 1995). For example, in the Caprivi region of Namibia, local attitudes towards elephants have changed concurrently with the increase in elephant numbers and expansion of agricultural areas (Lindeque, 1995; O'Connell-Rodwell et al., 2000).

Although local communities might regard problem elephants as pests, their attitudes could also reflect resistance and discontent over the negative impacts of protectionist conservation strategies (Gillingham and Lee, 1999; Sarker and Røskaft, 2011a). The tendency for respondents to exaggerate the negative effects of wildlife reflects important social dimensions in the human-wildlife conflict (Naughton-Treves, 1997). Exaggerated complaints can result from concerns over constraints on using resources imposed by conservation legislation, regional land-tenure systems or administrators (Madden, 2004). As a result, tensions between local farmers and wildlife authorities sometimes run high, something that politicians are quick to exploit by publicly blaming wildlife agencies for a constituency's problems. The media also use HEC incidents effectively to construct narratives of inequality, fear and political struggle. The combined pressure from politicians and the national media can force wildlife agencies to deviate from evidence-based elephant management in favour of visible but ineffective interventions designed to placate stakeholders.

Ultimately, negative attitudes generated by HEC are far more extreme than those created by common crop pests because wild elephants are simply terrifying to villagers. Incidents of human injury or death caused by elephants are traumatic but in some cases also economically catastrophic for the victim's family (Thirgood *et al.*, 2005). The overall potential cost of an attack is thus extremely high even when its probability is low (Røskaft *et al.*, 2007), with farmers feeling powerless to protect their crops from elephants because of personal fear (Osborn and Parker, 2002). It follows, unsurprisingly, that HEC drives negative local attitudes towards elephants, wildlife in general and conservation initiatives as a whole (de Boer and Baquete 1998; Røskaft *et al.*, 2007), and can sometimes even fuel political conflicts at the national level (Hart and O'Connell, 1998).

Conclusions and recommendations

Conflict with humans is a serious threat to the persistence of elephant populations in many parts of the African elephant's range. Whereas chronic HEC has always remained part of traditional life in some parts of Africa, it is re-emerging in others, such as in Botswana where the elephant population is still recovering from being decimated by 19th century ivory hunters. Facilitating the coexistence of humans and elephants is a complex problem that no single strategy can solve. A combined approach offers the greatest possibility of success. It is of fundamental importance, however, that communities affected by the problem are key participants in any attempts to reduce the conflict. In line with CBC principles, the most effective approach involves empowering the affected local communities to take responsibility for mitigating HEC in their own areas. Nevertheless, this requires technical support and incentive schemes to increase and sustain the benefits associated with accommodating elephants in community areas; the economic benefits must exceed the direct and indirect costs incurred by HEC. Likewise, the needs of people living with elephants must be recognised and respected at local, national, and international governance levels. A prerequisite is therefore having and implementing legislation that adequately recognises local people's roles and rights over wildlife.

As long as there are elephants to conserve, there are opportunities to derive benefits from them for local communities. The challenge is to reduce the associated costs.

Long-term land-use planning holds potential for minimising the human–elephant interface along which HEC can occur (Osborn and Parker, 2002). One example might be to incentivise the production of certain cash crops that are unpalatable to elephants (coffee, tea, hot peppers, etc.), especially in buffer zones adjacent to wildlife areas. Education and training of local communities are essential for disseminating innovative mitigation techniques, building local capacity in conflict resolution, and promoting an evidence-based understanding of HEC (Kideghesho *et al.*, 2007; Røskaft *et al.*, 2007). With time, such programmes can promote commitment to conservation and raise awareness of the essential role of wildlife in resilient social-ecological systems. Commitment, education, and empowerment are all keys to effective CBC. This requires effective governance at local, regional, and national levels, without which HEC will never be sufficiently mitigated to satisfy both community leaders and conservationists.

References

Adams, J.S. & McShane, T.O. 1996. *The Myth of Wild Africa: Conservation Without Illusion.* W.W. Norton & Company, New York, USA.

Ahlering, M.A., Millspaugh, J.J., Woods, R.J., Western, D. & Eggert, L.S. (2011) Elevated levels of stress hormones in crop-raiding male elephants. *Animal Conservation* 14, 124–130.

Barnes, R.F.W. (1983) Elephant behaviour in a semi-arid environment. *African Journal of Ecology* 21, 185–196.

Barnes, R.F.W. (1991) Man determines the distribution of elephants in the rainforests in north-eastern Gabon. *African Journal of Ecology* 29, 54–63.

Barnes, R.F.W. (1996) The conflict between humans and elephants in the central African forests. *Mammal Review* 26, 67–80.

Barnes, R.F.W. (2002) The problem of precision and trend detection posed by small elephant populations in West Africa. *African Journal of Ecology* 40, 179–185.

Barnes, R.F.W., Dubiure, U.F., Danquah, E., Boafo, Y., Nandjui, A., Hema, E.M. & Manford, M. (2007) Crop-raiding elephants and the moon. *African Journal of Ecology* 45, 112–115.

Bell, R.H.V. (1984) The man-animal interface: an assessment of crop damage and wildlife control. In: Bell, R.H.V. & McShane-Caluzi, E. (eds.) *Conservation and Wildlife Management in Africa.* U.S. Peace Corps Office of Training and Program Support, Malawi, pp. 387–416.

Blake, S., Strindberg, S., Boudjan, P., Makombo, C., Bila-Isia, I., Ilambu, O., Grossman, F., Bene-Bene, L., de Semboli, B., Mbenzo, V., S'hwa, D., Bayogo, R., Williamson, L., Fay, M., Hart, M. & Maisels, F. (2007) Forest elephant crisis in the Congo Basin. *PLoS Biol* 5(4), e111. doi:10.1371/journal.pbio.0050111

Blanc, J.J., Thouless, C.R., Hart, J.A., Dublin, H.T., Douglas-Hamilton, I., Craig, G.C. & Barnes, R.F.W. (2003) *African Elephant Status Report 2002: An Update from the African Elephant Database.* Occasional Paper of the IUCN Species Survival Commission No. 29, IUCN, Gland, Switzerland and Cambridge, UK.

Blanc J.J., Barnes R.F.W., Craig G.C., Dublin, H.T., Thouless, C.R., Douglas-Hamilton, I. & Hart, J.A. (2007) *African Elephant Status Report 2007. An Update from the African Elephant*

Database. Occasional Paper of the IUCN Species Survival Commission No. 33, IUCN, Gland, Switzerland.

Bonner, R. (1993) *At the Hand of Man: Peril and Hope for Africa's Wildlife*. Simon & Schuster, London, UK.

Burke, T., Page, B., Van Dyk, G., Millspaugh, J. & Slotow, R. (2008) Risk and ethical concerns of hunting male elephant: behavioural and physiological assays of the remaining elephants. *PLoS ONE* 3(6), e2417. doi:10.1371/journal.pone.0002417

Carrington, R. (1958) *Elephants: A Short Account of Their Natural History, Evolution and Influence on Mankind*. Chatto & Windus, London, UK.

Chamaillé-Jammes, S., Valeix, M. & Fritz, H. (2007) Managing heterogeneity in elephant distribution: interactions between elephant population density and surface-water availability. *Journal of Applied Ecology* 44, 625–633.

Chase, M. (2011) *Dry Season Fixed-wing Aerial Survey of Elephants and Wildlife in Northern Botswana*. Elephants Without Borders, Kasane, Botswana; Department of Wildlife and National Parks, Botswana; and Zoological Society of San Diego, USA. [online] http://www.elephantdatabase.org/population_submission_attachments/102

Chase, M.J. & Griffin, C.R. (2003) *Elephant Distribution and Abundance in the Caprivi Strip of Namibia, Results of an Aerial Survey in April 2003*. Conservation International, Maun, Botswana.

Child, B. (2006) Revenue distribution for empowerment and democratisation. *Participatory Learning and Action* 55, 20–29.

Chira, R.M. & Kinyamario, J.I. (2009) Growth response of woody species to elephant foraging in Mwea National Reserve, Kenya. *African Journal of Ecology* 47, 598–605.

Chiyo, P.I., Cochrane, E.P., Naughton, L. & Basuta, G.I. (2005) Temporal patterns of crop raiding by elephants: a response to changes in forage quality or crop availability? *African Journal of Ecology* 43, 48–55.

Chiyo, P.I., Moss, C., Archie, E.A., Hollister-Smith, J.A. & Alberts, S.C. (2011) Using molecular and observational techniques to estimate the number and raiding patterns of crop-raiding elephants. *Journal of Applied Ecology* 48, 788–796.

Cumming, D.H.M., du Toit, R.F & Stuart, S.N. (1990) *African Elephants and Rhinos: Status Survey and Conservation Action Plan*. IUCN/SSC, Gland, Switzerland.

Damiba, T.E. & Ables, E.D. (1993) Promising future for an elephant population: a case study in Burkina Faso, West Africa. *Oryx* 27, 97–103.

Davies, T.E., Wilson, S., Hazarika, N., Chakrabarty, J., Das, D., Hodgson, D.J. & Zimmerman, A. (2011) Effectiveness of intervention methods against crop-raiding elephants. *Conservation Letters* 4, 346–354.

de Boer, W.F. & Baquete, D.S. (1998) Natural resource use, crop damage and attitudes of rural people in the vicinity of the Maputo Elephant Reserve, Mozambique. *Environmental Conservation* 25, 208–218.

Department of Tourism. (2010) *Tourism Statistics 2006–2010*. Department of Tourism, Ministry of Wildlife, Environment and Tourism, Republic of Botswana, Gaborone, Botswana.

Diamond, J.M. (1988) The golden age that never was. *Discover* 9, 70–79.

Douglas-Hamilton, I. (1987) African elephants: population trends and their causes. *Oryx* 21, 11–24.

Douglas-Hamilton, I., Michelmore, F. & Inamdar, A. (1992) *African Elephant Database.* GEMS/GRID/UNEP, Nairobi, Kenya.

Douglas-Hamilton, I., Krink, T. & Vollrath, F. (2005) Movements and corridors of African elephants in relation to protected areas. *Naturwissenschaften* 92, 158–163.

Dunham, K.M., Ghiurghi, A., Cumbi, R. & Urbano, F. (2010) Human-wildlife conflict in Mozambique: a national perspective, with emphasis on wildlife attacks on humans. *Oryx* 44, 185–193.

Fernando, P., Leimgruber, P., Prasad, T. & Pastorini, J. (2012) Problem-elephant translocation: translocating the problem and the elephant? *PLoS ONE* 7(12), e50917. doi:10.1371/journal.pone. 0050917

Gadd, M.E. (2005) Conservation outside of parks: attitudes of local people in Laikipia, Kenya. *Environmental Conservation* 32, 50–63.

Gereta, E. (2010) The role of biodiversity conservation in development of the tourism industry in Tanzania. In: Gereta, E. & Røskaft, E. (eds.) *Conservation of Natural Resources: Some African & Asian Examples.* Tapir Academic Press, Trondheim, Norway, pp. 23–49.

Gillingham, S. & Lee, P.C. (1999) The impact of wildlife-related benefits on the conservation attitudes of local people around the Selous Game Reserve, Tanzania. *Environmental Conservation* 26, 218–228.

Hakansson, N.T. (2004) The human ecology of world systems in East Africa: the impact of the ivory trade. *Human Ecology* 32, 561–587.

Hart, L.A. & O'Connell, C.E. (1998) *Human Conflict with African and Asian Elephants and Associated Conservation Dilemmas.* Centre for Animals in Society, School for Veterinary Medicine and Ecology Graduate Group, University of California, Los Angeles.

Hill, C.M. (1997) Crop-raiding by wild vertebrates: the farmer's perspective in an agricultural community in Western Uganda. *International Journal of Pest Management* 43, 77–84.

Hoare, R. (1995) Options for the control of elephants in conflict with people. *Pachyderm* 19, 54–63.

Hoare, R. (2000) African elephants and humans in conflict: the outlook for co-existence. *Oryx* 34, 34–38.

Hoare, R. (2012) Lessons from 15 years of human-elephant conflict mitigation: management considerations involving biological, physical and governance issues in Africa. *Pachyderm* 51, 60–74.

Hoare, R.E. (1999) Determinants of human-elephant conflict in a land-use mosaic. *Journal of Applied Ecology* 36, 689–700.

Hoare, R.E. (2001) *A Decision Support System for Managing Human-Elephant Conflict Situations in Africa.* IUCN/SSC African Elephant Specialist Group, Nairobi, Kenya. [online] http://www.africanelephant.org/hec/pdfs/hecdssen.pdf

Hoare, R.E. & Du Toit, J.T. (1999) Coexistence between people and elephants in African savannas. *Conservation Biology* 13, 633–639.

Hofer, H., Campbell, K.L.I., East, M.L. & Huish, S.A. (2000) Modeling the spatial distribution of the economic costs and benefits of illegal game meat hunting in Serengeti. *Natural Resource Modeling* 13, 151–177.

Holmern, T. (2010) Bushmeat hunting in Western Serengeti: implications for community-based conservation. In: Gereta, E. & Røskaft, E. (eds.) *Conservation of Natural Resources: Some African & Asian Examples.* Tapir Academic Press, Trondheim, Norway, pp. 211–236.

Holmern, T., Røskaft, E., Mbaruka, J., Mkama, S.Y. & Muya, J. (2002) Uneconomical game cropping in a community-based conservation project outside the Serengeti National Park, Tanzania. *Oryx* 36, 364–372.

Hutton, J.M., Adams, W.M. & Murombedzi, J.C. (2005) Back to the barriers? Changing narratives in biodiversity conservation. *Forum for Development Studies* 2, 342–370.

Jackson, T.R., Mosojane, S., Ferreira, S.M. & van Aarde, R.J. (2008) Solutions for elephant *Loxodonta africana* crop raiding in northern Botswana: moving away from symptomatic approaches. *Oryx* 42, 83–91.

Jerozolimski, A. & Peres, C.A. (2003) Bringing home the biggest bacon: a cross-site analysis of the structure of hunter-kill profiles in Neotropical forests. *Biological Conservation* 111, 415–425.

Kangwana, K. (ed.) (1996) *Studying Elephants.* The African Wildlife Foundation, Nairobi, Kenya.

Kideghesho, J.R. (2006) *Wildlife conservation and local land use conflicts in Western Serengeti, Tanzania.* PhD Thesis, Department of Biology, Norwegian University of Science and Technology, Trondheim, Norway.

Kideghesho, J.R. (2010a) Wildlife conservation in Tanzania: whose interests matter? In: Gereta, E. & Røskaft, E. (eds.) *Conservation of Natural Resources: Some African & Asian Examples.* Tapir Academic Press, Trondheim, Norway, pp. 82–110.

Kideghesho, J.R. (2010b) Sustainable use and conservation of wildlife resources in Africa: What traditional cultural practices can offer? In: Gereta, E. & Røskaft, E. (eds.) *Conservation of Natural Resources: Some African & Asian Examples.* Tapir Academic Press, Trondheim, Norway, pp. 111–129.

Kideghesho, J.R., Røskaft, E. & Kaltenborn, B.P. (2007) Factors influencing conservation attitudes of local people in Western Serengeti, Tanzania. *Biodiversity and Conservation* 16, 2213–2230.

Kiiru, W. (1995) The current status of human-elephant conflict in Kenya. *Pachyderm* 19, 15–18.

King, L.E., Lawrence, A., Douglas-Hamilton, I. & Vollrath, F. (2009) Beehive fence deters crop-raiding elephants. *African Journal of Ecology* 47, 131–137.

King, L.E., Soltis, J., Douglas-Hamilton, I., Savage, A. & Vollrath, F. (2010) Bee threat elicits alarm call in African elephants. *PLoS ONE* 5(4), e10346. doi:10.1371/journal.pone.0010346

Laws, R.M., Parker, I.S.C. & Johnstone, R.C.B. (1975) *Elephants and Their Habitats: The Ecology of Elephants in North Bunyoro*, Uganda. Clarendon Press, Oxford, UK.

Lindeque, M. (1995) Conservation and management of elephants in Namibia. *Pachyderm* 19, 49–53.

Loarie, S.R., van Aarde, R.J. & Pimm, S.J. (2009) Elephant seasonal vegetation preferences across dry and wet savannas. *Biological Conservation* 142, 3099–3107.

Løe, J. & Røskaft, E. (2004) Large carnivores and human safety: a review. *Ambio* 33, 283–288.

Madden, F. (2004) Creating coexistence between humans and wildlife: global perspectives on local efforts to address human-wildlife conflict. *Human Dimensions of Wildlife* 9, 247–259.

Marchais, J.E.S. (2008) Field notes about human-elephant conflict in the southern buffer zone of the Okavango Delta, Botswana. *Pachyderm* 14, 96–97.

Mbaiwa, J.E. (2003) The socio-economic and environmental impacts of tourism development on the Okavango Delta, north-western Botswana. *Journal of Arid Environments* 54, 447–467.

Mfunda, I.M. (2010) Benefit and cost sharing in collaborative wildlife management in eastern and southern Africa: country experiences, lessons and challenges. In: Gereta, E. & Røskaft, E. (eds.) *Conservation of Natural Resources: Some African & Asian Examples.* Tapir Academic Press, Trondheim, Norway, pp. 166–185.

Naughton-Treves, L. (1997) Farming the forest edge: vulnerable places and people around Kibale National Park, Uganda. *The Geographical Review* 87, 27–46.

Njumbi, S.J., Waithaka, J.M., Gachago, S.W., Sakwa, J.S., Mwathe, K.M., Mugai, P., Mulama, M.S., Mutinda, H.S., Ormondi, P.O.M. & Litoroh, M.W. (1996) Translocation of elephants: the Kenya experience. *Pachyderm* 22, 61–65.

Nyahongo, J.W. (2010) Community participation in management and sustainable use of wildlife: Advantages and disadvantages. In: Gereta, E. & Røskaft, E. (eds.) *Conservation of Natural Resources: Some African & Asian Examples.* Tapir Academic Press, Trondheim, Norway, pp. 155–165.

Nyhus, P.J., Tilson, R. & Sumianto, P. (2000) Crop-raiding elephants and conservation implications at Way Kambas National Park, Sumatra, Indonesia. *Oryx* 34, 262–274.

O'Brien, C. (2002) Bees buzz elephant. *New Scientist* 176, 8.

O'Connell-Rodwell, C.E., Rodwell, T., Rice, M. & Hart, L.A. (2000) Living with the modern conservation paradigm: can agricultural communities co-exist with elephants? A five-year case study in East Caprivi, Namibia. *Biological Conservation* 93, 381–391.

Osborn, F.V. (2004) Seasonal variation of feeding patterns and food selection by crop-raiding elephants in Zimbabwe. *African Journal of Ecology* 42, 322–327.

Osborn, L. & Parker, G.E. (2002) *Community-based methods to reduce crop loss to elephants: experiments in the communal lands of Zimbabwe Pachyderm* 33, 33–38.

Osborn, L. & Rasmussen, L.E.L. (1995) Evidence for the effectiveness of an oleo-resin capsicum aerosol as a repellent against wild elephants in Zimbabwe. *Pachyderm* 20, 55–64.

Packer, C., Ikanda, D., Kissui, B. & Kushnir, H. (2005) Lion attacks on humans in Tanzania. Understanding the timing and distribution of attacks on rural communities will help to prevent them. *Nature* 436, 927–928.

Pamo, E.T. & Tchamba, M.N. (2001) Elephants and vegetation change in the Sahelo-Soudanian region of Cameroon. *Journal of Arid Environments* 48, 245–253.

Parker, I.S.C. & Graham, A.D. (1989a) Elephant decline: downward trends in African elephant distribution and numbers (Part I). *International Journal of Environmental Studies* 34, 287–305.

Parker, I.S.C. & Graham, A.D. (1989b) Elephant decline: downward trends in African elephant distribution and numbers (Part II). *International Journal of Environmental Studies* 35, 13–26.

Remis, M.J. & Kpanou, J.B. (2011) Primate and ungulate abundance in response to multi-use zoning and human extractive activities in a Central African Reserve. *African Journal of Ecology* 49, 70–80.

Romañach, S.S., Lindsey, P.A. & Woodroffe, R. (2007) Determinants of attitudes towards predators in central Kenya and suggestions for increasing tolerance in livestock dominated landscapes. *Oryx* 41, 185–195.

Røskaft, E., Händel, B., Bjerke, T. & Kaltenborn, B.P. (2007) Human attitudes towards large carnivores in Norway. *Wildlife Biology* 13, 172–185.

Rozemeijer, N. (2002) *Network Who? The Impact of Network on the Participation of Community-based Natural Resources Management (CBNRM) in Botswana.* Ninth Biennial Conference of the International Association for Study of Common Property (LASCP), Victoria Falls, Zimbabwe. [online] http://www.cbnrm.net/index.html

Said, M.Y., Chunge, R., Craig, G.C., Thouless, C.R., Barnes, R.F. & Dublin, H.T. (1995) *African Elephant Data Base 1995.* IUCN, Gland, Switzerland.

Sarker, A.H.M.R. (2010) Human-wildlife conflict: a comparison between Asia and Africa with special reference to elephants. In: Gereta, E. & Røskaft, E. (eds.) *Conservation of Natural Resources: Some African & Asian Examples.* Tapir Academic Press, Trondheim, Norway, pp. 186–210.

Sarker, A.H.M.R. & Røskaft, E. (2010) Human attitudes towards conservation of Asian elephants (*Elephas maximus*) in Bangladesh. *International Journal of Biodiversity and Conservation* 2, 316–327.

Sarker, A.H.M.R. & Røskaft, E. (2011a) Human-wildlife conflicts and management options in Bangladesh with special reference to Asian elephants (*Elephas maximus*). *International Journal of Biodiversity Science, Ecosystem Services & Management* 6, 164–175.

Sarker, A.H.M.R. & Røskaft, E. (2011b) Human attitudes towards the conservation of protected areas: a case study from four protected areas in Bangladesh. *Oryx* 45, 391–400.

Sitati, N.W., Walpole, M.J., Smith, R.J. & Leader-Williams, N. (2003) Predicting spatial aspects of human-elephant conflict. *Journal of Applied Ecology* 40, 667–677.

Skarpe, C., Aarrestad, P.A., Andreassen, H.P., Dhillion, S., Dimakatso, T., du Toit, J.T., Halley, D.J., Hytteborn, H., Makhabu, S., Mari, M., Marokane, W., Masunga, G., Modise, D., Moe, S.R., Mojaphoko, R., Mosugelo, D., Motsumi, S., Neo-Mahupeleng, G., Ramotadima, M., Rutina, L., Sechele, L., Sejoe, T.B., Stokke, S., Swenson, J.E., Taolo, C., Vandewalle, M., & Wegge, P. (2004) The return of the giants: ecological effects of an increasing elephant population. *Ambio* 33, 276–282.

Spinage, C.A. (1994) *Elephants.* T. & A.D. Poyser Ltd, London, UK.

Stokes, E.J., Strindberg, S., Bakabana, P.C., Elkan, P.W., Iyenguet, F.C., Madzoké, B., Malanda, G.A.F., Mowawa, B.S., Moukoumbou, C., Ouakabadio, F.K. & Rainey, H.J. (2010) Monitoring great ape and elephant abundance at large spatial scales: measuring effectiveness of a conservation landscape. *PLoS ONE* 5(4), e10294. doi:10.1371/journal.pone.0010294.

Tchamba, M.N. & Elkan, P. (1995) Status and trends of some large mammals and ostriches in Waza National Park, Cameroon. *African Journal of Ecology* 33, 366–376.

Tchamba, M.N. (1996) History and present status of the human elephant conflict in the Waza-Logone Region, Cameroon, West Africa. *Biological Conservation* 75, 35–41.

Thirgood, S., Polasky, S. & Mlingwa, C. (2005) Who pays for conservation? Current and future financing scenarios for the Serengeti ecosystem. In: Sinclair, A.R.E., Packer, C., Coughenour, M.B., Galvin, K. & Mduma, S. (eds.) *Serengeti III: Biodiversity and*

Biocomplexity in a Human-influenced Ecosystem. University of Chicago Press, Chicago, pp. 443–470.

Thouless, C.R. (1994) Conflict between humans and elephants on private land in northern Kenya. *Oryx* 28, 119–127.

Tingvold, H.G., Fyumagwa, R., Baardsen, L.F., Rosenlund, H., Bech, C. & Røskaft, E. (2013) Determining adrenocortical activity as a measure of stress in African elephants (*Loxodonta africana*) in relation to human activities in Serengeti ecosystem. *African Journal of Ecology.* [e-publication ahead of print] doi:10.1111/aje.12069.

van Aarde, R.J. & Jackson, T.P. (2007) Megaparks for metapopulations: addressing the causes of locally high elephant numbers in southern Africa. *Biological Conservation* 134, 289–297.

Van De Vijver, C.A.D.M., Foley, C.A. & Olff, H. (1999) Changes in the woody component of an East African savanna during 25 years. *Journal of Tropical Ecology* 15, 545–564.

Vollrath, F. & Douglas-Hamilton, I. (2002) African bees to control African elephants. *Naturwissenschaften* 89, 508–511.

Waitaka, J. (1993) The elephant menace. *Wildlife Conservation* 96, 62–65.

Warren, Y., Buba, B. & Ross, C. (2007) Patterns of crop-raiding by wild and domestic animals near Gashaka Gumti National Park, Nigeria. *International Journal of Pest Management* 53, 207–216.

Elephants and Heterogeneity in Savanna Landscapes

Johan T. du Toit[1], Christina Skarpe[2] and Stein R. Moe[3]

[1] Department of Wildland Resources, Utah State University, USA
[2] Faculty of Applied Ecology and Agricultural Sciences, Hedmark University College, Norway
[3] Department of Ecology and Natural Resource Management, Norwegian University of Life Sciences, Norway

Ecosystem scientists are interested in measuring and understanding how parts of ecosystems interact and change across space and time, whereas managers are concerned with inducing those parts to remain within, or return to, desired states. The innumerable differences within an ecosystem – such as between here and there and now and then – all contribute to its heterogeneity at multiple scales. Such heterogeneity can be intractable unless contextualised by a spatially and temporally explicit focus on some process, pattern or assemblage of organisms and their resources. For example, the aim of wildlife management is typically to conserve biodiversity at the landscape scale for the benefit of present and future generations. In this context ecological heterogeneity, which is the basis for biodiversity, can be analysed within a conceptual framework of agents, substrates, controllers and responders (Pickett *et al.*, 2003), as described in Chapter 1. Even then, all this heterogeneity makes it difficult to formulate science-based policies that managers can adhere to because the temporal component of heterogeneity means that almost everything is nearly always changing within the ecosystem. Distribution and abundance patterns of plants, animals, soils and water supplies are in constant flux at rates that vary from fast (e.g. some herbivory effects) to slow (e.g. plant community succession), whereas the temporal and spatial scales of the superimposed management domain, and the manner of its functioning,

Elephants and Savanna Woodland Ecosystems: A Study from Chobe National Park, Botswana,
First Edition. Edited by Christina Skarpe, Johan T. du Toit and Stein R. Moe.
© 2014 John Wiley & Sons, Ltd. Published 2014 by John Wiley & Sons, Ltd.

are mismatched with those of the ecosystem (Cumming *et al.*, 2006). This mismatch is inevitable because park managers have to operate within the framework of their institutions and at the spatiotemporal scales prescribed by their parks' boundaries, budgets, human resources, political mandates, and so forth.

A fundamental rule of ecosystem science is that temporal heterogeneity is destabilising whereas spatial heterogeneity is stabilising (May, 1974). This means that a rich mosaic of habitats should support a comparatively stable community, in terms of the persistence of species, because some habitat patches (refugia) provide dispersers to repopulate areas depleted by periodic disturbances. Consequently the system is stable across this dynamic space-time mosaic at the landscape scale and above, but within each habitat patch the system is unstable. Now, with park boundaries cutting across ecosystems, and human land use transforming the matrix outside, the stability of the enclosed portion of the ecosystem becomes a function of its scale. Such considerations of scale are important for effective conservation in general (du Toit, 2010), and are increasingly recognised as crucial for the management of African elephant, *Loxodonta africana*, populations in particular (Scholes and Mennell, 2008).

In recovering after the end of the commercial ivory-hunting era in the late 19th century, most southern African elephant populations have become impressive agents of change in the parks to which they are now mostly confined. It has been variously argued that such changes are natural and should be allowed (Lewin, 1986), or that uncontrolled elephant populations threaten biodiversity (Cumming *et al.*, 1997) and represent management failures through which parks have become 'overpopulated animal slums' (Child, 2004: 18). This chapter reviews a body of evidence from Chobe and savannas elsewhere in Africa that indicates how elephant populations can both generate, and be regulated by, heterogeneity at the landscape scale. At the outset, however, we must make it clear that these ecosystem-level feedbacks are increasingly unlikely to occur. That is because the biology of elephants is evolutionarily adapted for spatiotemporal scales that are incompatible with the space and time limitations imposed by modern humans.

Elephant drinking sites, transition zones, and population regulation

Elephants are water-dependent, needing to drink at intervals of no more than 3 days (Chapter 8). Visits to water involve social interactions, wallowing, and play behaviour that result in herds congregating near water for longer than simply to drink. In most areas of their range (excluding desert-adapted populations) savanna elephants have close access to multiple drinking sites along drainage systems and across the landscape in ephemeral pans during the wet season. In the dry season, however, the availability of surface water can become restricted to the extent that some populations depend on digging for water beneath the sand in dry riverbeds, whereas others, such as in northern Botswana, congregate close to major rivers. During the dry season in Chobe, elephant family units usually remain within 3–4 km from perennial water, in contrast to the wet season when they are usually found 5–9 km away (Stokke and du Toit, 2002). Furthermore, in the dry season, elephants are browsers whereas in the wet season they are grazers, so their effects on woody vegetation structure occur mostly in the dry season within a distinct zone around perennial water. Within this zone, elephants are the primary agents of vegetation change, from a wooded savanna on the outer periphery to a denuded scrubland near water, with a graded transition in-between. Tourists, and therefore wildlife managers, concentrate their activities in this transition zone because it is where wildlife viewing opportunities are usually optimal. As a result, they see elephant-disturbed woodland at its 'worst'. They then compare such areas to riparian woodland at its 'best' when they return to their riverside hotels outside the park where elephants are excluded. The contrast is dramatic and so it is quite understandable when public concerns are expressed about elephant overpopulation and ineffective park management. But can – or should – anything be done?

Compare the phenomenon of elephants congregating at water in the dry season to the throng of people in a city pub. The closest pub within walking distance of a football stadium (e.g.) is packed after a game. Even if attendance at games could be greatly reduced – by restricting ticket sales (instead of culling!) by as much as 50% – there would still be no empty stools at the bar after a game. In the same way, elephant density would remain high near water in the dry season even if total population size was substantially reduced. Culling elephants in northern Botswana would not alleviate their disturbance of vegetation along the Chobe riverfront unless the entire population was decimated to the extent that ivory hunters achieved in the latter part of the 19th century (Chapter 6). In addition, the devastating effects of the rinderpest panzootic on the ungulate community (Chapter 6) would have to be locally replicated and maintained through culling, to allow tree seedlings to establish and the woodland to regenerate (Chapter 10). That is obviously not an acceptable management option, which leaves park authorities with the responsibility of educating tourists and stakeholders in the tourist industry to understand and accept elephant-caused transition zones around perennial water.

Although the striking spectacle of elephant 'damage' to woody vegetation in a transition zone might worry stakeholders and tourists it does not necessarily indicate damage to regional biodiversity and ecosystem processes (Chapters 5, 12, 13). This is as long as perennial drinking sites are widely dispersed at the scale of elephant home ranges. Furthermore, for elephant populations occurring at or near saturation density (whatever that might be) transition zones could be part of a mechanism of population regulation. From vegetation changes that have occurred near the Chobe riverfront during the period of elephant population recovery (Mosugelo *et al.*, 2002), together with monitoring of elephant locations in wet and dry seasons (Stokke and du Toit, 2002), it seems that a transition zone of about 5 km width or radius is to be expected around perennial water. This matches measured daily displacement distances for satellite-tracked elephant cows in 13 conservation areas across southern Africa (Young and van Aarde, 2010). As the dry season progresses, elephant herds have to move to the outer edge of this zone and back to water every day or two, with travel time reducing feeding time. As a result, herds speed up when making drinking trips (Chapter 8). Neonates and weaned calves suffer increased mortality during droughts, apparently due to exhaustion, thirst and predation by lions (Dudley et al., 2001; Loveridge et al., 2006; Foley et al., 2008; Young and van Aarde, 2010). These factors are compounded when family units have to commute farther and farther between feeding areas and water, as indicated by two findings: (i) there is a negative relationship between weaned calf survival and daily displacement distance (Young and van Aarde, 2010); and (ii) calf loss is highest among the youngest and therefore least experienced mothers (Foley et al., 2008). Presumably the youngest mothers are least successful in maintaining contact with their calves during fast commuting trips to and from water. Being low in the social hierarchy, these individuals also cannot control their herds' movements.

There is no conclusive evidence yet that elephant populations can be self-regulated by density-dependent processes (van Aarde et al., 2008) although the population density within Chobe has been stable at 2.5–3.0 animals km^{-2} for the past decade (Chase, 2011), with the probable mechanism being density-dependent dispersal (Junker et al., 2008). For a closed population the mean annual mortality of juveniles (younger than 12 years) would have to be around 17% to constrain growth of that population (Woolley et al., 2008). This is much higher than has ever been recorded in southern Africa, although in the Tarangire area of northern Tanzania 20% of monitored calves (younger than 8 years) died within nine months during a severe drought in 1993 (Foley et al., 2008). Non-senescent adult mortality is low because adult elephants are immune from non-human predators and can forage over large distances for food, including low-quality tree bark and roots. Nevertheless, chronic under-nutrition could reduce fertility. Demographic modelling indicates that if the intercalving interval is prolonged and age of first parturition delayed, then elephant population growth should eventually be reduced (Owen-Smith et al., 2006). But calf mortality is expected to be the main factor limiting the growth of elephant populations in effectively protected areas, with the bulk of those deaths occurring in heavily browsed vegetation within 5 km of water during the late dry season, especially in years of below-average rainfall.

Introducing artificial waterpoints (e.g. pumped boreholes or dams) to divert elephants away from heavily disturbed areas is unlikely to be a successful management solution (Owen-Smith, 1996). Elephants will quickly begin using the new water supply and transform the vegetation in a piosphere around that site. Furthermore, unless the new waterpoint is farther than 10 km from natural sources of perennial water, the same elephants will still frequent these natural water sources for drinking, wallowing and socialising. Conversely, closing an artificial waterpoint will not lead to the local recovery of elephant-disturbed vegetation if it is within about 5 km of either a natural perennial water source or another artificial, but still functioning, waterpoint. With access to surface water in the dry season being the likely limiting factor inducing density-dependent population regulation in unmanaged elephant populations (Chapter 8), it follows that artificial waterpoints would simply enable larger population sizes.

Available area, distance between waterpoints, and elephant overpopulation

The mean home range area of elephant breeding herds in southern Africa is 1678 km^2 and the maximum is 10,738 km^2 (van Aarde *et al.*, 2008), which indicates the spatial scale of elephant population ecology. Elephants confined to small reserves such as Addo Elephant National Park (currently about 250 km^2) can be expected to affect all accessible vegetation types and be agents responsible for homogenising landscapes, threatening some animal and plant populations, and even some endemic plant species, unless intensively managed (Kerley and Landman, 2006). For a minimum viable population of elephants (let us say 500–1000 animals), the level of management concern arising from the genuine risk of long-term damage to the ecosystem can be expected to be influenced by both reserve size and mean distance between dry season waterpoints. The two variables are autocorrelated in small reserves but not in large ones. For example, the vast southern African transfrontier conservation areas (TFCAs) such as Kavango-Zambezi TFCA, which includes Chobe and Hwange, and Great Limpopo TFCA, which includes Kruger, provide ample space for their elephant populations to continue growing until they become self-regulated through density-dependent processes, whenever that might happen. Within these TFCAs, however, some management units, such as Kruger, have reduced the mean distance between dry-season waterpoints. This statistic is the cause of greater management concern than the total area available to the elephant population. In the dry season more than 90% of the area inside Kruger (20,000 km^2) is within 5 km of surface water (Redfern *et al.*, 2003), largely because of artificial waterpoints. The contrast between this situation in Kruger and that in Chobe illustrates the need to clarify what is actually meant by 'the elephant problem' when it is invoked for a particular area. It also throws into question the concept of 'overpopulation' as a basis for management action.

Overpopulation is a value judgment (Caughley, 1981; Côté *et al.*, 2004) made by people comparing a population, or its undesirable effects, against some benchmark that is perceived to be 'right', a concept termed 'aesthetic carrying capacity'. Overpopulation by elephants causes what, for the past 50 years, has been called 'the elephant problem' (Glover, 1963), where an elephant population exceeds its aesthetic carrying capacity at the local scale, rather than its ecological carrying capacity-assuming that concept can be applied to elephants-at the regional scale (Cumming, 1981). The notion of overpopulation, however, is not useful because it implies an intrinsic problem of dysfunctional population regulation, whereas the problem is extrinsic, associated with humans and not necessarily with elephant population density or size. A comparatively low density of elephants can include crop-raiders (Chapter 16), and small populations of elephants can degrade scenic landscapes and cause species extinctions if they are confined to small areas (Kerley and Landman, 2006). Population ecology offers no solutions for 'the elephant problem', which should be more accurately termed 'the human-elephant problem'. Its management requires a facilitated process involving local communities, government agencies, investors, special interest groups and politicians, with technical support being provided by applied social scientists, resource economists and wildlife managers. Prior to that, scientists need to have clarified the ecological interactions because changes ascribed to elephant herbivory could be transitions between stable states involving other agents or controllers of change. For example, in the Serengeti-Mara woodlands the agent driving the transition from woodland to grassland is in fact fire, with elephants controlling the reverse transition (Dublin *et al.*, 1990). In Chobe the agent behind the transition from riparian woodland to scrubland is indeed elephants, but here impala, *Aepyceros melampus*, control the reverse transition (Chapter 10).

Are elephants agents of landscape heterogeneity or homogeneity?

The Chobe riverfront presently has the appearance of a conventional warfare battleground in the dry season, becoming a verdant thicket of coppicing trees and regenerating shrubs in the wet season. Nevertheless, within this highly disturbed zone, fragments of riparian forest remain (e.g. near Serondela) in small-scale (10–1000 m^2) refugia that elephants seldom visit. These vary from single trees to small stands and always occur above steep gradients, which elephants cannot negotiate, caused by bank erosion on the outer bends of the meandering river. It was trees in these refugia that must have supplied the seeds of the 'decreaser' species that were found in the seedbank (Chapter 13) and which germinated into a naturally regenerating community inside replicated exclosures along the riverfront (Moe *et al.*, 2009; Chapter 10). We suggest that source-sink seed dispersal from these refugia maintains a supply of propagules of riparian forest trees, thus buffering tree-species richness even when elephant-driven disturbance along the riverfront is high. Another example of the importance of refugia is in northern Kruger, where baobab trees, *Adansonia digitata*, are vulnerable to bark-stripping except in rocky outcrops, which are inaccessible to elephants (Edkins *et al.*, 2008). Further research is needed to test the implication that, because of refugia associated with fluvial geomorphology, transition zones along rivers should retain a richer woody plant community than transition zones in piospheres around pumped waterholes in flat terrain.

Consider a hypothetical scenario in which an elephant population self-regulates through density-dependent limitation caused mainly by juvenile mortality within a 5 km-wide transition zone on either side of a perennial river, or around pools along its course. The vegetation in the transition zone grades from a wooded savanna on the outer margin to coppicing scrubland near water, except for clumps of mature trees in refugia associated with topographical features that make them inaccessible or undesirable feeding sites for elephants. Rainfall alternates in wet and dry interannual cycles, with some parts of the landscape within years receiving more or less rainfall than the regional average. Extended droughts occur with intervals of a few decades, sometimes associated with the El Niño-Southern Oscillation (ENSO) and sometimes not (Richard *et al.*, 2001). Rivers flood and run dry, pools fill and empty; sedimentation and erosion cause some pools to disappear and new ones to form. Elephant herds shift their feeding areas around the landscape; ungulate populations fluctuate. The components of the heterogeneity framework (Pickett *et al.*, 2003) would all be operating with the relative strengths of the disturbance of the vegetation (substrate) caused by elephants (agents), impalas (controllers of woody regeneration; Chapter 10), and buffalo, *Syncerus caffer* (controllers of grass abundance; Chapter 11), continually changing across space and time. In all of this, pockets of vegetation remain comparatively undisturbed in refugia. Some of these refugia, such as rocky outcrops, would be more permanent than others, such as riverbank cliffs. This scenario remains hypothetical because no elephant population has yet been found conclusively to self-regulate through density-dependent processes. It is nevertheless the probable scenario that prevailed across African savannas prior to the episode of overhunting for ivory in the 19th century. Before then, as indeed now, indigenous humans hunted elephants (Chapter 6) and used all methods at their disposal to frighten them from their crops and villages (Chapter 16). Humans in traditional societies, by continually modifying the 'landscape of fear', are thus integral to the suite of natural factors maintaining spatiotemporal variation in the effects of elephants on savanna vegetation.

Now consider an alternative scenario in which elephant herds are constrained within park boundaries by habitat loss and barriers of manmade infrastructure. Park boundaries exclude many refugia and practically all traditional human settlements. River flow regimes are regulated by dams, and modified by abstraction and intensive land use upstream. Artificial waterpoints are scattered across the landscape so that virtually any location is within 5 km of permanent water. This is not hypothetical. It describes the actual situation causing concern in South Africa, where chronic disturbance by elephants might eradicate the baobab population in northern Kruger (Edkins *et al.*, 2008) and various endemic plant species in Addo (Kerley and Landman, 2006). It also broadly describes the situation across most of the remaining fragments of the African savanna elephant's range, such as the Zambezi Valley in Zimbabwe (Cumming *et al.*, 1997) and Amboseli in Kenya (Western, 2006). In such cases elephants are agents creating landscape homogeneity, because humans constrain the spatial and temporal scales implicit in the first scenario, in which elephants are agents promoting heterogeneity.

Conclusion

Whether elephants operate as agents creating heterogeneity or homogeneity in savanna landscapes depends on the spatiotemporal constraints imposed upon them by humans. It is only in large protected areas with minimal artificial provisioning of water that elephant populations can interact with their environment in ways that could be considered characteristic of a 'natural' savanna ecosystem. That scenario could return – if wildlife managers are allowed to let it happen – in Chobe and across the vast new TFCAs in southern Africa. Even so, wildlife managers, tourists and other stakeholders must understand and accept there will be areas near water that get transformed and there will be times when young elephants die in large numbers. As for the mounting human-elephant problems arising from the under-provision of habitat or the over-provision of waterpoints, their solutions are contingent upon human values and are appropriately being debated in other forums.

References

Caughley, G. (1981) Overpopulation. In: Jewell, P.A. & Holt, S. (eds.) *Problems in Management of Locally Abundant Wild Mammals*. Academic Press, New York, pp. 7–20.

Chase, M. (2011) *Dry Season Fixed-wing Aerial Survey of Elephants and Wildlife in Northern Botswana*. Elephants Without Borders, Kasane, Botswana; Department of Wildlife and National Parks, Botswana; and Zoological Society of San Diego, USA. [online] http://www.elephantdatabase.org/population_submission_attachments/102

Child, G. (2004) Growth of modern nature conservation in southern Africa. In: Child, B. (ed.) *Parks in Transition: Biodiversity, Rural Development and the Bottom Line*. Earthscan, London, pp. 7–27.

Côté, S.D., Rooney, T.P., Tremblay, J.P., Dussault, C. & Walker, D.M. (2004) Ecological impacts of deer overabundance. *Annual Review of Ecology, Evolution, and Systematics* 35, 113–147.

Cumming, D.H.M. (1981) The management of elephant and other large mammals in Zimbabwe. In: Jewel, P.J. & Holt, S. (eds.) *Problems in Management of Locally Abundant Wild Animals*. Academic Press, New York, pp. 91–118.

Cumming, D.H.M., Fenton, M.B., Rautenbach, I.L., Taylor, R.D., Cumming, G.S., Cumming, M.S., Dunlop, J.M., Ford, G.S., Hovorka, M.D., Johnston, D.S., Kalcounis, M.C., Mahlanga, Z. & Portfors, C.V. (1997) Elephants, woodlands and biodiversity in southern Africa. *South African Journal of Science* 93, 231–236.

Cumming, G.S., Cumming, D.H.M. & Redman, C.L. (2006) Scale mismatches in socio-ecological systems: causes, consequences, and solutions. *Ecology and Society* 11, 14 [online] http://www.ecologyandsociety.org/vol11/iss1/art14/

Dublin, H.T., Sinclair, A.R.E. & McGlade, J. (1990) Elephants and fire as causes of multiple stable states in the Serengeti-Mara woodlands. *Journal of Animal Ecology* 59, 1147–1164.

Dudley, J.P., Craig, G.C., Gibson, D. StC., Haynes, G. & Klimowicz, J. (2001) Drought mortality of bush elephants in Hwange National Park, Zimbabwe. *African Journal of Ecology* 39, 187–194.

du Toit, J.T. (2010) Considerations of scale in biodiversity conservation. *Animal Conservation* 13, 229–236.

Edkins, M.T., Kruger, L.M., Harris, K. & Midgley, J.J. (2008) Baobabs and elephants in Kruger National Park: nowhere to hide. *African Journal of Ecology* 46, 119–125.

Foley, C., Pettorelli, N. & Foley, L. (2008) Severe drought and calf survival in elephants. *Biology Letters* 4, 541–544.

Glover, J. (1963) The elephant problem at Tsavo. *African Journal of Ecology* 1, 30–39.

Junker, J., van Aarde, R.J. & Ferreira, S.M. (2008) Temporal trends in elephant *Loxodonta africana* numbers and densities in northern Botswana: is the population really increasing? *Oryx* 42, 58–65.

Kerley, G.I.H. & Landman, M. (2006) The impacts of elephants on biodiversity in the Eastern Cape Subtropical Thickets. *South African Journal of Science* 102, 395–402.

Lewin, R. (1986) In ecology, change brings stability. *Science* 234, 1071–1073.

Loveridge, A.J., Hunt, J.E., Murindagomo, F. & Macdonald, D.W. (2006) Influence of drought on predation of elephant (*Loxodonta africana*) calves by lion (*Panthera leo*) in an African wooded savannah. *Journal of Zoology* 270, 523–530.

May, R.M. (1974) Ecosystem patterns in randomly fluctuating environments. In: Rosen, R. & Snell, F.M. (eds.) *Progress in Theoretical Biology*, vol. 3. Academic Press, New York, pp. 1–50.

Moe, S.R., Rutina, L.P., Hytteborn, H. & du Toit, J.T. (2009) What controls woodland regeneration after elephants have killed the big trees? *Journal of Applied Ecology* 46, 223–230.

Mosugelo, D.K., Moe, S.R., Ringrose, S. & Nellemann, C. (2002) Vegetation changes during a 36-year period in northern Chobe National Park, Botswana. *African Journal of Ecology* 40, 232–240.

Owen-Smith, N. (1996) Ecological guidelines for waterpoints in extensive protected areas. *South African Journal of Wildlife Research* 26, 107–112.

Owen-Smith, N., Kerley, G.I.H., Page, B., Slotow, R. & van Aarde, R.J. (2006) A scientific perspective on the management of elephants in the Kruger National Park and elsewhere. *South African Journal of Science* 102, 389–394.

Pickett, S.T.A., Cadenasso, M.L. & Benning, T.L. (2003) Biotic and abiotic variability as key determinants of savanna heterogeneity at multiple spatiotemporal scales. In: du Toit, J.T., Rogers, K.H. & Biggs H.C. (eds.) *The Kruger Experience: Ecology and Management of Savanna Heterogeneity*. Island Press, Washington DC, pp. 22–40.

Redfern, J.V., Grant, R., Biggs, H. & Getz, W.M. (2003) Surface-water constraints on herbivore foraging in the Kruger National Park, South Africa. *Ecology* 84, 2092–2107.

Richard, Y., Fauchereau, N., Poccard, I., Rouault, M. & Trzaska, S. (2001) 20th Century droughts in southern Africa: spatial and temporal variability, teleconnections with oceanic and atmospheric conditions. *International Journal of Climatology* 21, 873–885.

Scholes, R.J. & Mennell, K.G. (eds.) (2008) *Elephant Management: A Scientific Assessment for South Africa*. Wits University Press, Johannesburg, South Africa.

Stokke, S. & du Toit, J.T. (2002) Sexual segregation in habitat use by elephants in Chobe National Park, Botswana. *African Journal of Ecology* 40, 360–371.

van Aarde R., Ferreira, S. Jackson, T., Page, B., de Beer, Y., Gough, K., Guldemond, R., Junker, J., Olivier, P., Ott, T. & Trimble, M. (2008) Elephant population biology and ecology. In: Scholes, R.J. & Mennell, K.G. (eds.) *Elephant Management: A Scientific Assessment for South Africa*. Wits University Press, Johannesburg, South Africa, pp. 84–145.

Western, D. (2006) A half century of habitat change in Amboseli National Park, Kenya. *African Journal of Ecology* 45, 302–310.

Woolley, L-A., Mackey, R.L., Page, B.R. & Slotow, R. (2008) Modelling the effect of age-specific mortality on elephant *Loxodonta africana* populations: can natural mortality provide regulation? *Oryx* 42, 49–57.

Young, K.D. & van Aarde, R.J. (2010) Density as an explanatory variable of movements and calf survival in savanna elephants across southern Africa. *Journal of Animal Ecology* 79, 662–673.

Index

Elephants and Savanna Woodland Ecosystems: A Study from Chobe National Park, Botswana,
First Edition. Edited by Christina Skarpe, Johan T. du Toit and Stein R. Moe.
© 2014 John Wiley & Sons, Ltd. Published 2014 by John Wiley & Sons, Ltd.